STRUCTURAL, CIVIL, AND PIPE DRAFTING FOR CAD TECHNICIANS

STRUCTURAL, CIVIL, AND PIPE DRAFTING FOR CAD TECHNICIANS

DAVID L. GOETSCH

THOMSON
™
DELMAR LEARNING

Australia • Canada • Mexico • Singapore • Spain • United Kingdom • United States

Structural, Civil, and Pipe Drafting for CAD Technicians

David L. Goetsch

Vice President, Technology and Trades SBU:
Alar Elken

Executive Director:
Sandy Clark

Senior Acquisitions Editor:
James Devoe

Senior Development Editor:
John Fisher

Editorial Assistant:
Mary Ellen Martino

Marketing Director:
Cyndi Eichelman

Production Director:
Mary Ellen Black

Channel Manager:
Fair Huntoon

Production Manager:
Andrew Crouth

Production Editor:
Stacy Masucci

Art and Design Specialist:
Mary Beth Vought

Marketing Coordinator:
Sarena Douglass

Cover Image:
Brand X Pictures

Cover Designer:
Scott Keidong

Library of Congress Cataloging-in-Publication Data

Goetsch, David L.
  Structural, civil, and pipe drafting for CAD technicians
    p. cm.
  1-4018-9656-1
    1. Engineering drawings. 2. Mechanical drawing. I. Title.
TA175.G64 2003
624—dc22   2003053372

NOTICE TO THE READER

Dedicated to:
The Drafting and Design Program of Okaloosa–Walton
Community College, past, present, and future

CONTENTS

SECTION I

OVERVIEW OF STRUCTURAL DRAFTING   1

SECTION II

STRUCTURAL STEEL DRAFTING

65

SECTION III

STRUCTURAL PRECAST CONCRETE DRAFTING

143

Unit 15—Precast Concrete Fabrication Details   192

Unit 16—Precast Concrete Bills of Materials 214

SECTION IV

STRUCTURAL POURED-IN-PLACE CONCRETE   229

Unit 17—Poured-in-Place Concrete Foundations   230

Unit 18—Poured-in-Place Concrete Walls and Columns   241

Unit 19—Poured-in-Place Concrete Floor Systems   254

Unit 20—Poured-in-Place Stairs and Ramps 261

SECTION V

STRUCTURAL WOOD DRAFTING   267

Unit 21—Structural Wood Floor Systems 268

SECTION VII

EMPLOYMENT IN DRAFTING — **385**

PREFACE

Structural, civil, and pipe drafting are specialized drafting fields. The old adage "it must be drawn before it can be built" is particularly true in these fields. This book is designed to assist the drafting student in developing the knowledge and skills necessary to begin a career in these drafting fields at a productive level.

Students completing all of the reinforcement activities contained in this textbook will have developed skills equivalent to those possessed by a CAD technician with at least one full year of work experience. The text is divided into seven sections with a total of thirty-one units of instruction. The first five units provide background information pertinent to a study of drafting. Unit 6 through Unit 28 are knowledge and skills development lessons. Unit 29 through Unit 31 teach the student how to find, get, and keep a job in drafting after acquiring the necessary skills.

The book is designed for self-directed, individualized instruction so that any course relying on it as the required textbook can be offered on a multilevel basis with other drafting courses. Each unit contains performance objectives, a presentation of the material, original drawings prepared by the author, a summary, review questions, and CAD activities.

The CD in back of the book contains a student version of the Structural Engineering Visual Encyclopedia (SEVE); this supplement was originally developed by Robert M. Henry at the University of New Hampshire with the support of an NSF grant. Dr. Henry is an Associate Professor of Civil Engineering and a registered professional structural engineer. For the past 25 years he has been actively engaged in the development of interactive structural engineering software and virtual reality modeling of structural systems. His wife, Nancy, and two sons, Mark and Brian, have provided technical support during the development of SEVE. In 1998, the UNH version of SEVE won the Premiere Award for Excellence in Engineering Education Courseware.

The author, Dr. David L. Goetsch, has worked in drafting for thirty years. He has owned his own drafting and design consulting service and taught drafting at the high school, vocational school, and community college level for over twenty-seven years. He is currently Provost at Okaloosa-Walton Community College in Niceville, Florida. His drafting program was selected for inclusion in the Florida Department of Education's Catalogue of Innovative Programs and maintains a 95 percent successful placement record of its graduates. In 1984, his program was selected as "Outstanding Technical Program" in Region IV of the United States by the U.S. Secretary of Education and, as a result, received the U.S. Secretary of Education's Award for technical programs.

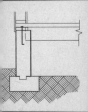

ACKNOWLEDGMENTS

CONTRIBUTORS OF PHOTOGRAPHS AND TECHNICAL MATERIAL

The author wishes to thank those people, organizations, and companies that provided materials for use in this book. Sources of such materials are:

American Concrete Institute
American Institute of Steel Construction: Appendix A
Bethlehem Steel Corporation
Design-Build Systems
Goodheart-Wilcox Company, Inc.

National Forest Products Association: Appendix B
Southern Prestressed Concrete Company, Inc.
Weyerhaeuser Company
Southern Prestressed, Inc.

Charles D. Willis

Billy Queysen, Sherrell Helm, Tom Benton, Clark Williams, Dieter Maucher, Roger Tonnessen, and Doug Smith of Southern Prestressed Concrete, Inc.

Mark Barrett of Design-Build Systems, Inc.

Linden Lyons of Mallet Engineering Company

The author also wishes to thank his wife Deb for providing valuable photographic and illustration assistance and Roxane Bennett for her help with the index.

CLASSROOM TESTING

The material in this text was classroom tested at Okaloosa-Walton Community College, Niceville, Florida.

REVIEWERS

A technical review of the text was completed by Jeff Levy of Pulaski County Technical Center, Dublin, VA.

A special acknowledgment is due to the following instructors who reviewed the chapters in detail: Larry Lamont, Moraine Park Technical College, Fond du Lac, WI; Art Leonard, Universal Technical Institute, Phoenix, AZ; James Overton, Southeast College of Technology, Memphis, TN; Malcolm Roberts, Southeast College, Mobile, AL; and Keven Standiford, Arkansas State Teacher's Retirement System, Little Rock, AR.

INTRODUCTION

OVERVIEW OF STRUCTURAL, CIVIL, AND PIPE DRAFTING

This book prepares students for careers in three closely related fields of drafting: structural, civil, and pipe. Drafting in these fields is the act of documenting design and planning information so that something can be built, laid out, remodeled, recorded for legal purposes, or fabricated. CAD technicians in these fields work with engineering personnel in developing plans for buildings, bridges, parking decks, stadiums, and other structures; roads, canals, highways, utility lines, sewer lines, water lines, and other entities that involve cutting into the earth; and piping systems that process, transport, and store liquids and gases. Figure I.1 through Figure I.8 are examples of the types of work that require comprehensive design and drawing packages.

A typical engineering team in the modern workplace includes one or more engineers, CAD technicians, and experienced CAD technicians who serve as checkers. A checker is usually a CAD technician who has gained sufficient experience to be able to check the work of other drafting personnel to ensure that it complies with the engineer's instructions and all applicable codes, regulations, and standards.

Although most CAD technicians eventually work exclusively in one of the three fields covered in this book—structural, civil, or pipe drafting—it is wise to study all three fields. A knowledge of one field will broaden students' understanding of the others and will also increase their employability (the ability to secure a good job). Many consulting engineering firms have departments that correspond to all three of these fields. Consequently, versatility on the part of a job applicant is viewed positively by those who do the hiring in such firms. In addition, a knowledge of all three fields will help students make a more informed decision when deciding which field they like best or which has the best career potential.

Structural CAD technicians typically work in such settings as consulting engineering firms that have a structural engineering department, prestressed and precast concrete companies, structural steel companies, and firms that specialize in the construction of prefabricated metal buildings. Civil CAD technicians typically work in such settings as consulting engineering firms that have a civil engineering department, surveying companies, government engineering departments (e.g., county road departments, city or county engineering departments, and state road departments), and local property appraisers' offices as well as full-service architectural and engineering firms. Pipe CAD technicians typically work in consulting engineering firms that have a mechanical engineering department that specializes in pipe, construction companies that specialize in building piping systems, processing companies, and water and sewer departments of local governments.

CAD IN STRUCTURAL, CIVIL, AND PIPE DRAFTING

Drafting is the process through which the documentation of work performed by engineers, designers, architects, surveyors, and other technical professionals is prepared. The need for drafting is found in the old adage, *a picture is worth a thousand words*. Drafting has evolved over the years from being a process accomplished using ink pens and linen to a more efficient process using pencils and vellum to yet an even more efficient process using computers.

Using the computer as the basic tool in drafting came to be known as *computer-aided drafting,* or CAD. Although CAD is a radically different way to document the design process, creating documentation is still what drafting is about. CAD is not a different drafting field. Rather, it is a different way of doing drafting. Before something can be constructed, manufactured, processed, or laid out on the ground, it must first be documented. CAD is now the normal way of creating that documentation for most engineering, architectural, and design applications including the following: predesign, cost estimation, site analysis, project design, product design, construction development, bid preparation, contract negotiations, construction management, and project follow-up.

This book is designed to help students studying to be CAD technicians learn to develop the types of documentation needed in structural, civil, and pipe drafting settings. It is not a book on developing CAD skills per se, but a book that will help students who have basic CAD skills learn how to apply them to structural, civil, and pipe drafting. This book assumes that students have basic CAD skills as well as basic mechanical drawing skills (e.g., orthographic projection, sectioning, and dimensioning).

FIGURE I.1 ■ The plans for this college library were developed by structural CAD technicians.

FIGURE I.2 ■ The plans for this water tower were developed by structural CAD technicians.

FIGURE I.3 ■ The plans for this bridge were developed by structural CAD technicians.

FIGURE I.4 ■ The plans for this road that is being constructed were developed by civil CAD technicians.

FIGURE I.5 ■ The plot plan for this house was developed by civil CAD technicians.

FIGURE I.6 ■ The plans for this parking lot were developed by civil CAD technicians.

FIGURE I.7 ■ The plans for this process piping system for natural gas were developed by pipe CAD technicians.

FIGURE I.8 ■ The plans for this water and sewer pipe system were developed by pipe CAD technicians.

SECTION I

Overview of Structural Drafting

1 UNIT

Introduction to Structural Drafting

OBJECTIVES

Upon completion of this unit, the student will be able to:

- Define structural drafting.
- Identify the different types of structural drawings.
- List the most common employers of structural CAD technicians.
- Demonstrate proper structural drafting techniques in the areas of linework, lettering, and scale use.
- Explain the use of CAD in structural drafting.

STRUCTURAL DRAFTING DEFINED

In heavy construction, anything composed of parts is called a *structure*. All products of the heavy construction industry—bridges, buildings, towers, and countless other possibilities—are composed of parts, making them structures. In order for the various parts to be designed, manufactured, and put together to form a completed structure, drawings must be made. This need for drawing forms the foundation for drafting as an important occupation in the heavy construction industry. Figure 1.1 through Figure 1.6 show examples of typical structures that were built according to plans drawn by structural CAD technicians.

TYPES OF STRUCTURAL DRAWINGS

Structural CAD technicians are called upon to prepare two separate types of drawings: engineering drawings and shop drawings. **Engineering drawings** are used to provide an overall picture of a job for sales, marketing, estimating, or engineering purposes. **Shop drawings** are much more detailed. They are used for designing, fabricating, manufacturing, and erecting the structural products that go into a job. Engineering and shop drawings may be combined into one set to form **working drawings**.

FIGURE 1.1 ■ Intercity bridge over Columbia River. (*Prestressed Concrete Institute*)

FIGURE 1.2 ■ Private home. (*Prestressed Concrete Institute*)

FIGURE 1.3 ■ Cleveland State Office Building. (*Bethlehem Steel Corporation*)

FIGURE 1.4 ■ Picnic and concession shelters. (*Prestressed Concrete Institute*)

FIGURE 1.5 ■ Pedestrian walkway. (*Bethlehem Steel Corporation*)

FIGURE 1.6 ■ Parking structure. (*Photo by Author*)

Engineering drawings for heavy construction jobs are usually prepared by structural CAD technicians employed by architects, engineers, contractors, or sales engineers. Information found on engineering drawings includes the following:

■ Locations of major structural components such as columns, beams, and girders

■ Basic dimensions

■ Typical sections

■ Notes to clarify complicated situations

Figure 1.7 and Figure 1.8 show examples of structural engineering drawings.

Shop drawings for heavy construction jobs are usually prepared by structural CAD technicians employed by companies that actually manufacture structural products. For the most

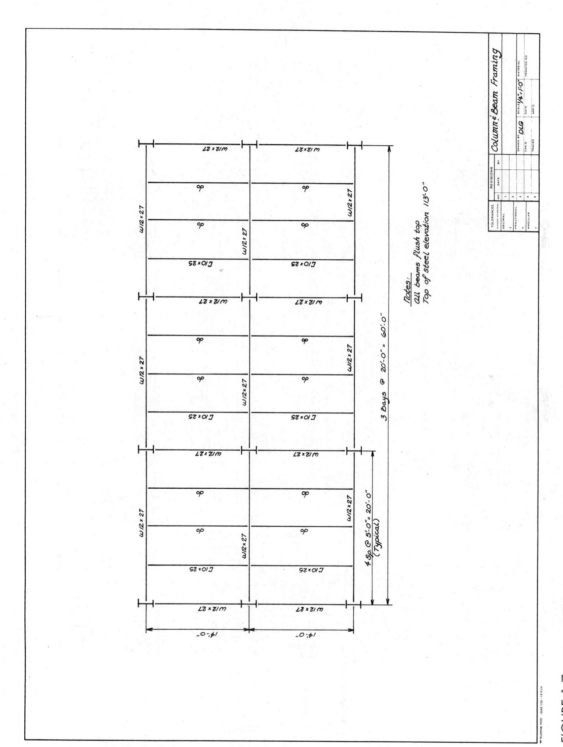

FIGURE 1.7 ■ Structural steel engineering drawing.

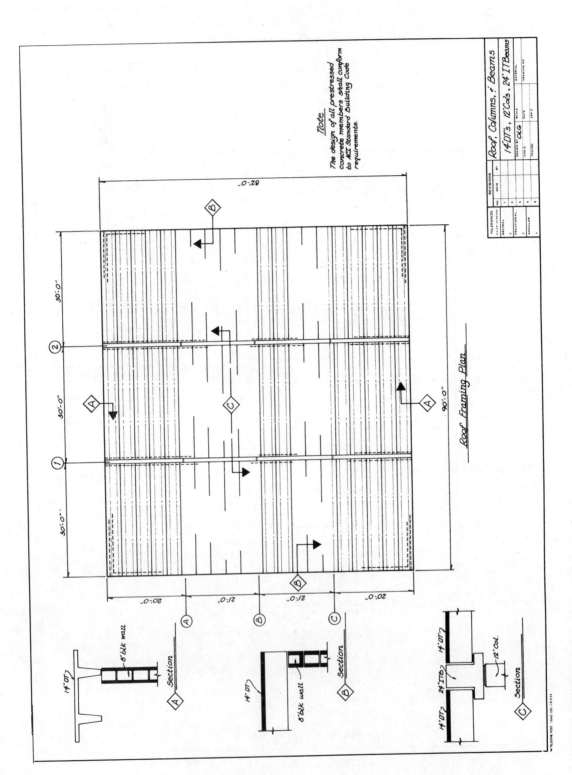

FIGURE 1.8 ■ Precast concrete engineering drawing.

part, these are structural steel fabrication companies and precast concrete manufacturers. Shop drawings include the following:

■ Precise connection details

■ Fabrication details for every structural member in a job

■ Details of miscellaneous metal connectors required

■ Bills of material listing every item that will be used during fabrication of the products and erection of the final structure

Figure 1.9 and Figure 1.10 show examples of structural fabrication details, which are the primary component of a set of shop drawings.

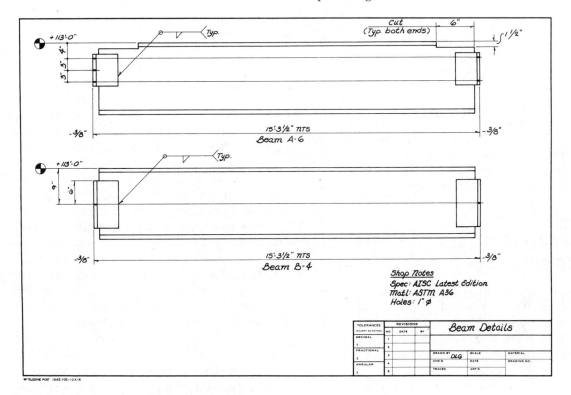

FIGURE 1.9 ■ Structural steel fabrication detail.

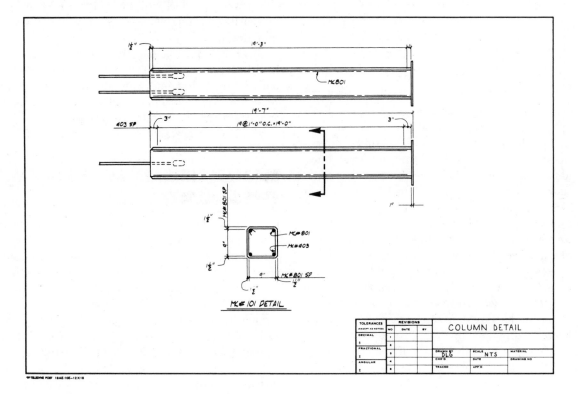

FIGURE 1.10 ■ Precast concrete fabrication detail.

EMPLOYERS OF STRUCTURAL CAD TECHNICIANS

Structural CAD technicians are usually employed in one of two ways. The first is preparing engineering and shop drawings of wood, concrete, or steel structures for structural consulting engineering firms or for full-service architectural/engineering firms. The second is preparing shop drawings for structural steel or precast concrete manufacturers.

The CAD student preparing to enter the world of structural engineering should develop knowledge and skills in several areas. These areas include structural steel, precast concrete, poured-in-place concrete, and structural wood drafting. In certain employment situations, such as with a consulting structural engineering firm, the structural CAD technician is often called upon to prepare plans for structures involving all four types of structural products.

REVIEW OF DRAFTING/CAD TECHNIQUES

Before developing specific knowledge and skills in the areas of structural steel, precast concrete, and poured-in-place concrete, the student must learn several general items that apply to all areas of structural drafting. These *structural drafting/CAD techniques* include the following:

- Linework
- Lettering
- Scale use
- Paper sizes
- Title blocks and borders

Linework

As is the case in all types of drafting, structural drafting has a set of line types that are commonly used. This set includes object lines, hidden lines, phantom lines, centerlines, dimension lines, extension lines, cutting plane lines, and break lines. Figure 1.11 illustrates each of these types of lines drawn to an acceptable width. It is evident from Figure 1.11 that object lines may be drawn to one of several widths. Actually, in structural drafting, this is true of all lines. The examples provided in Figure 1.11 are meant to serve as guidelines for reference when preparing structural drawings. However, note that though the shapes of the lines should agree with the examples provided, the widths may actually vary slightly in use. In application, line widths are sometimes varied. This is done to emphasize one aspect of a drawing or de-emphasize another aspect. The width of a line is determined by how the line is to be used and the individual circumstances of the drawing. With computer-aided drafting (CAD), linework is taken care of by the software and plotter.

Line Types. The most commonly used line is the *object line.* An object line is a continuous solid line used to show the outline of the object being drawn. Object lines are drawn in slightly varied widths depending upon the amount of emphasis desired by the CAD technician (Figure 1.12).

Hidden lines are used almost as frequently as object lines. They show the edges or outlines of structural components that are important but could not be seen by an actual viewer of a given part of a structure. Hidden lines are composed of a number of very short line segments. These line segments should be drawn with sharp, crisp pencil strokes to achieve consistency and uniformity (Figure 1.13).

Phantom lines are used to show parts of a structure that do not actually appear in a given view for the purpose of clarity. For example, phantom lines may be used to show the outline of a roof that is actually above a framing plan or the outline of a wall that is not included on a framing plan (Figure 1.13).

Centerlines are used frequently in structural drafting to locate centers of columns on framing plans or holes on fabrication details. Centerlines are thin lines broken by a short dash. Centerlines are often extended beyond the object being drawn and used as extension lines to enclose dimension lines. Figure 1.13 illustrates how centerlines are used on structural plans.

Dimension lines are enclosed by extension lines and are frequently used in structural drafting to indicate size and distance. Dimension lines are thin, continuous lines that end in any one of a variety of arrowheads or arrowhead substitutes (Figure 1.14).

Cutting plane lines are thick lines that are used to, in a sense, cut through an object for the purpose of clarification. A cutting plane line indicates the area that has been sliced through and a direction for the viewer's sight. A sectional view is then drawn showing the viewer what would actually be seen if the object were sliced as indicated. Figure 1.14 shows two examples of how cutting plane lines are commonly used in structural drafting.

Break lines may be constructed mechanically or freehand. In either case, they are used to cut out unnecessary or lengthy portions of a drawing. This allows the CAD technician to show only those portions of a detail that are needed to convey an idea, thereby economizing on space and drafting time. Figure 1.13 shows an example of a freehand break line.

ENGINEERING LETTERING AND CAD

Historically, the lettering used in structural, civil, and pipe drafting has been an issue. On the one hand, it had to be legible and clear to people reading the plans; on the other hand, it also reflected the style of the individual creating it. Figure 1.15 is an example of some of the styles of lettering historically found on structural, civil, and pipe drawings.

One of the many benefits of CAD is that it simplified the issue of engineering lettering. Lettering in CAD amounts to typing in the necessary characters. Modern CAD systems have a range of lettering options available to CAD technicians. Many systems can match the various styles shown in Figure 1.15 and

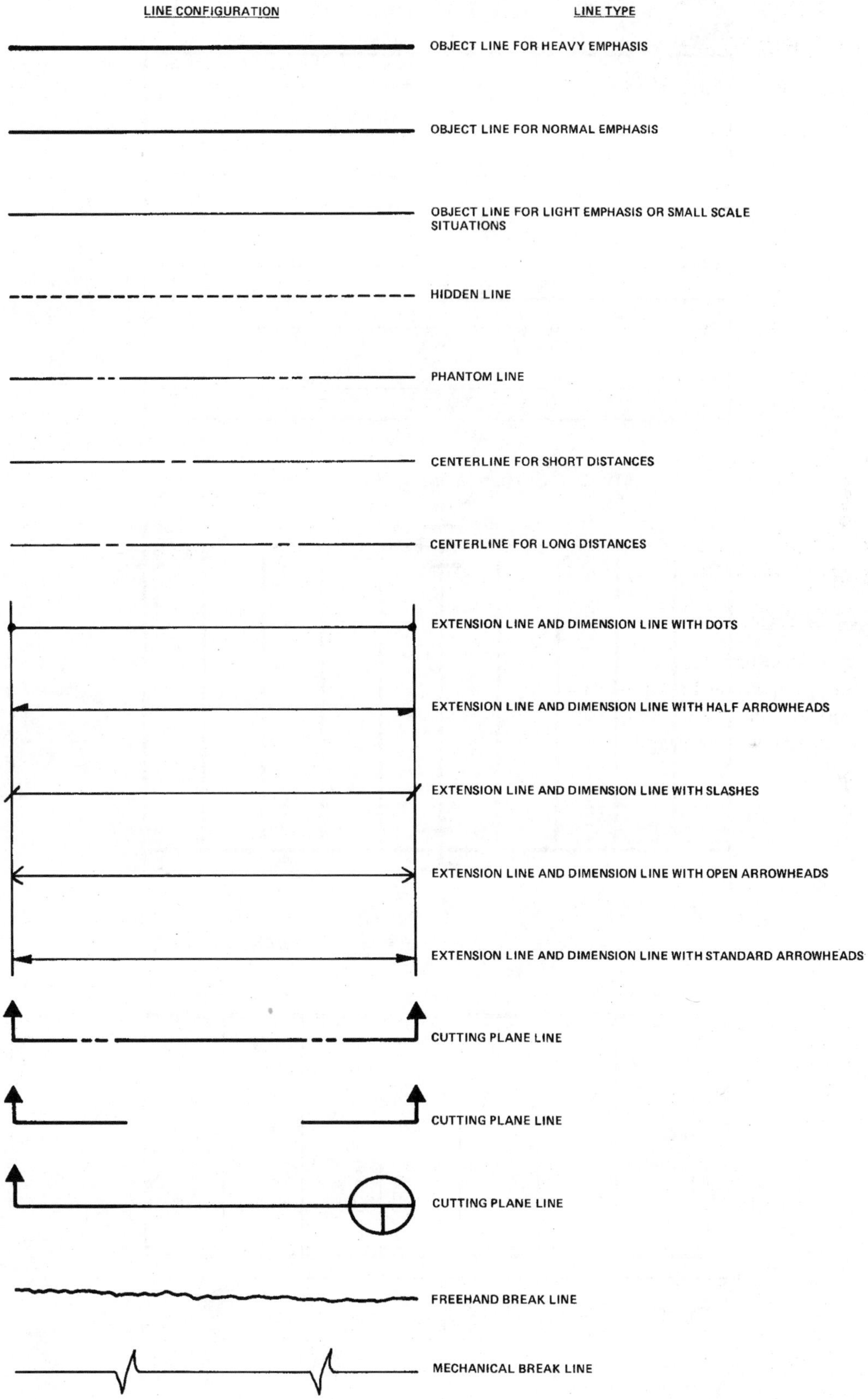

LINE CONFIGURATION   LINE TYPE

———————————————————— OBJECT LINE FOR HEAVY EMPHASIS

———————————————————— OBJECT LINE FOR NORMAL EMPHASIS

———————————————————— OBJECT LINE FOR LIGHT EMPHASIS OR SMALL SCALE
SITUATIONS

– – – – – – – – – – – – – – – HIDDEN LINE

——— – – ——— – – ——— PHANTOM LINE

——— – ——— – ——— CENTERLINE FOR SHORT DISTANCES

——— – – ——— – – ——— CENTERLINE FOR LONG DISTANCES

EXTENSION LINE AND DIMENSION LINE WITH DOTS

EXTENSION LINE AND DIMENSION LINE WITH HALF ARROWHEADS

EXTENSION LINE AND DIMENSION LINE WITH SLASHES

EXTENSION LINE AND DIMENSION LINE WITH OPEN ARROWHEADS

EXTENSION LINE AND DIMENSION LINE WITH STANDARD ARROWHEADS

CUTTING PLANE LINE

CUTTING PLANE LINE

CUTTING PLANE LINE

FREEHAND BREAK LINE

MECHANICAL BREAK LINE

FIGURE 1.11 ■ Commonly used line types in drafting/CAD.

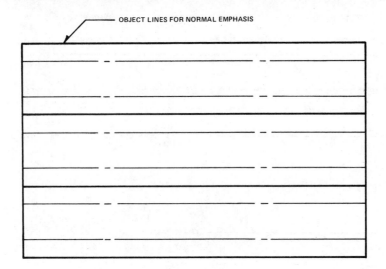

OBJECT LINES FOR NORMAL EMPHASIS

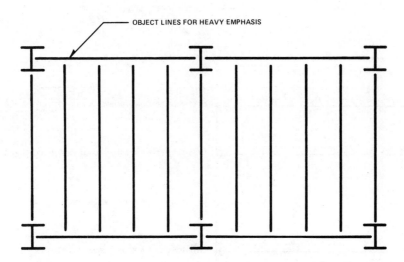

OBJECT LINES FOR HEAVY EMPHASIS

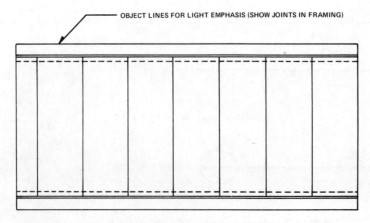

OBJECT LINES FOR LIGHT EMPHASIS (SHOW JOINTS IN FRAMING)

FIGURE 1.12 ■ Object line samples.

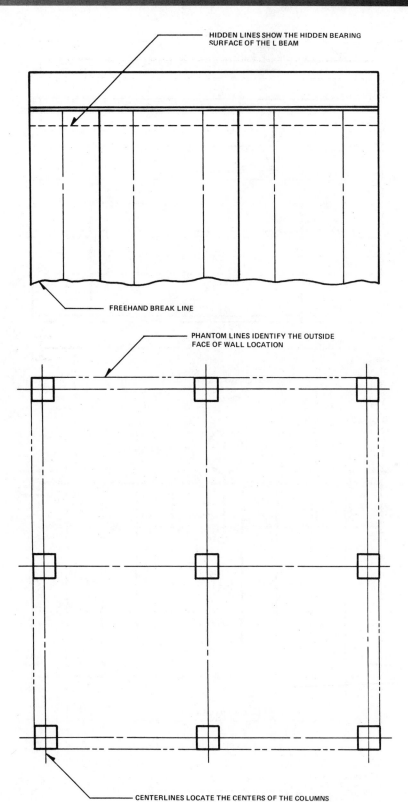

HIDDEN LINES SHOW THE HIDDEN BEARING SURFACE OF THE L BEAM

FREEHAND BREAK LINE

PHANTOM LINES IDENTIFY THE OUTSIDE FACE OF WALL LOCATION

CENTERLINES LOCATE THE CENTERS OF THE COLUMNS

FIGURE 1.13 ■ Hidden, phantom, freehand break, and centerline samples.

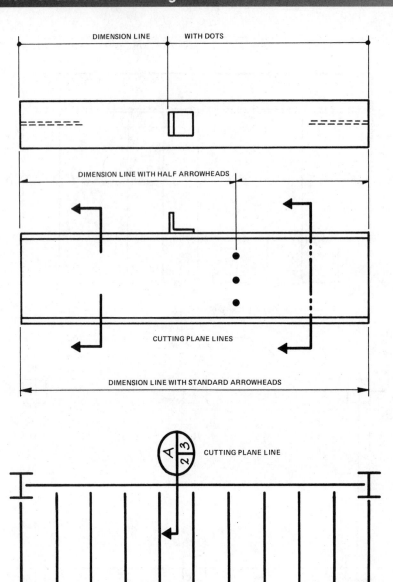

FIGURE 1.14 ■ Dimension and cutting plane lines.

STRUCTURAL DRAFTING LETTERING SAMPLE

STRUCTURAL DRAFTING LETTERING SAMPLE

STRUCTURAL DRAFTING LETTERING SAMPLE

STRUCTURAL DRAFTING LETTERING SAMPLE

Structural Drafting Lettering Sample

Structural Drafting Lettering Sample

Structural Drafting Lettering Sample

Structural Drafting Lettering Sample

A A A A A A a a a a
B B B B B B b b b b
C C C C C C c c c c
D D D D D D d d d d
E E E E E E e e e e
F F F F F F f f f f
G G G G G G g g g g
H H H H H H h h h h
J J J J J J j j j j
K K K K K K k k k k
M M M M M M m m m m
N N N N N N n n n n
P P P P P P p p p p
Q Q Q Q Q Q q q q q
R R R R R R r r r r
S S S S S S s s s s
T T T T T T t t t t
U U U U U U u u u u
W W W W W W w w w w
Y Y Y Y Y Y y y y y
1 2 3 4 5 6 7 8 9 0
1 2 3 4 5 6 7 8 9 0

FIGURE 1.15 ■ Examples of lettering styles that might be used in structural engineering settings.

provide even more options. Some companies standardize the lettering style to be used for all drawings, while others allow CAD technicians to select from among the various options available in their systems. The student is encouraged to experiment with various lettering style options available in his CAD system unless the instructor assigns a given style.

Review of Manual Scale Use

Structural, civil, and pipe drafting involve drawing the plans for very large structures or plats on relatively small sheets of paper. Historically, this task was accomplished using scales. The advent of CAD changed how the task of scaling drawings is actually accomplished. However, in order to fully understand how to scale drawings in CAD, it is necessary—or at least helpful—to understand the old-fashioned manual way presented in this section. The drawings presented in each unit indicate the appropriate manual scales for each respective type of drawing. These scales and their CAD equivalents should be used in all drawing assignments. To make a large building or piece of land fit on a small sheet of paper, the engineer or technician uses a scale. The two scales commonly used are the architect's scale and the engineer's scale.

The *architect's scale* is divided into ten separate scales that set off multiples of an inch and fractions of an inch equal to one foot. An architect's scale contains various-sized scales ranging from the smallest 3/32″ = 1′-0″ to 3″ = 1′-0″ (Figure 1.16). Two additional scales may be obtained from the architect's scale, full scale and 1/16″ = 1′-0″. The full scale is a normal ruler-type scale and is not often used. The scale, 1/16″ = 1′0″, may be obtained by using the full scale and setting every 1/16″ off as being 1 foot. This would be done in the event that an extreme-

3/32″ = 1′-0″	3/16″ = 1′-0″
1/8″ = 1′-0″	1/4″ = 1′-0″
3/8″ = 1′-0″	3/4″ = 1′-0″
1/2″ = 1′-0″	1″ = 1′-0″
1 1/2″ = 1′-0″	3″ = 1′-0″

FULL SIZE

FIGURE 1.16 ■ Architect's scale.

ly large structure was required to fit on one small sheet of paper. The architect's scale is read by placing it along the line to be measured, aligning the nearest full foot mark with the end of the line, and reading any leftover inches and fractions on the additional foot that is provided just beyond the zero point on each scale (Figure 1.17).

The *engineer's scale* is used when a site plan, plot plan, or property boundaries must be drawn. It is a decimal scale that ranges from 1″ = 10′ to 1″ = 60′ (Figure 1.18). The scales listed in Figure 1.18 may also be increased by tens since the engineer's scale is a decimal scale. For example, the scale 1″ = 10′ may also be used as 1″ = 100′, 1000′, and so on. The engineer's scale is read by placing it along the line that is to be measured, reading the number of full feet off of the scale, and estimating the decimals of a foot when the end of the line falls between full foot marks (Figure 1.19).

Review of CAD Scales. When creating a drawing manually, the scale is selected before beginning the drawing. The image on the piece of paper is drawn to the scale chosen. This is not

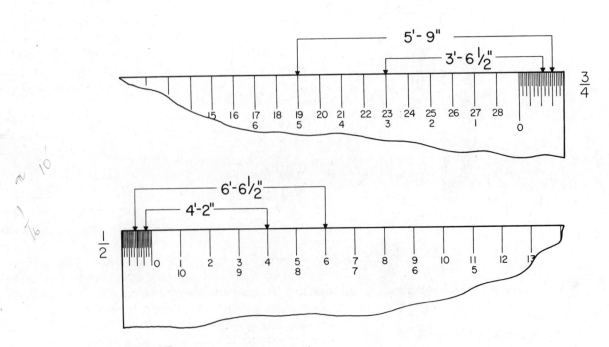

FIGURE 1.17 ■ Reading the architect's scale.

1″ = 10′	. . . .	1″ = 20′	
1″ = 30′	. . . .	1″ = 40′	
1″ = 50′	. . . .	1″ = 60′	
1″ = 100′	. . . .	1″ = 200′	
1″ = 300′	. . . .	1″ = 400′	
1″ = 500′	. . . .	1″ = 600′	

FIGURE 1.18 ■ Engineer's scale.

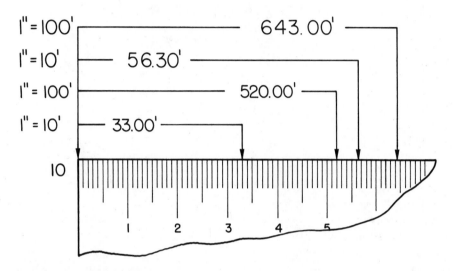

FIGURE 1.19 ■ Reading the engineer's scale.

the case with CAD. When creating a drawing on a CAD system, the image is theoretically created life size. In other words, if drawing a lot line that is actually 50 feet long on the ground, the computer theoretically draws it 50 feet long. Scaling takes place during the printing/plotting of the drawing. If the drawing is to be plotted or printed at a scale of 1 inch equals 100 feet, that instruction is sent to the plotter or printer, which then applies the desired scale to the drawing.

To review, the relationship between manual scales and CAD is one of proportion. There is a set reference scale in CAD for each manual drawing scale. That reference scale may be determined by cross multiplying the manual scale times 12 (number of inches in a foot). For example, the reference scale for the manual scale of 1/8″ = 1′-0″ is 96, as follows:

$$\frac{1}{8} \times \frac{12}{1} = \frac{96}{1} = 96$$

Frequently used manual scales and their corresponding reference scales are:

$$1/8'' = 96$$
$$1/4'' = 48$$
$$1/2'' = 24$$
$$1'' = 12$$

Paper Sizes

Most structural engineering firms use paper that has been pre-cut to standard sizes. Sizes B, C, and D are the most commonly used sizes of paper in structural drafting. B-size paper is 12″ wide × 18″ long, C-size paper is 18″ wide × 24″ long, and D-size paper is 24″ wide × 36″ long. These figures include the area outside of the border.

Figure 1.20 contains examples of B-size and C-size paper with 1/2″ borders. These formats should be used along with the title block shown in Figure 1.21 for all drafting activities requiring borders and a title block.

Title Blocks and Borders

Many structural engineering firms purchase precut paper with proper borders and a company title block printed on the paper. Others build templates in their CAD system and print the title block and borders. In the structural drafting classroom, the student may be required to apply his own borders and title blocks. The border lines should be drawn to approximate the width of an object line for either normal or heavy emphasis. Figure 1.21 contains a sample of a title block that may be used by students on B, C, or D paper.

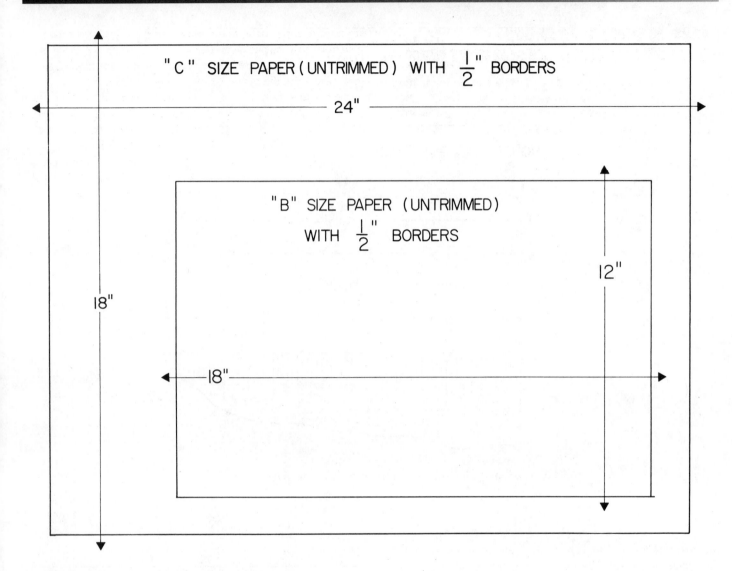

FIGURE 1.20 ■ Sample paper sizes with border formats.

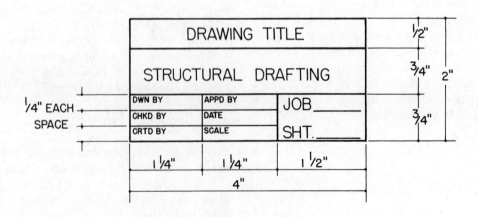

FIGURE 1.21 ■ Sample title block.

CAD SYSTEMS IN STRUCTURAL DRAFTING

Computer-aided drafting, or CAD, is now the norm in the modern structural engineering department. It has simplified the drawing, checking, correcting, and revising processes (Unit 3) significantly. Figure 1.22 is an example of a typical CAD workstation that might be found in a contemporary structural engineering department.

This workstation has all of the components and devices one would expect to find in a modern CAD system. These devices are used to input data, manipulate data, and interact with the system's software package. The devices include a CPU, CRT, keyborad, scanner, and plotter. Some systems use flatbed plotters, others use the drum variety as well as laser printers.

Computer-aided drafting has had the same impact on the quality of structural engineering drawings that it has had in other drafting disciplines (i.e., architectural, mechanical, civil). Generally speaking, drawing quality has improved. Figure 1.23, Figure 1.24, and Figure 1.25 are examples of shop drawings that were developed on a modern CAD system. Notice the consistent quality of the linework and lettering.

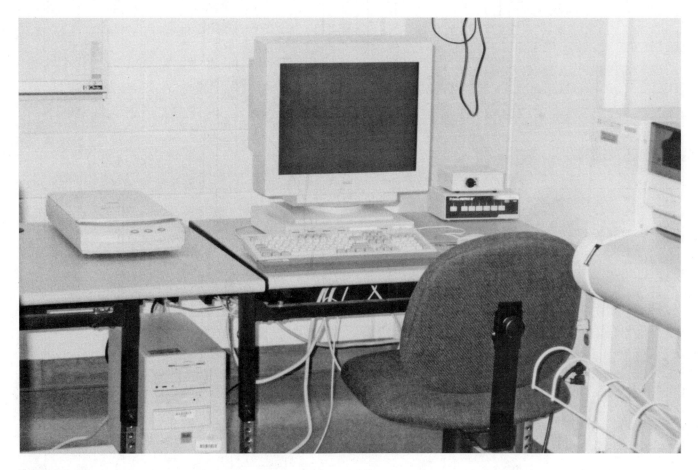

FIGURE 1.22 ■ An example of a CAD workstation. (Photo by Author)

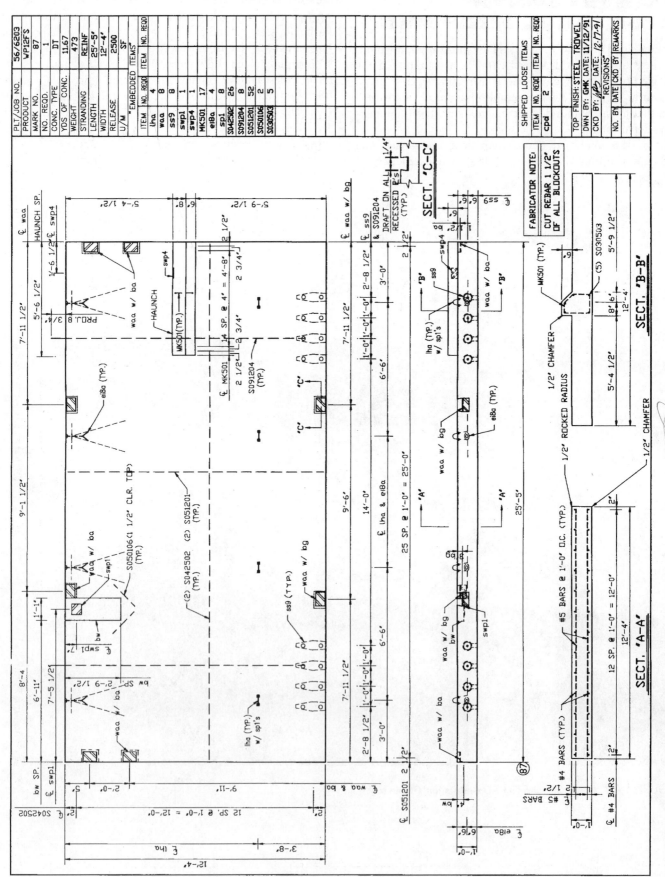

FIGURE 1.23 ■ Figure 1.23 and Figure 1.24 show how CAD can be used to produce intricate shop drawings with consistent line weight and lettering style.

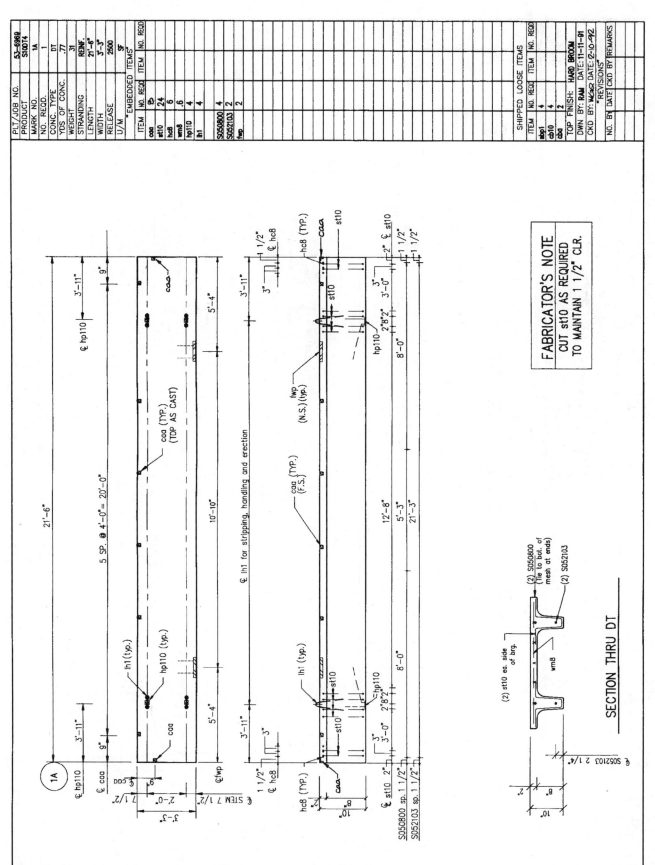

FIGURE 1.24 ■ Figure 1.23 and Figure 1.24 show how CAD can be used to produce intricate shop drawings with consistent line weight and lettering style.

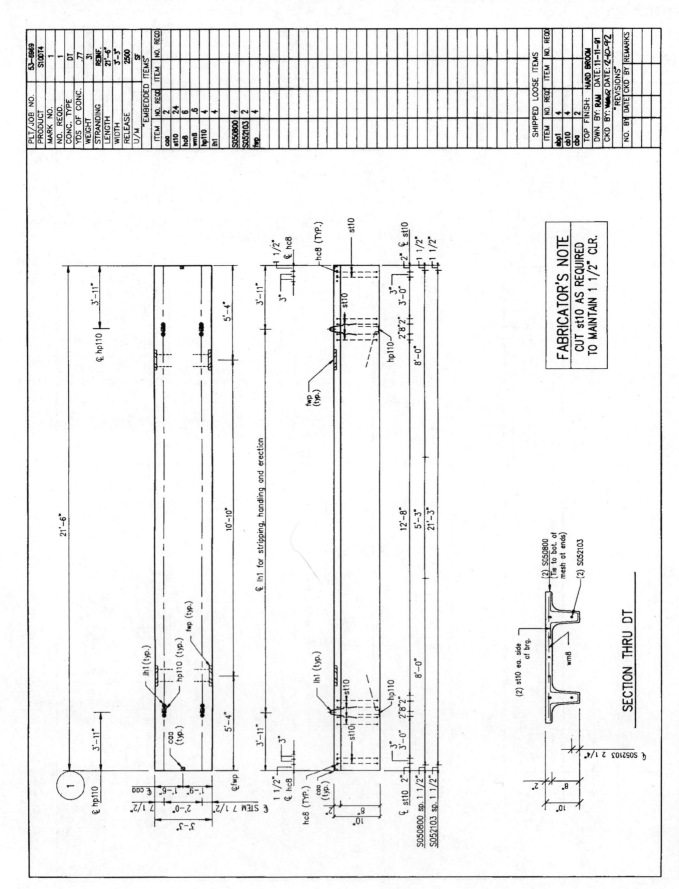

FIGURE 1.25 ■ Shop drawing created on a CAD system.

SUMMARY

- The need for structural drafting is based on the requirement to have drawings in order to properly design, manufacture, and erect structures.

- There are two types of structural drawings: engineering drawings and shop drawings.

- Engineering drawings and shop drawings may be combined into one set to form structural working drawings.

- Engineering drawings are prepared by structural CAD technicians employed by architects, engineers, contractors, and sales personnel and contain a minimum of detail.

- Shop drawings are prepared by structural drafters employed by companies that actually manufacture steel and concrete products.

- Structural CAD technicians are most commonly employed by consulting engineering firms and manufacturers of structural products.

- Linework in structural drafting may vary in width according to the needs of the job but must be sharp, clear, and rapidly done.

- Line types most commonly used in structural drafting are object, hidden, phantom, center, dimension, and cutting plane lines.

- Structural CAD technicians must be able to use the architect's and the engineer's scales as well as CAD scaling techniques.

- Structural CAD technicians use B, C, and D sizes of paper.

- Title blocks for drawings may be prepared according to specifications set forth by the instructor, a structural drafting firm, or samples in this textbook.

- The advent of CAD has simplified the drawing, checking, correcting, and revising processes in structural drafting. CAD has improved drawing quality in structural drafting.

REVIEW QUESTIONS

1. Why are structural drawings needed in the heavy construction industry?

2. Name two types of structural drawings.

3. What do engineering drawings and shop drawings form when combined into one set?

4. Name four different situations in which a structural CAD technician would be called upon to prepare engineering drawings.

5. Shop drawings are prepared by what types of companies?

6. Name the two most common employers of structural CAD technicians.

7. Sketch an example of an object line, a hidden line, and a centerline.

8. List the three most commonly used sizes of paper in structural drafting.

9. Explain the impact CAD has had on structural drafting.

10. List the input and output devices typically found in a modern CAD system.

Typical Structural CAD Department

OBJECTIVES

Upon completion of this unit, the student will be able to:

- Sketch an organizational chart for a typical structural CAD department.
- Write a job description for an entry-level structural CAD technician.
- List the primary duties of a junior CAD technician, a CAD technician, a senior CAD technician, a checker, and a drafting manager in a typical structural CAD department.

CAD DEPARTMENT ORGANIZATION

One of the main complaints of students entering their first structural CAD position is that they do not know what to expect on the job. Most have the technical skills required. However, many do not know exactly how or where they fit in. The structural CAD student needs to understand the structure of a typical CAD department. The inconsistency on the part of employers in position descriptions and position titles has further complicated the problem. However, enough consistency exists that similarities can be drawn and generalities stated.

The typical CAD department in a company involved in structural engineering includes a drafting clerk, junior CAD technicians, CAD technicians, senior CAD technicians, checkers, and a drafting manager. Although the position titles may vary from company to company, the positions themselves as well as the responsibilities and tasks required remain constant (Figure 2.1).

Drafting Clerk

Persons employed as *drafting clerks* are sometimes also referred to as CAD/engineer secretaries. The following are the primary responsibilities of persons in this position:

- Running printers and plotters (see Figure 2.2 for typical copy request)
- Typing internal and external correspondence
- Filing

- Running errands
- Maintaining the supply room

The structural CAD student, having completed the instruction in this textbook, usually starts at a higher level than drafting clerk. However, this position should not be overlooked as a possible stepping stone to a career in CAD. Many successful structural CAD technicians have started in the field as a drafting clerk.

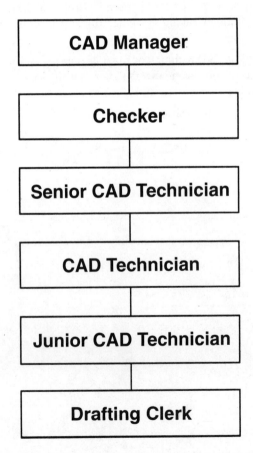

FIGURE 2.1 ■ Organizational chart—typical structural drafting department.

```
┌─────────────────────────────────────────────────────────────┐
│                                                               │
│               REQUEST FOR COPIES OF DRAWINGS                  │
│                                                               │
│   JOB NUMBER _____  DATE _____     │
│                                                               │
│   SHEET NUMBER _____      │
│                                                               │
│   NUMBER OF PRINTS REQUIRED _____      │
│                                                               │
│                      STAMPS REQUIRED                          │
│                                                               │
│   _____  DATE ONLY                                       │
│                                                               │
│   _____  FOR YOUR APPROVAL                               │
│                                                               │
│   _____  RESUBMITTED                                     │
│                                                               │
│   _____  REVISED                                         │
│                                                               │
│   _____  APPROVED FOR PRODUCTION/FABRICATION             │
│                                                               │
│   _____  FOR FIELD USE                                   │
│                                                               │
│   _____  PRELIMINARY                                     │
│                                                               │
│   _____  INCOMPLETE                                      │
│                                                               │
│   REMARKS                                                     │
│                                                               │
│                                                               │
└─────────────────────────────────────────────────────────────┘
```

FIGURE 2.2 ■ A work order for the drafting clerk.

Junior CAD Technician

Persons employed as *junior CAD technicians* are sometimes referred to as detailers, apprentices, trainees, or beginning CAD technicians. The primary responsibilities of persons in this position are as follows:

■ Monitoring plotters when the drafting clerk is overloaded (see Figure 2.3)

■ Preparing elementary engineering and shop drawings

■ Drawing details

■ Performing revisions and corrections

■ Preparing bills of materials

■ Maintaining disk files for CAD systems

Most structural CAD students, having completed a course of study in structural drafting but lacking work experience in the field, begin their careers as junior CAD technicians.

CAD Technician

There is very little variation in position titles for persons employed in structural drafting at this level. The term *CAD technician* and the previously used term *drafting technician* are

FIGURE 2.3 ■ Plotters should be monitored. *(Photo by Author)*

FIGURE 2.4 ■ A senior drafter at work at his CAD system. *(Photo Courtesy of Hewlett-Packard Company)*

almost universal. The primary responsibilities of persons employed as CAD technicians are the following:

■ Preparing engineering and shop drawings in accordance with architectural plans, contractor's sketches, engineer's sketches, sales personnel sketches, or verbal instructions from any of these sources

■ Assisting junior CAD technicians assigned to cooperative projects and teams

■ Adhering to projected timetables and work schedules

Senior CAD Technician

Senior CAD technicians are sometimes referred to as designers or CAD team leaders. The primary responsibilities of persons in this position include the following:

■ Preparing more complicated engineering and shop drawings in accordance with standard drafting procedures and the raw data available for a job

■ Supervising CAD technicians and junior CAD technicians assigned to a job or CAD team (Figure 2.4)

■ Ensuring that projected timetables and work schedules are met

■ Performing minor checking duties

■ Acting as liaison between the checker and project engineer assigned to a job (Figure 2.5)

Checker

Checkers are sometimes referred to as senior designers and are persons of much experience and expertise in structural CAD. Responsibilities of persons in this position include checking:

■ Engineering and shop drawings for dimensional accuracy

■ For adherence to company drafting procedures

■ For adherence to information presented in the raw data for a job

FIGURE 2.5 ■ Project engineers provide the structural design data for drafters. (*Deborah M. Goetsch*)

FIGURE 2.6 ■ New jobs are reviewed thoroughly before work is assigned to drafters. (*Photo Courtesy of Hewlett-Packard Company*)

■ General CAD techniques

■ Bills of material

In addition to these, the checker is sometimes required to help drafters with less experience to interpret engineering calculations and design sketches.

CAD Manager

The *CAD manager* is sometimes referred to as the chief drafter. Persons serving in this position are the administrators or managers of CAD departments. The following are the primary responsibilities of the CAD manager.

■ Supervising all CAD department personnel

■ Scheduling and assigning work, ensuring that all functions of the CAD department are carried out properly and on time

■ Reviewing new projects and estimating the amount of time that will be required to complete drawings (Figure 2.6)

■ Requesting supplies for the CAD department

■ Conducting interviews of prospective department employees

SUMMARY

■ The structural CAD student needs to understand the organizational structure of a typical CAD department.

■ Position titles vary somewhat from company to company, but most companies have positions that equate to drafting clerk, junior CAD technician, CAD technician, senior CAD technician, checker, and drafting manager.

■ The primary responsibilities of the drafting clerk are making copies of drawings, preparing and distributing correspondence, filing, and running errands.

■ The primary responsibilities of the junior CAD technician are performing corrections and revisions, preparing elementary drawings, and assisting the drafting clerk when the work load is heavy.

■ The primary responsibility of the CAD technician is preparing engineering and shop drawings from architectural, engineering, contractor, or sales sketches.

■ The primary responsibilities of the senior CAD technician are preparing complicated engineering and shop drawings, coordinating the efforts of the CAD team, and assisting less-experienced CAD technicians in meeting timetables and work schedules.

■ The primary responsibilities of the checker are checking engineering and shop drawings and helping CAD technicians with less experience to interpret design calculations.

■ The primary responsibilities of the CAD manager are maintaining overall supervision and operation of the CAD department, examining new jobs to determine how much time and how many people will be required to complete drawings, and interviewing prospective employees.

REVIEW QUESTIONS

1. What positions are usually found in a typical structural CAD department?

2. List the primary responsibilities of the drafting clerk.

3. List the primary responsibilities of the junior CAD technician.

4. List the primary responsibilities of the CAD technician.

5. List the primary responsibilities of the senior CAD technician.

6. List the primary responsibilities of the checker.

7. List the primary responsibilities of the CAD manager.

CAD ACTIVITIES

GENERAL INSTRUCTIONS

The following activities may be completed on any CAD system. Before reading the *specific instructions* for each activity (below), go through each step in the following planning checklist. The checklist applies to any CAD system and will help ensure the optimum use of your time and resources.

1. Analyze the problem carefully. Decide exactly what you are being asked to do.

2. Determine what resources and references you will need in order to complete the problem and collect them.

3. Decide if any particular standards apply to the project and have those standards available.

4. Determine what types of views will be required and how many of each.

5. Determine what the final plotted scale of the drawing will need to be, and select the appropriate paper size for plotting/printing (make sure the appropriate paper size is available).

6. Plan your drawing sequence. In what order will you develop the drawing (i.e., lines, features, dimension lines, leaders, dimensions, notes, etc.)?

7. Review the various CAD commands you will have to use in order to develop the drawing.

8. Examine your CAD system to ensure that everything is in working order, then begin the project.

SPECIFIC INSTRUCTIONS

Activity 2.1—Convert the contractor's sketch in Figure 2.7 into a finished drawing. Save your drawing as *Activity 2.1*.

Activity 2.2—Convert the engineer's sketch in Figure 2.8 into a finished drawing. Save your drawing as *Activity 2.2*.

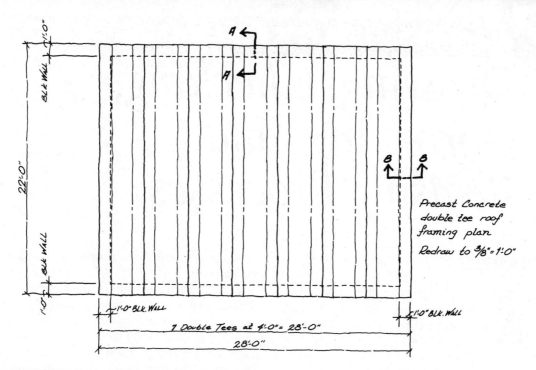

FIGURE 2.7 ■ Contractor's sketch for CAD Activity 2.1.

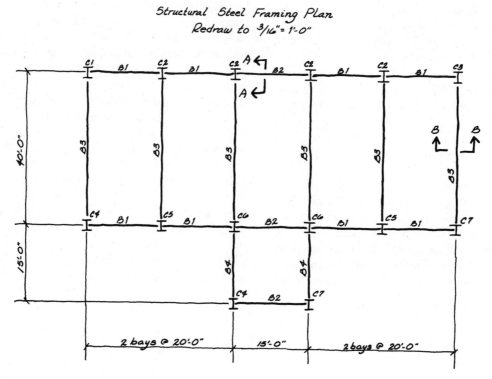

FIGURE 2.8 ■ Engineer's sketch for CAD Activity 2.2.

3 UNIT

Drawing, Checking, Correcting, and Revising Processes

OBJECTIVES

Upon completion of this unit, the student will be able to:

■ Explain the original drawing process, the checking process, the correcting process, and the revising process in structural CAD.

ORIGINAL DRAWING

Structural CAD technicians prepare original engineering and shop drawings based on raw data that may be supplied by a number of different sources. Some of the more common sources are architects, engineers, contractors, and sales personnel. The raw data may consist of a complete set of thoroughly prepared architectural plans with comprehensive specifications. On the other hand, the raw data may consist of as little as a freehand sketch made in the field.

The process that takes place in the CAD drafting department in preparing original drawings is the same. The drafting manager collects, examines, and analyzes all of the data that are available for a job. A CAD team is then selected by the drafting manager to prepare the original drawings file. The team is given a projected timetable for completion.

The duties of each team member vary according to each person's experience and abilities and the job size. In a large job, the bulk of the drawing is assigned to CAD technicians and senior CAD technicians. The senior CAD technicians perform the most complicated tasks. Figure 3.1 shows an architectural drawing of the floor plan for a small bank. Such drawings are commonly provided as the raw data. From this, a structural CAD technician must develop the structural draw-ings that are needed to manufacture products and construct the building.

Figure 3.2 contains a sample prestressed concrete engineering drawing that was prepared based on the architect's drawing in Figure 3.1. It is a roof framing plan of structural prestressed concrete Lin Tee members. Figure 3.3 is an example of a structural engineering drawing involving a combination of poured-in-place concrete and structural steel. The drawing consists of the layout information for concrete columns and beams and steel bar joists. It was prepared based on the architect's drawing in Figure 3.1.

CHECKING

Once an original drawing file has been completed, it must be checked. A checker is assigned to each job that is brought into the CAD department. A checker ensures that the drawings have been properly prepared before they are turned over to the originator of the job for approval. The checker checks all drawings against standard company procedures, accepted practices, and the raw data used in preparing the original drawings.

A common checking practice is to mark through incorrect information and write in the correct information. This is done on a hard copy but may be done electronically. Once a drawing has been completely checked, it is coded *checkprint #1* or *revision #1* and returned to be corrected. This process continues with every sheet in a set of drawings until all mistakes have been identified and corrected. This might involve many checkprints for each sheet in the job or as few as one. When there is more than one checkprint, each successive print is numbered and kept in order and on file until the job is completed and the building has been constructed.

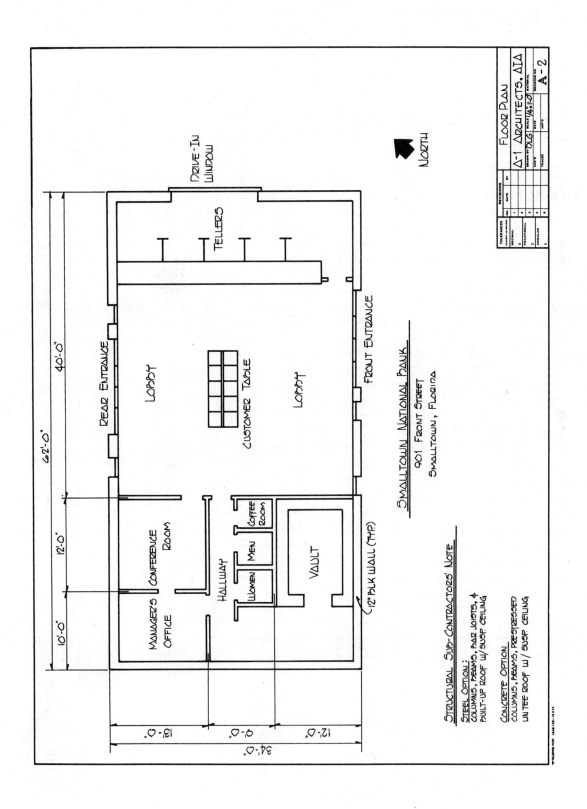

FIGURE 3.1 ■ Architect's floor plan for a small bank.

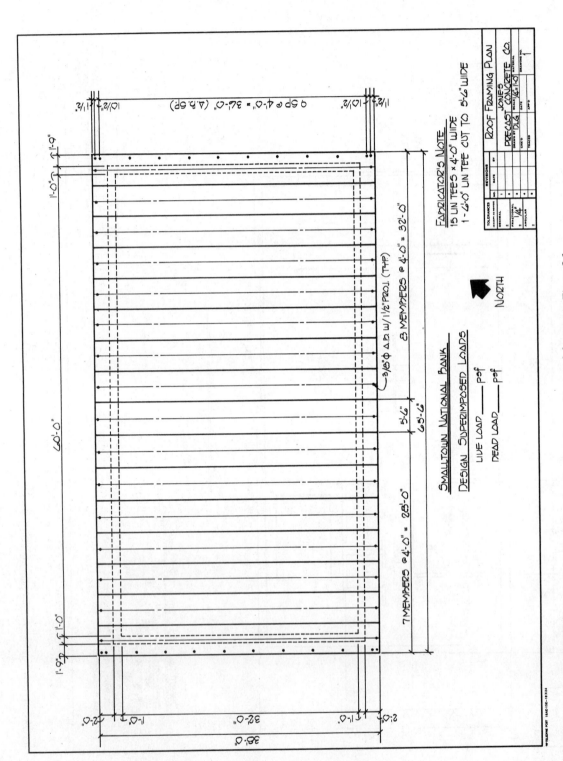

FIGURE 3.2 ■ Prestressed concrete engineering drawing based on architectural drawing in Figure 3.1.

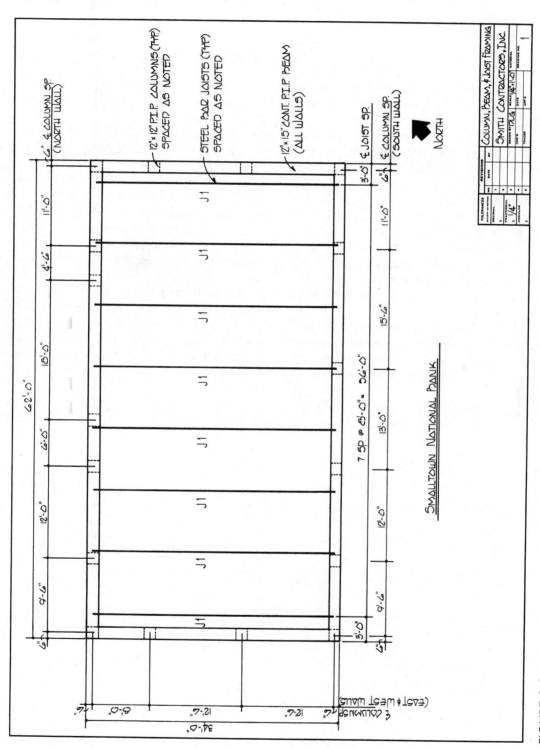

FIGURE 3.3 ■ Structural engineering drawing based on architectural drawing in Figure 3.1.

Figure 3.4 and Figure 3.5 are examples of checkprints that have been marked by a checker and are ready to be corrected. Both samples are part of a complete set of structural steel drawings.

CORRECTING

Correcting drawings is often the job of the junior CAD technicians. By performing corrections, junior CAD technicians become familiar with structural drawings and CAD practices. This familiarity makes it easier for them to develop more advanced skills and achieve CAD technician positions.

Junior CAD technicians receive checkprints and any necessary instructions from the checker before they begin corrections. Once they have been instructed, they call up the appropriate drawings from the files and make the corrections indicated on the checkprints. The junior CAD technician crosses off the corrections as they are made, runs a new copy of the corrected original, and returns the checkprints to the checker. The checker then ensures that all corrections were properly made. When satisfied that a drawing is completely correct, the checker initials the title block on a hard copy. This signifies that it has been checked and corrected and is now ready to be approved.

When all drawings for a job are ready for approval, the drafting manager instructs the drafting clerk to make copies and stamp them *for your approval.* These hard copies are forwarded to the architect, contractor, or engineer who originated the job to be examined and approved. The originator may approve the plans, disapprove the plans, or approve the plans with revisions. Figure 3.6 shows a sample structural drawing that has been checked, corrected, and prepared for approval of the originator. Notice the checker's initials in the title block and the approval stamp.

REVISING

The originator of a job may approve, disapprove, or, as is often the case, approve the structural drawings with revisions. A *revision* is a change in design, configuration, or plan that was made after the drawings were developed. Revisions are caused by the originator of a job, not the company preparing the structural drawings.

Revisions are marked on the set of approval plans by the originator. The plans are then marked *approved with revisions* and returned to the company that prepared the original drawings to be corrected. Note the distinction between revisions and corrections. Corrections are caused by mistakes made in preparing the drawings. Revisions are the result of changes by the originator of a job in her design, plan, or approach. Time spent in performing revisions to the original drawings is usually charged to the person causing the revisions.

Figure 3.7 contains an example of an original drawing that has been revised by a structural CAD technician. Note the revision triangles that identify the number of each revision and the revision notes that explain each revision.

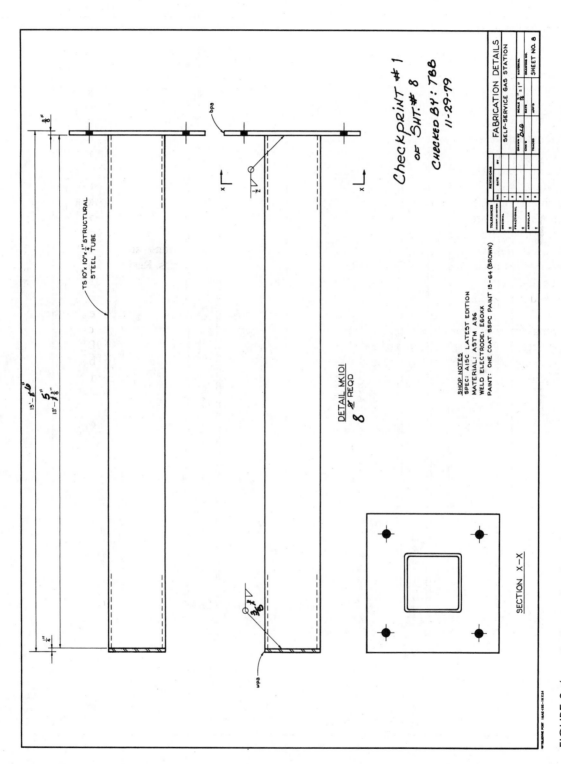

FIGURE 3.4   ■   Checkprint of a column detail.

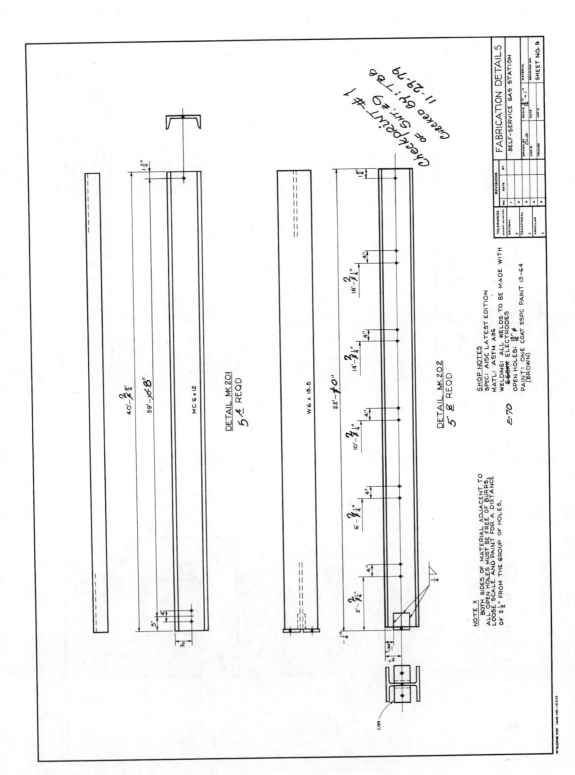

FIGURE 3.5 ■ Checkprint of a beam detail.

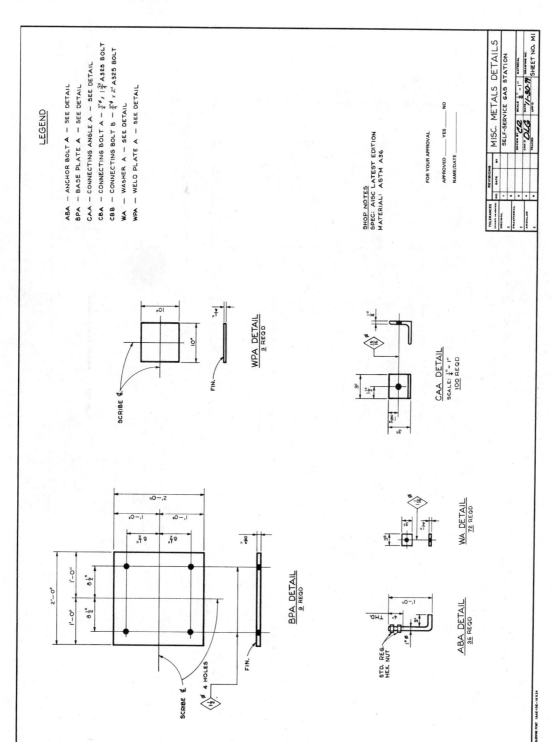

FIGURE 3.6 ■ Structural drawing ready to be submitted for approval.

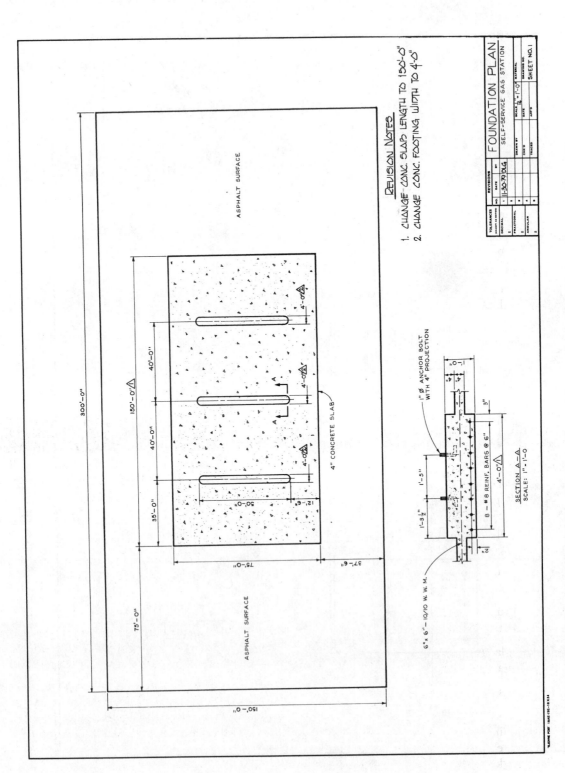

FIGURE 3.7 ■ Revised original drawing for a poured-in-place concrete job.

SUMMARY

■ Raw data for preparing structural drawings come from architects, engineers, contractors, or sales personnel.

■ Structural drawings might be produced based on very complete architectural plans or on very quickly prepared freehand sketches.

■ Every original structural drawing must be checked for accuracy, conformance to raw data specifications, and proper drafting practices.

■ Checkers make their marks on prints of original drawings called checkprints.

■ Corrections are usually performed by junior CAD technicians.

■ Once drawings are corrected, they must be printed and forwarded to the originator for approval.

■ The originator of a structural job must check the drawings and approve them, disapprove them, or approve them with revisions.

■ A revision differs from a correction in that it is caused by a change in plans rather than by a mistake.

REVIEW QUESTIONS

1. Explain where structural CAD technicians get the raw data they use in preparing the structural drawings for a job.

2. Who must approve structural drawings once they are complete?

3. Explain how a correction and a revision differ.

4. What step immediately follows the preparation of the original drawings?
   a. checking
   b. revisions
   c. approval
   d. product fabrication

5. Checkers make their marks on:
   a. original drawings
   b. specifications
   c. checkprints
   d. file copies

6. Corrections to original drawings are often performed by:
   a. CAD technicians
   b. junior CAD technicians
   c. drafting managers
   d. checkers

CAD ACTIVITIES

GENERAL INSTRUCTIONS

The following activities may be completed on any CAD system. Before reading the *specific instructions* for each activity (below), go through each step in the following planning checklist. The checklist applies to any CAD system and will help ensure the optimum use of your time and resources.

1. Analyze the problem carefully. Decide exactly what you are being asked to do.

2. Determine what resources and references you will need in order to complete the problem and collect them.

3. Decide if any particular standards apply to the project and have those standards available.

4. Determine what types of views will be required and how many of each.

5. Determine what the final plotted scale of the drawing will need to be, and select the appropriate paper size for plotting/printing (make sure the appropriate paper size is available).

6. Plan your drawing sequence. In what order will you develop the drawing (i.e., lines, features, dimension lines, leaders, dimensions, notes, etc.)?

7. Review the various CAD commands you will have to use in order to develop the drawing.

8. Examine your CAD system to ensure that everything is in working order, then begin the project.

SPECIFIC INSTRUCTIONS

Activity 3.1—Begin to understand the original drawing and correcting process by redrawing and correcting the marked-up drawing in Figure 3.8. Save your drawing as *Activity 3.1*.

Activity 3.2—Begin to understand the revision process by redrawing the revised drawing in Figure 3.9. Identify each revision with a number inside a revision triangle and explain each with a set of revision notes. Save your drawing as *Activity 3.2*.

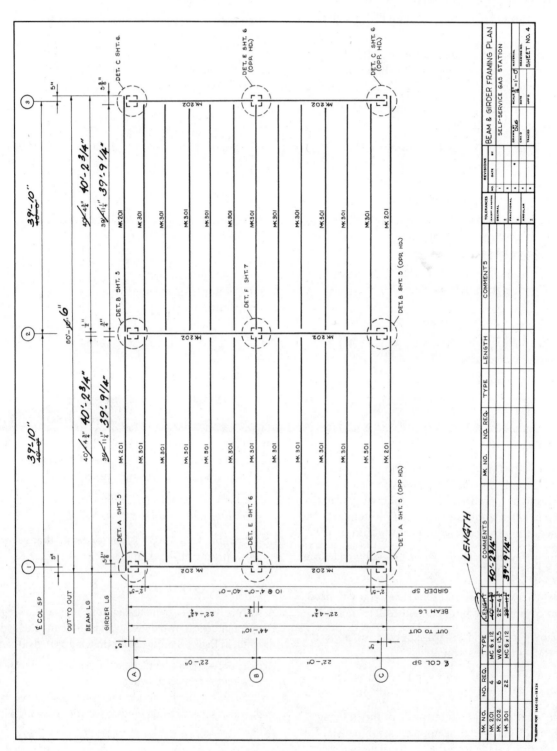

FIGURE 3.8 ■ Drafting problem for CAD Activity 3.1.

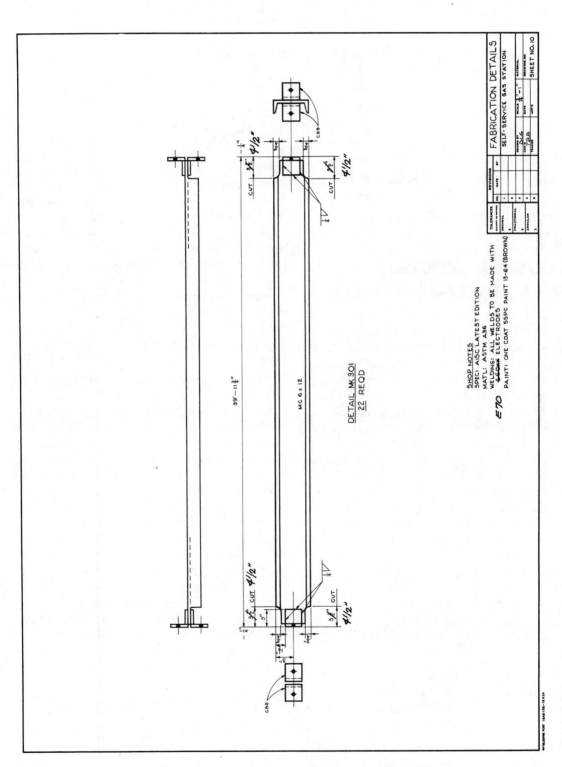

FIGURE 3.9 ■ Drafting problem for CAD Activity 3.2.

4 UNIT

Product Fabrication and Shipping

OBJECTIVES

Upon completion of this unit, the student will be able to:

■ Explain the product fabrication processes for structural steel, precast concrete, and poured-in-place concrete.

■ Explain how structural steel and precast concrete products are shipped to the jobsite.

STRUCTURAL STEEL PRODUCT FABRICATION

Structural steel is produced in standard shapes in rolling mills (Figure 4.1, Figure 4.2, and Figure 4.3). From there, it is shipped to steel construction companies to be altered and used according to the needs of each individual job the company contracts. Dimensions for detailing and properties for designing steel components of a structure are listed in the *Manual of Steel Construction,* which is put out by the American Institute of Steel Construction (AISC). It is the primary reference source for structural steel CAD technicians and engineers.

Excerpts from the *Manual of Steel Construction* have been included in the appendix of this textbook. It should be referred to by students when determining the dimensions of steel shapes contained in drawings. For example, if the width and depth of a W14 × 426 steel beam were required, the information could be found by using the appendix in this text or the section in the *Manual of Steel Construction* labeled *W Shapes—Dimensions for Detailing.* The notation W14 × 426 appears under the heading *Designation.* The *Depth Column* lists the desired depth as 18 5/8″, and the *Width Column* lists the desired width as 16 3/4″.

A structural steel shape must go through a number of steps during fabrication. The following are the most common steps:

■ Handling and cutting
■ Punching and drilling
■ Straightening
■ Bending
■ Bolting
■ Riveting
■ Welding
■ Finishing

Not every steel product undergoes all of these processes. Only those processes needed to prepare the product for delivery to and erection at the jobsite are used.

Handling and Cutting

Structural steel products are very heavy. Therefore, they require special methods for handling, cutting, and transporting. Special handling is accomplished by forklifts, cranes, overhead hoists, or straddle carriers (Figure 4.4).

Cutting steel products can be done in one of several ways depending on the product shape and size. Thin, flat shapes can be cut on a shear. Medium-sized members such as beams and columns are often cut with a special hot cutting saw (Figure 4.5), while thicker structural shapes are usually cut with a flame cutting torch.

Punching and Drilling

Structural steel shapes are often connected at the jobsite by bolting or less frequently by riveting. The holes required in a structural member may be drilled or punched. Punching is confined to the thinner structural members. When punching is not possible, the desired holes may be made by drilling.

Single holes are punched in structural members by a machine called a *detail punch.* Several holes may be punched in a piece of steel at the same time by a machine called a *multiple punch.* Punch machines are versatile and fast. However, they do occasionally bend the structural member being punched. The thicker the material, the more prone the punch is to disfigure it. This makes drilling the more practical process for putting holes in thick products.

Several different types of drills are used in structural steel fabrication shops. Some of these are the drill press, the radial arm drill, multiple spindle drills, drills on jibs, and gantry drills. In modern steel fabrication shops, much of the drilling is automated. This makes the drilling fast and accurate and thereby cuts down on waste due to human error.

Straightening and Bending

Steel members that have been punched or mishandled often become bent and must be straightened before they can be used. The most common machine used for straightening structural steel shapes is the *bend press* or *gag press.* Structural steel shapes

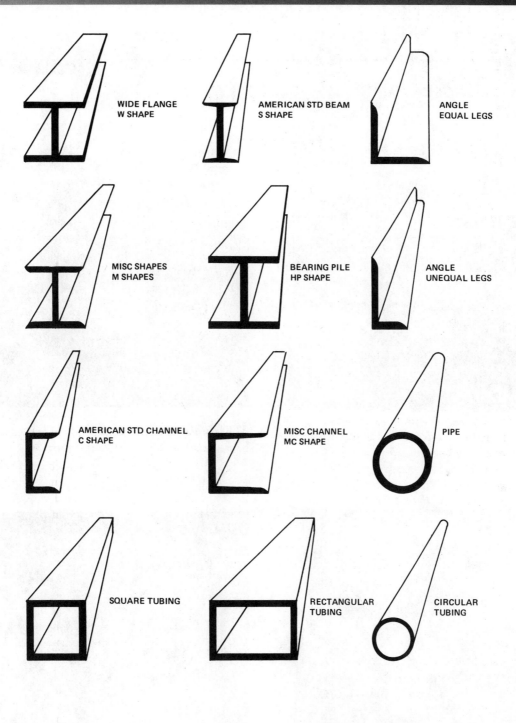

FIGURE 4.1 ■ Standard structural steel shapes.

FIGURE 4.2 ■ Steel ingots are reduced to standard sizes and shapes. *(Bethlehem Steel Corporation)*

FIGURE 4.4 ■ Heavy steel members are stacked with an overhead hoist. *(Bethlehem Steel Corporation)*

FIGURE 4.3 ■ White-hot ingots are moved through the rolling mill on conveyor lines. *(Bethlehem Steel Corporation)*

FIGURE 4.5 ■ Steel shapes are cut to desired lengths by hot sawing. *(Bethlehem Steel Corporation)*

are straightened by resting one edge of the member against a rigid retainer and then applying pressure with a high-strength plunger apparatus. This machine is also used for bending members when long radius curves are required.

Bolting, Riveting, and Welding

Three basic methods are used in fastening structural steel shapes together to form a completed structure: bolting, riveting, and welding (Unit 5). Bolting and welding are frequently used alone or in combination, but riveting is no longer considered a major connecting process.

Bolting of structural steel shapes is a common fastening method, especially for connections that are made in the field. Bolts may be applied by hand or, as is more often the case, with power wrenches. Most structural steel applications in heavy construction require special high-strength bolts that are tightened by a compressed-air device known as an *impact wrench*.

Welding is a common fastening method that is particularly useful in making shop connections and permanent field connections. Structural steel shapes that are to be welded are carefully marked off or layed out to ensure that all welds are accurately applied. Welding processes may be performed by hand or they may be automated. Most modern shops have cut down on the time and waste from human error by switching from hand welding to automated welding.

Until the year 1950, riveting was the primary method for connecting structural members. However, due to improvements in welding processes and the arrival of reliable, high-strength bolting processes, riveting is used less and less. It is no longer considered a major connecting process; but, since many riveted structures are still in use, the drafting student should be aware of this process. All three of the preceding fastening processes are discussed in depth in Unit 5.

Finishing

The final process that structural steel members must undergo during fabrication is finishing. The finishing process involves smoothing rough edges or surfaces to ensure the proper flatness for bearing purposes or fit. Ends of beams that have been cut, edges and tops of bearing plates, and ends of columns that will rest on a baseplate are commonly finished. Finishing is done by any one of the following methods: sawing, milling, filing, high-pressure blasting, or various other means.

PRECAST CONCRETE FABRICATION

Like structural steel, **precast concrete products** are manufactured in standard shapes that may be altered to suit the needs of the individual job (Figure 4.6 through Figure 4.13). Precast concrete can be broken into two separate categories: prestressed products and reinforced products.

Prestressed concrete products are poured in steel beds or forms through which high-strength steel strands have been passed. Before the concrete is poured into the bed, the strands

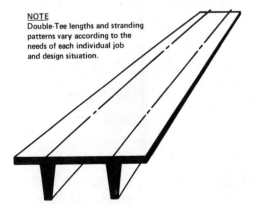

Double Tees

NOTE
Double-Tee lengths and stranding patterns vary according to the needs of each individual job and design situation.

Typical Uses

Roof and floor systems for commercial and industrial buildings; Wall panels for commercial and industrial buildings; Pier decks, tank covers, catwalks, tunnel covers, and conveyor trestle decks

Standard Sizes

10″ deep by 4′ wide
14″ deep by 4′ wide
16″ deep by 4′ wide
24″ deep by 8′ wide

FIGURE 4.6 ■ Standard precast concrete double-tee members.

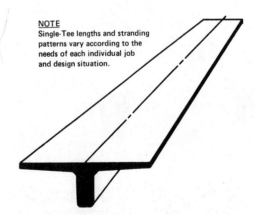

Single Tees

NOTE
Single-Tee lengths and stranding patterns vary according to the needs of each individual job and design situation.

Typical Uses

Roof and floor systems for commercial and industrial buildings; Wall panels for commercial and industrial buildings; Bridges, conveyor trestle decks, recreational facilities

Standard Sizes

Depths: From 12″ deep to 36″ deep
Widths: Widths of 6′ and 8′ are standard. Other widths can be manufactured to meet job and design needs.

FIGURE 4.7 ■ Standard precast concrete single-tee members.

Flat Slabs

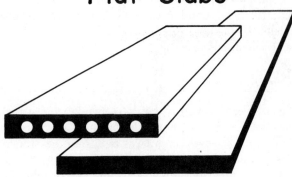

Typical Uses

Roof or floor systems where a flat ceiling or minimum structural depth is particularly desired. Also used for wall panels, tank covers, and tunnel covers.

NOTE
Flat-slab lengths and stranding patterns vary according to the needs of each individual job and design situation.

Standard Sizes

Solid slabs: 3″ to 6″ deep and widths to 10′-0″
Cored slabs: 6″, 8″, and 10″ by 10′-0″ wide

FIGURE 4.8 ■ Standard precast concrete flat-slab members.

Building Beams

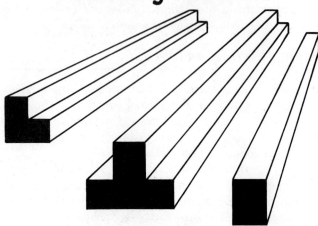

Typical Uses

Primary structural beams for all types of roof and floor systems and other structural framing systems

Standard Sizes

Sizes and shapes of all beams can be adjusted to meet the design requirements of each individual job.

FIGURE 4.10 ■ Standard precast concrete building beams.

Joists

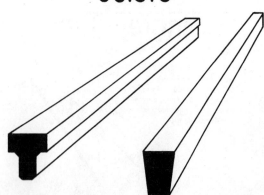

Typical Uses

Roof and floor systems with prefabricated or poured decks; Also used as fins, frames, posts, and columns.

NOTE
Joist lengths and stranding patterns vary according to the needs of each individual job and design situation.

Standard Sizes

8″ and 12″ deep in the keystone shapes.
16″ deep in the tee shapes.

FIGURE 4.9 ■ Standard precast concrete joist members.

Piles And Columns

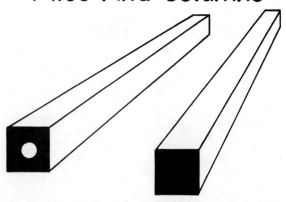

Typical Uses

Bridge and pier bearing piles; Foundation piles and building columns

NOTE
Pile and column lengths and reinforcing or stranding vary according to the needs of each individual job and design situation.

Standard Sizes

Sizes and shapes of piles and columns may be manufactured to meet the needs of the individual job and design situation.

FIGURE 4.11 ■ Standard precast concrete piles and columns.

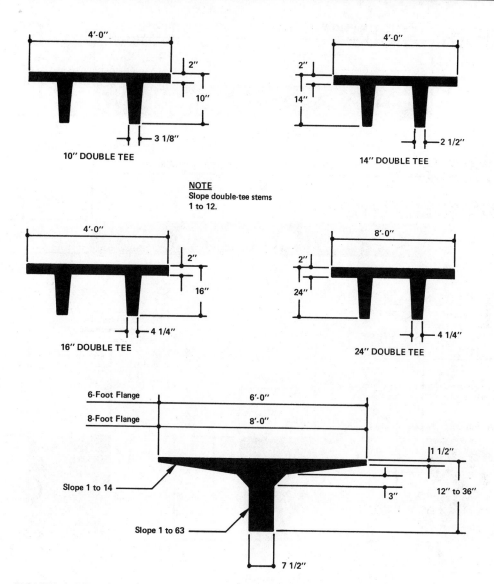

FIGURE 4.12 ■ Standard precast concrete products—dimensions for detailing.

are stretched to a predetermined amount of tension. The concrete is then poured into the bed around the strands and allowed to harden. Once the concrete has sufficiently hardened, the tension on the strands is released and the concrete member is placed in compression or prestressed. The principle is commonly illustrated by the attempt to lift a row of books on a shelf by pressing against each end and lifting (Figure 4.14). If enough pressure is exerted inward from both ends, there is sufficient compression to lift all of the books at once.

Prestressed concrete members are further strengthened by the addition of *nonstressed reinforcing bars* (rebars) or corners, around openings, or in other high-stress situations. Rebars are designated by numbers that correspond with their diameters measured in eighths of an inch. For example, a #3 bar is 3/8″ in diameter, a #4 bar is 4/8″ or 1/2″ in diameter, a #8 bar is 8/8″ or 1″ in diameter, and so on.

Reinforced concrete products are also poured in beds or forms, but without the prestressing strands. Instead, they are reinforced solely with wire mesh and reinforcing bars. Prestressed and reinforced concrete products are removed from the beds after hardening by cranes or overhead lifts, stacked on wooden blocks for storage, and shipped to the jobsite by truck or train. Figure 4.15 through Figure 4.23 illustrate both categories of precast concrete fabrication.

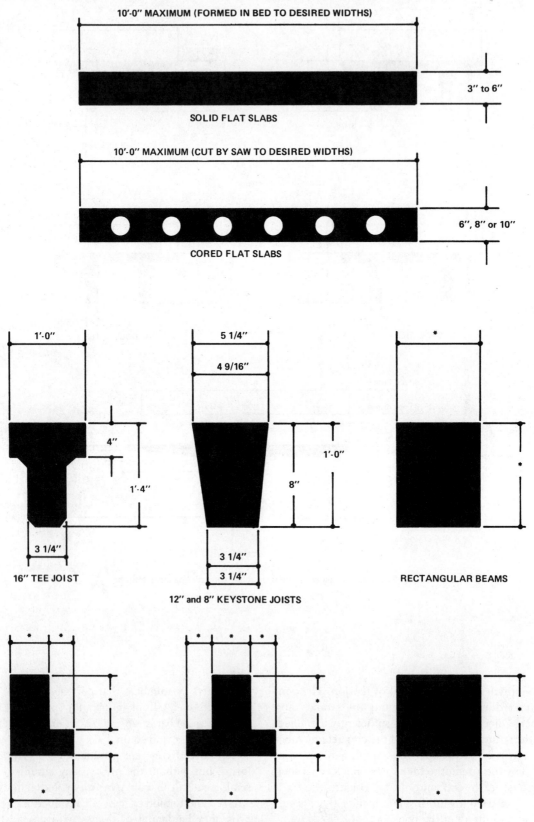

FIGURE 4.13 ■ Standard precast concrete products—dimensions for detailing.

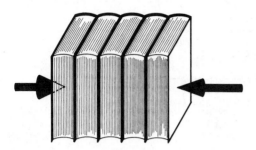

FIGURE 4.14 ■ The prestressing principle.

FIGURE 4.17 ■ Precast concrete beam form. (*Deborah M. Goetsch*)

FIGURE 4.15 ■ Steel abutment for pulling prestressing strands.
(*Deborah M. Goetsch*)

FIGURE 4.18 ■ Reinforcing bars stacked for use in a precast
concrete job. (*Deborah M. Goetsch*)

FIGURE 4.16 ■ Precast concrete double-tee form. (*Deborah M.
Goetsch*)

FIGURE 4.19 ■ Precast concrete column. (*Deborah M. Goetsch*)

FIGURE 4.20 ■ Precast concrete flat-slab forms. *(Deborah M. Goetsch)*

FIGURE 4.21 ■ Precast double-tee members are stacked awaiting shipping. *(Deborah M. Goetsch)*

FIGURE 4.22 ■ Precast concrete flat slabs stacked and labeled for shipping. *(Deborah M. Goetsch)*

FIGURE 4.23 ■ High-strength steel prestressing strands. *(Deborah M. Goetsch)*

POURED-IN-PLACE CONCRETE FABRICATION

Poured-in-place concrete is similar to the reinforced concrete already discussed; in fact, it is reinforced concrete. The difference between precast reinforced concrete and poured-in-place reinforced concrete is that the latter is poured at the jobsite. Precast reinforced concrete is poured at a plant or shop and shipped to the jobsite for erection. Poured-in-place reinforced concrete is usually poured in wooden forms as a part of the structure. These wooden forms are built in place at the jobsite and torn down after the concrete has hardened. Figure 4.24 and Figure 4.25 show examples of wooden forms.

In structural steel and precast concrete construction, the general contractor works with a steel or concrete subcontractor. In poured-in-place concrete construction, there is no concrete or steel subcontractor. Using shop drawings, the contrac-tor's workers build the necessary forms, place the prebent reinforcing bar cages into the forms, and pour the concrete. Figure 4.24 through Figure 4.27 show examples of poured-in-place concrete fabrication.

GLUED-LAMINATED WOOD FABRICATION

Several separate pieces of wood may be pressed together to form a laminated structural member that is stronger than its individual parts. Laminated wood members such as columns, beams, and arches are usually made of kiln-dried lumber. The moisture content of this lumber is less than 15 percent. Special high-strength, waterproof glues are used under pressure to permanently seal individual pieces of lumber into one solid piece. These single pieces may be bent or shaped to the desired form with special equipment in a fabrication shop (Figure 4.28).

FIGURE 4.24 ■ Wooden form for a poured-in-place concrete beam.

FIGURE 4.26 ■ Reinforcing bars to be used in a poured-in-place concrete job. (*Deborah M. Goetsch*)

FIGURE 4.25 ■ Wooden form for a poured-in-place column base. (*Deborah M. Goetsch*)

FIGURE 4.27 ■ Wooden form for poured-in-place concrete bank vault. (*Deborah M. Goetsch*)

FIGURE 4.28 ■ Fabrication of glued-laminated structural wood members. (*Weyerhaeuser Company*)

SHIPPING PRODUCTS TO THE JOBSITE

In steel and precast concrete and heavy timber construction, the structural products are shipped to the jobsite after being manufactured or fabricated in a shop. Most companies have a shipping deck or yard. From this deck, products that are ready for delivery are loaded onto trucks or trains by cranes or overhead hoists. Due to the size of the products, loading equipment, and vehicles used for shipping, this load area must be very large.

Shipping is an important phase in the heavy construction industry and sometimes involves the drafter. If a large structural member is to be shipped to the jobsite by truck, the state road department or highway patrol will sometimes request sketches. These sketches assist them in recommending the best possible route for the truck. Maximum clearances under bridges or tunnels, as well as the truck's minimum turning radius, must be considered before shipment. In such cases, the drafter will be required to prepare a basic drawing of the truck showing length, width, and height dimensions. These dimensions must take into consideration the product that is being shipped (Figure 4.29).

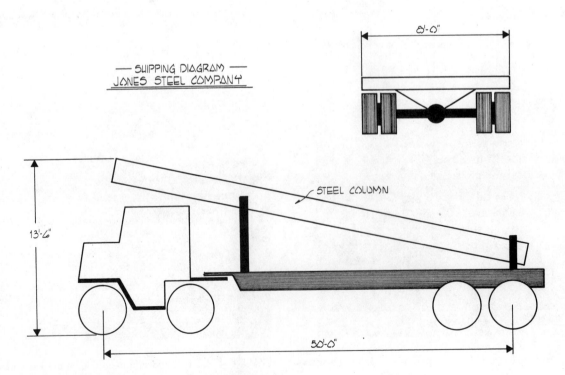

FIGURE 4.29 ■ CAD technician's shipping diagram.

SUMMARY

■ Structural steel is produced in standard shapes, including: W shapes; S shapes; M shapes; HP shapes; angles; C shapes; MC shapes; pipe; square tubing; rectangular tubing; circular tubing; and structural tees cut from W, S, and M shapes.

■ The most common structural steel fabrication processes are: handling, cutting, punching, drilling, straightening, bending, bolting, riveting, welding, and finishing.

■ Handling of structural steel products is accomplished by forklifts, cranes, overhead hoists, or straddle carriers.

■ Cutting of structural steel products is accomplished by shearing, by hot sawing, or by using a cutting torch.

■ Holes are placed in structural steel products by punching (for thin members) or drilling (for thick members).

■ Structural steel members may be straightened or bent on a machine known as a bend press or gag press.

■ There are three basic methods for fastening structural steel members: bolting, riveting, and welding. Bolting and welding are the most frequently used.

■ The final phase in fabricating steel products is finishing, which is accomplished by: sawing, milling, filing, high-pressure blasting, and various other means.

■ Precast concrete products are manufactured in two categories: prestressed concrete and reinforced concrete.

■ Prestressed concrete products are manufactured in steel beds and contain high-stressed steel strands that place the concrete member in compression.

■ Reinforced concrete products are poured in steel or wooden forms and are reinforced with steel reinforcing bars but are not prestressed.

■ Common prestressed and reinforced concrete products include: columns, beams, double tees, single tees, flat slabs, joists, bridge girders, and stadium seats.

■ Poured-in-place concrete members are reinforced concrete products that are poured, usually in temporary wooden forms, at the jobsite.

■ Structural wood members are either solid timber or built-up glu-lam members. Common glued-laminated products are beams, columns, and arches.

■ Drafters are sometimes called upon to make shipping diagrams so that the highway patrol and state road department can map out special routes when large structural products are to be shipped by truck.

REVIEW QUESTIONS

1. Sketch the basic configuration of the following structural steel shapes:
   a. W shape
   b. S shape
   c. M shape
   d. HP shape
   e. C shape
   f. MC shape
   g. pipe
   h. square tubing

2. List the ten most common structural steel fabrication processes.

3. List three methods of cutting structural steel products.

4. List the three basic methods for fastening structural steel members.

5. List four different methods that might be used during the finishing process.

6. List the two categories of precast concrete products.

7. Sketch the basic configuration of the following precast concrete products:
   a. double tee
   b. single tee
   c. square column
   d. rectangular beam
   e. flat slab
   f. cored flat slab
   g. keystone joist
   h. bridge girder

8. Explain the difference between precast reinforced concrete fabrication and poured-in-place reinforced concrete fabrication.

9. Explain how CAD technicians sometimes become involved in the shipping process.

10. Handling of heavy structural products is accomplished by:
    a. forklifts
    b. cranes
    c. overhead hoists
    d. straddle carriers
    e. all of the above

11. Holes are placed in thin structural steel members by:
    a. drilling
    b. shearing
    c. punching
    d. torching

12. The machine used for bending or straightening structural steel members is:
    a. a gag press
    b. a bend press
    c. an offset press
    d. a vise press
    e. *a* and *b*
    f. *c* and *d*

CAD ACTIVITIES

GENERAL INSTRUCTIONS

The following activities may be completed on any CAD system. Before reading the *specific instructions* for each activity (below), go through each step in the following planning checklist. The checklist applies to any CAD system and will help ensure the optimum use of your time and resources.

1. Analyze the problem carefully. Decide exactly what you are being asked to do.

2. Determine what resources and references you will need in order to complete the problem and collect them.

3. Decide if any particular standards apply to the project and have those standards available.

4. Determine what types of views will be required and how many of each.

5. Determine what the final plotted scale of the drawing will need to be, and select the appropriate paper size for plotting/printing (make sure the appropriate paper size is available).

6. Plan your drawing sequence. In what order will you develop the drawing (i.e., lines, features, dimension lines, leaders, dimensions, notes, etc.)?

7. Review the various CAD commands you will have to use in order to develop the drawing.

8. Examine your CAD system to ensure that everything is in working order, then begin the project.

SPECIFIC INSTRUCTIONS

Activity 4.1—Build your keyboarding speed and CAD skills by copying the notes in Figure 4.30. Save your notes as *Activity 4.1*.

Activity 4.2—Continue to develop your CAD skills by drawing the precast concrete section shown in Figure 4.31. Save your drawing as *Activity 4.2*.

Activity 4.3—Continue to develop your CAD skills by drawing the structural steel shop drawing shown in Figure 4.32. Save your drawing as *Activity 4.3*.

Activity 4.4—Begin to develop skills in the use of the AISC *Manual of Steel Construction* by completing the following activity. Construct and dimension a detail similar to the example in Figure 4.33 for each of the following steel shapes:

$$W\ 36 \times 300$$
$$W\ 30 \times 132$$
$$W\ 24 \times\ 62$$
$$W\ 14 \times 117$$
$$W\ \ 8 \times\ 36$$

Save your drawing as *Activity 4.4*.

GENERAL NOTES

1. GENERAL CONTRACTOR SHALL FIELD-CHECK AND VERIFY ALL DIMENSIONS AND CONDITIONS AT JOBSITE.
2. ERECTION INCLUDES PLACING MEMBERS AND MAKING MEMBER CONNECTIONS ONLY.
3. ERECTION BY OTHERS, PRODUCTS F.O.B. TRUCKS JOBSITE.
4. NO GROUTING, POINTING, OR FIELD-POURED CONCRETE BY JONES PRECAST CONCRETE COMPANY.
5. RELEASE STRENGTH: 3500 PSI (UNLESS OTHER-WISE NOTED)
6. CEILING FINISH: SUSPENDED
7. BLOCKOUTS SMALLER THAN 10" x 10" TO BE CUT IN THE FIELD BY PROPER TRADES (EXCEPT AS NOTED ON THE SHOP DRAWINGS).
8. WALL PANEL FINISHES: INTERIOR PANELS — EXPOSED
   EXTERIOR PANELS — CRUSHED GRAVEL
9. THE FOLLOWING ITEMS ARE TO BE SHIPPED LOOSE TO THE JOBSITE:
   24 WAA, 15 WAB, 7 WAC

LEGEND

ba - BLOCKOUT - a - 24" x 2" x 5'-0 3/4"
301 - NUMBER 3 BAR BY 16'-6" long (STRAIGHT)
302 - NUMBER 3 BAR BY 5'-7" long (BENT) SEE DETAIL ON SHEET M-1
WAA - WELD ANGLE - A - SEE DETAIL ON SHEET M-1
WAB - WELD ANGLE - B - SEE DETAIL ON SHEET M-1
WAC - WELD ANGLE - C - SEE DETAIL ON SHEET M-1

DIRECTIONS FOR THE REMAINDER OF THE ACTIVITY

USE THE SPACE REMAINING ON THE SHEET TO IMPROVE YOUR LETTERING SPEED. TAKE THE ENTRIES IN THE GENERAL NOTES ABOVE AND LETTER THEM AT A NORMAL PACE MAKING NOTE OF HOW LONG IT TAKES TO LETTER EACH ENTRY. THEN REPEAT EACH ENTRY MAKING A SPECIAL EFFORT TO DO IT FASTER. MAKE NOTE OF THE TIME AND CALCULATE HOW MUCH FASTER YOU COMPLETED THE LETTERING ON THE SECOND TRY. SPEED IS IMPORTANT, BUT REMEMBER, YOU CANNOT SACRIFICE QUALITY TO ACHIEVE IT. WORK ON YOUR LETTERING UNTIL IT IS NEAT AND FAST!

FIGURE 4.30 ■ Key boarding activity for CAD Activity 4.1.

PRECAST CONCRETE SECTION

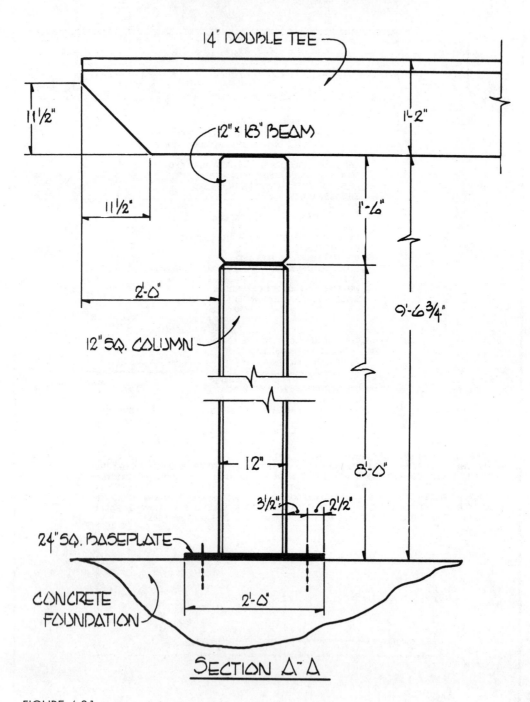

14" DOUBLE TEE

11½"

1'-2"

12" × 18" BEAM

11½"

1'-6"

2'-0"

12" SQ. COLUMN

9'-6¾"

12"

8'-0"

3½" 2½"

24" SQ. BASEPLATE

CONCRETE
FOUNDATION

2'-0"

SECTION A-A

FIGURE 4.31 ■ Example for CAD Activity 4.2.

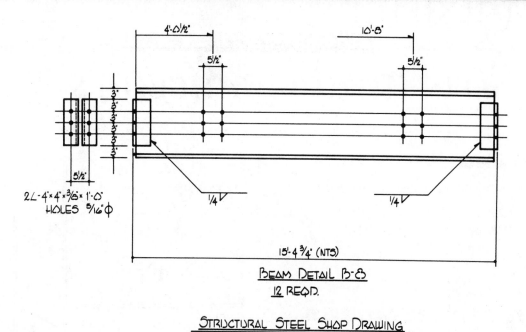

FIGURE 4.32 ■ Example for CAD Activity 4.3.

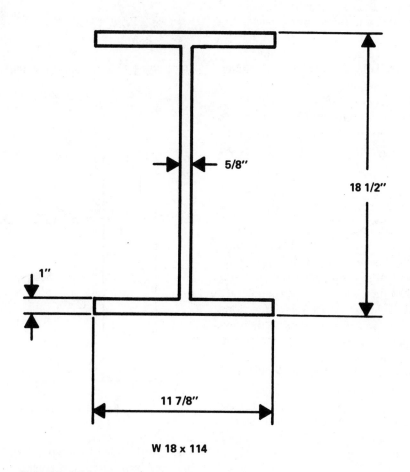

W 18 x 114

FIGURE 4.33 ■ Sample for CAD Activity 4.4.

5 UNIT

Structural Connectors

OBJECTIVES

Upon completion of this unit, the student will be able to:

■ Explain the application of bolted, welded, riveted, split ring and shear plate connections in heavy construction.

■ Interpret common welding symbols.

CONNECTING STRUCTURAL MEMBERS

Heavy construction with steel and precast concrete members requires numerous field connections during the erection process. The reason for this is that the structural members are manufactured in a plant and shipped to the jobsite in pieces. Columns, beams, girders, joists, floor/roof members, and wall panels are all prepared for field connections during the fabrication process.

Structural members in heavy construction are connected by any one of three basic methods: bolting, welding, and riveting. Most large jobs involve a combination of at least two of these methods. Designing connections is the job of an engineer or designer. However, CAD technicians must also be familiar with structural connections in order to convert engineering calculations and sketches into finished connection details (Figure 5.1 and Figure 5.2).

Bolted Connections

Two categories of bolts are used for field connections in heavy construction: common bolts and high-strength bolts. In addition, threaded rods of various diameters and lengths are used for bolting large, precast concrete members together (Figure 5.3).

Common bolts, identified by their square heads, are the least-expensive type of bolt available for structural connections. However, their applications are limited. This is due to their low carbon-steel content and their inability to be tightened firmly enough for use in high-stress situations.

Most bolted connections in heavy construction require high-strength bolts. *High-strength bolts,* classified as A325 and A490, are made of a special steel. This special steel gives them a much greater tightening capacity and allows them to be used in places that would otherwise require hot-driven rivets. This is important because bolted connections have several advantages over riveted connections.

Due to the greater thicknesses of precast concrete members, threaded rods are sometimes used for making bolted connections. *A threaded rod* is a steel rod of specified diameter with

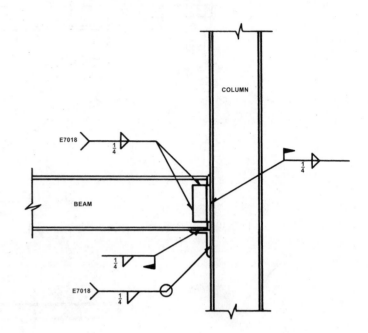

FIGURE 5.1 ■ Structural steel connection detail.

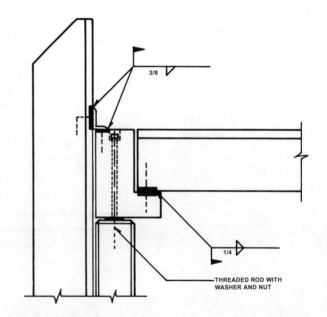

FIGURE 5.2 ■ Structural concrete connection detail.

one end threaded to accept a bolt (Figure 5.3). Threaded rods are used most often in precast concrete beams to column connections. Figure 5.4 explains how to interpret thread notes commonly used with bolts and threaded rods in structural drafting.

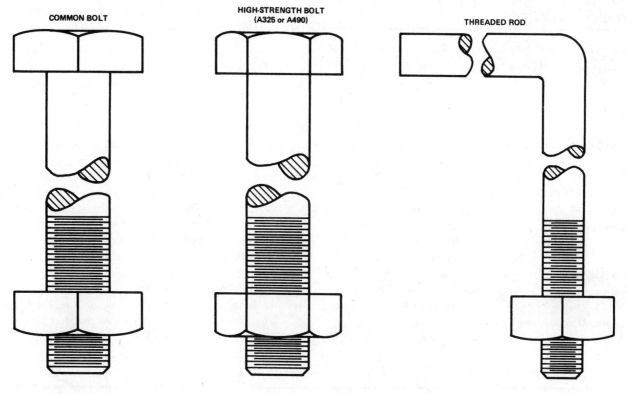

FIGURE 5.3 ■ Bolted connectors.

Split-Ring Connectors and Shear Plates. The applications of heavy timber and glued-laminated wood members are increased through the use of special metal connectors. The most common wooden fasteners are split-ring connectors and shear plates. *Split-ring connectors* are placed in grooves that have been cut in two members so they align when placed together for fastening. The joint between the two wood mem-

bers is fastened with a bolt passed through a hole in the center of the split ring (Figure 5.5). In situations where a wooden joint may have to be disassembled and reassembled, shear plates are especially useful. *Shear plates* are commonly used in fabricating and erecting glued-laminated members. As in split-ring connectors, bolts are the fastening devices used with shear plates (Figure 5.5).

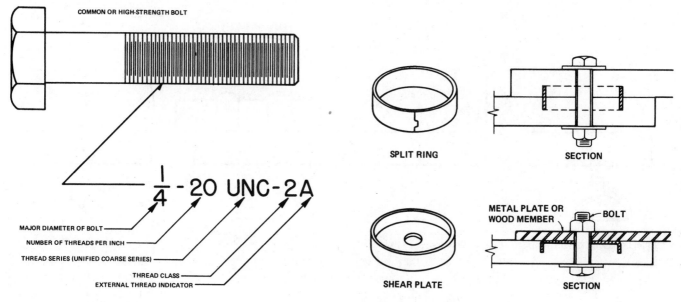

FIGURE 5.4 ■ Interpreting thread notes.

FIGURE 5.5 ■ Metal fasteners for wood connections.

Welded Connections

Arc welding, the fusion of metal by an electric arc, is the most common type of connection in heavy construction. Specifications for electrodes and standards of practice adopted by the American Society for Testing Materials (ASTM) and the American Welding Society (AWS) have eliminated problems that were associated with welding in the past and upgraded it to the most important type of structural connection process. Welding electrodes commonly used in heavy construction connections are of the E60 and E70 series.

Weld Types and Symbols. Welds are classified according to the type of joint on which they are used. Four common types of welds are used in heavy construction connections. These are the back weld, the fillet weld, the plug or slot weld, and the groove weld (Figure 5.6). Groove welds are subdivided into several subcategories. These subcategories are based on the shape of the groove to be welded: square groove, V groove, bevel groove, U groove, and flare bevel groove.

Welds are indicated on drawings by weld symbols. A completed weld symbol consists of the following:

■ A horizontal reference line with a connected sloping line ending in an arrowhead

■ A basic symbol indicating the type and/or size of weld

■ A supplementary symbol further explaining the desired welding specification (Figure 5.7)

The basic symbols used for the various types of welds are contained in Figure 5.8. Supplementary symbols that are commonly used by structural CAD technicians are shown in Figure 5.9.

The structural CAD technician must be familiar not only with welding symbols but also with certain rules governing their use. The arrow on the weld symbol shown in Figure 5.7 may point either to the right or to the left. However, this arrow must always form an angle with the horizontal reference line. The V-shaped tail of this same welding symbol is included when specifications concerning the weld are included at the tail of the symbol. If no specifications are included here, the V may be left off. When using a basic weld symbol that has a perpendicular

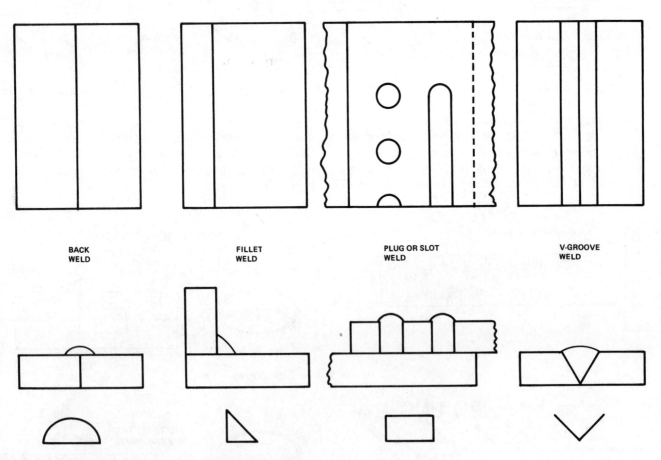

BACK WELD   FILLET WELD   PLUG OR SLOT WELD   V-GROOVE WELD

FIGURE 5.6 ■ Common structural weld types with symbols.

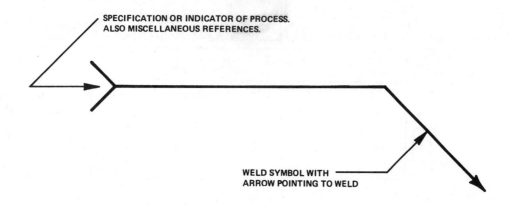

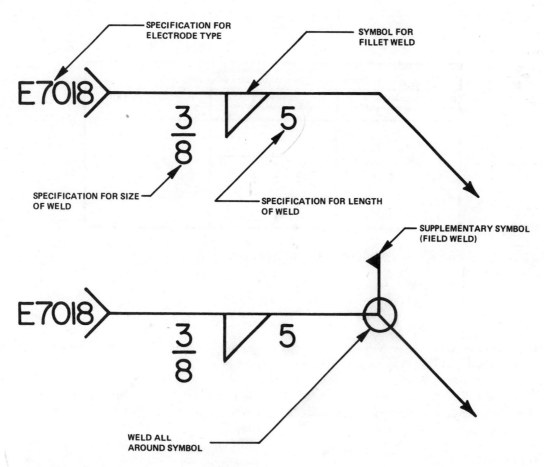

FIGURE 5.7 ■ Explanation of typical weld symbols.

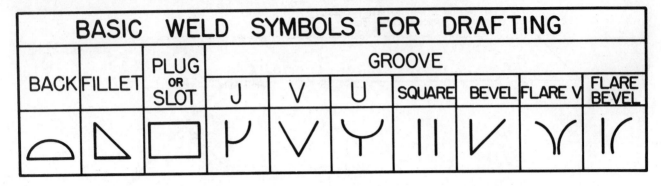

FIGURE 5.8 ■ Basic weld symbols for structural drafting.

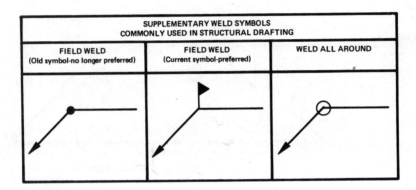

FIGURE 5.9 ■ Supplementary weld symbols.

leg, such as the fillet weld and certain groove welds (Figure 5.8), the leg is always drawn on the left-hand side as facing the symbol. Figure 5.10 illustrates several symbols, completely annotated to assist the student in learning to interpret welding symbols. Figure 5.11 shows examples of how welding symbols are used on structural drawings.

Riveted Connections

Riveting is no longer considered a major method of making structural connections in heavy construction. However, until around 1950, riveting was the only major connecting method. Because of its extensive applications in the past, the structural drafting student should be familiar with this connection process.

In heavy structural connections, hot rivets are driven through holes prepared in two structural steel members. The riveting process involves placing a heated rivet, with a button head on one end, into a prepared hole and hammering on the other end with a riveter or pneumatic hammer. This hammering secures the rivet firmly in the hole and fastens the structural members together.

Rivets are strong, long-lasting connectors. However, with the advent of high-strength bolts and vast improvements in welding materials and processes, riveted connections are no longer widely used.

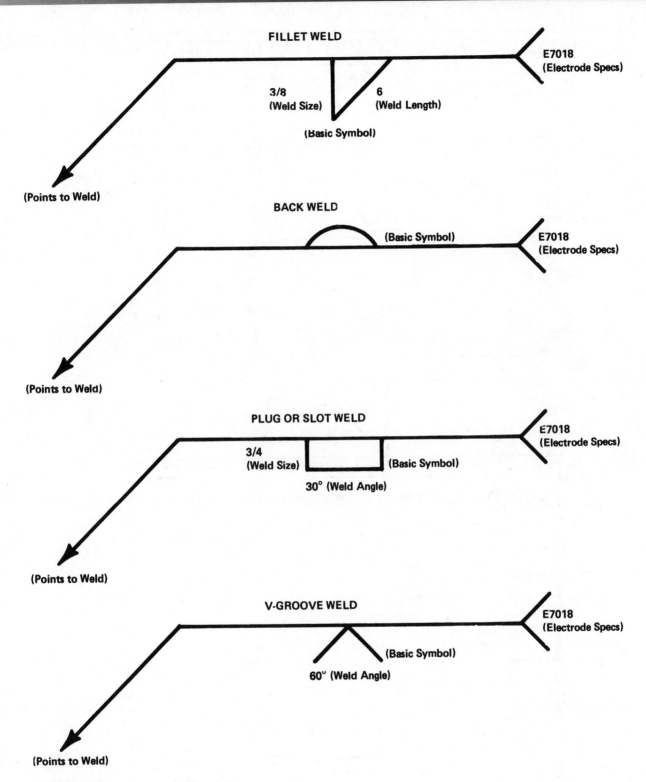

FILLET WELD

E7018
(Electrode Specs)

3/8
(Weld Size)

6
(Weld Length)

(Basic Symbol)

(Points to Weld)

BACK WELD

(Basic Symbol)

E7018
(Electrode Specs)

(Points to Weld)

PLUG OR SLOT WELD

E7018
(Electrode Specs)

3/4
(Weld Size)

(Basic Symbol)

30° (Weld Angle)

(Points to Weld)

V-GROOVE WELD

E7018
(Electrode Specs)

(Basic Symbol)

60° (Weld Angle)

(Points to Weld)

FIGURE 5.10 ■ Annotated weld symbols.

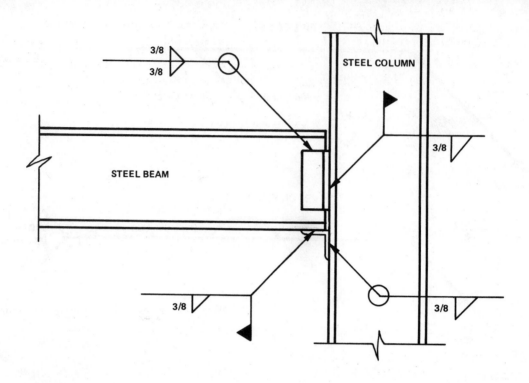

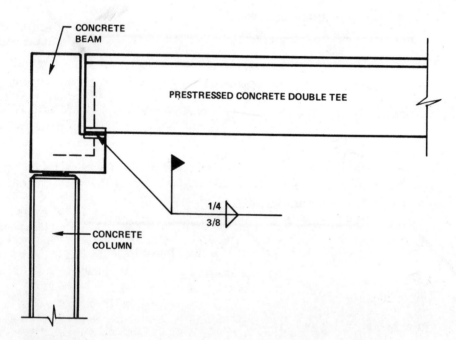

FIGURE 5.11 ■ Sample applications of weld symbols in structural drafting.

SUMMARY

■ Structural members in heavy construction are connected by bolting, welding, or riveting.

■ Structural connections are designed by engineers or designers and detailed by CAD technicians.

■ There are two categories of bolts used for field connections in heavy construction: common bolts and high-strength bolts.

■ Common bolts have limited applications and cannot be used in high-stress situations.

■ High-strength bolts used in heavy construction connections are classified as A325 and A490.

■ Structural wood connections are commonly made with split-ring connectors and shear plates.

■ Threaded rods are commonly used for precast concrete bolted connections.

■ Welding is the most common type of structural connection used in heavy construction.

■ The four most common types of welds in heavy construction connections are: back, fillet, plug or slot, and groove welds.

■ Groove welds are subdivided into several categories: square groove, V groove, bevel groove, U groove, and flare bevel groove.

■ Riveting is no longer considered a major structural connection method.

REVIEW QUESTIONS

1. List the three major types of structural connections used in heavy construction.

2. List the two categories of bolts used for field connections in heavy construction.

3. What type of bolts are most often used in heavy construction bolted connections?

4. List the two classifications of high-strength bolts used in structural connections.

5. How are thick, precast concrete columns and beams bolted together?

6. What is the most common type of structural connection used in heavy construction?

7. List the four most common types of welds used in heavy construction connections.

8. List five types of groove welds.

9. Sketch an example of a structural wood connection using a split-ring connector and one using a shear plate.

10. Which of the following is no longer considered a major structural connection process?
    a. welding
    b. riveting
    c. nailing
    d. bolting

11. Which of the following groups is responsible for the design of structural connections?
    a. junior CAD technicians
    b. contractors
    c. architects
    d. engineers

CAD ACTIVITIES

GENERAL INSTRUCTIONS

The following activities may be completed on any CAD system. Before reading the *specific instructions* for each activity (below), go through each step in the following planning checklist. The checklist applies to any CAD system and will help ensure the optimum use of your time and resources.

1. Analyze the problem carefully. Decide exactly what you are being asked to do.

2. Determine what resources and references you will need in order to complete the problem and collect them.

3. Decide if any particular standards apply to the project and have those standards available.

4. Determine what types of views will be required and how many of each.

5. Determine what the final plotted scale of the drawing will need to be, and select the appropriate paper size for plotting/printing (make sure the appropriate paper size is available).

6. Plan your drawing sequence. In what order will you develop the drawing (i.e., lines, features, dimension lines, leaders, dimensions, notes, etc.)?

7. Review the various CAD commands you will have to use in order to develop the drawing.

8. Examine your CAD system to ensure that everything is in working order, then begin the project.

SPECIFIC INSTRUCTIONS

Activity 5.1—Begin to develop skills in the understanding and use of weld symbols by drawing weld symbols for each of the following sets of specifications. Save your drawing as *Activity 5.1*.
   a. back weld with E7018 electrodes to be field welded
   b. fillet weld with E7018 electrodes 1/4″ × 6″ long
   c. slot weld 3/4″ with a 30-degree weld angle
   d. V-groove weld at 60 degrees with E7018 electrodes
   e. fillet weld to be field welded all around by 3/8″.

Activity 5.2—Learn to apply weld symbols on structural drawings by completing the following activity. Redraw the connection details in Figure 5.12. For each numbered situation requiring a weld, substitute the proper weld symbol. Also, size all weld plates and angles based on their proportions in Figure 5.12 and the needs of the situation. Save your drawing as *Activity 5.2*.

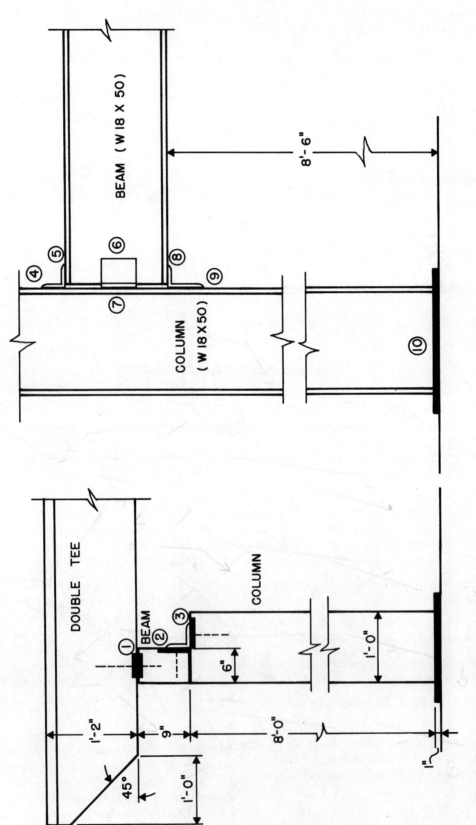

FIGURE 5.12 ■ Connection details for CAD Activity 5.2.

SECTION II

Structural Steel Drafting

6 UNIT

Structural Steel Framing Plans

OBJECTIVES

Upon completion of this unit, the student will be able to:

- Distinguish between engineering drawings and shop drawings.

- Describe, designate, and illustrate the various structural steel products used in framing plans.

- Properly use the American Institute of Steel Construction's *Manual of Steel Construction* for determining structural steel product designations and dimensions.

- Properly construct structural steel framing plans according to engineering specifications.

STRUCTURAL STEEL DRAWINGS

In Unit 1, it was learned that structural CAD technicians prepare two basic types of drawings: engineering drawings and shop drawings. Engineering drawings are used to provide general information for sales, marketing, engineering, and erection purposes. Shop drawings are used to provide more detailed information for fabrication purposes.

In structural steel drafting, engineering drawings are sometimes referred to as **erection drawings**, depending on how they are to be used. Engineering drawings are prepared by CAD technicians from sketches provided by structural engineers. The drawings include **framing plans** and **sections**, which are symbolic representations of all steel members used in a structure (Figure 6.1).

This unit deals with the preparation of structural steel framing plans. In order to prepare structural steel framing plans, the CAD technician must first become familiar with the structural steel products used in framing plans.

STRUCTURAL STEEL FRAMING PRODUCTS

Rolled steel products are classified as being either a plate, a bar, or a shape. **Plates** are flat pieces of steel of various thicknesses. They are used as a framing product only for making changes to other framing members. Common uses are as stiffeners, gusset plates, and in making built-up girders. Plates are called out on

drawings according to their thickness (in inches), their width (in inches), and their length (in feet and inches) (Figure 6.2).

Bars are the smallest structural steel products and may have round, square, rectangular, or hexagonal cross-sectional configurations (Figure 6.3). Bars are not a structural steel framing product, but they can be used in modifying other steel framing products.

Shapes consist of W shapes, M shapes, S shapes, angles, channels, structural tees, structural tubing, and pipe and are the most important structural steel framing products.

W, S, and M Shapes

W, S, and M shapes are the new designations, set forth by the American Institute of Steel Construction (AISC), for shapes that previously were designated as WF, I, and M or Jr shapes. It is important to learn to use the new and proper designations when calling out structural steel shapes on drawings. The proper callout designations, dimensions for detailing, and properties for designing structural steel shapes are provided in the previously mentioned manual by the AISC. For the student's convenience, a portion of this manual has been reproduced in the appendix of this book. Figure 6.4 contains a list of examples showing the old and the new designations for selected structural steel products.

S shapes in cross section have an I configuration with narrow flanges that slope in a manner similar to channel flanges (Figure 6.5). W shapes also have an I configuration in cross section, but their flanges are wider than those on S or M shapes and have a constant thickness (Figure 6.5). M shapes are miscellaneous shapes and include all rolled shapes with an I cross-sectional configuration that cannot be classified as W or S shapes (Figure 6.5).

Angles

Angles are properly designated as L shapes and are of two types: equal angles and unequal angles (Figure 6.6). In both types, the legs have the same thickness, even though the legs of unequal angles differ in length. Examples of angles properly called out on drawings are: L 6 × 5 × 1/2 and L 4 × 4 × 1/4. The first example denotes an unequal angle in which one of the legs is 6″ long, the other is 5″ long, and both legs are 1/2″ thick. The

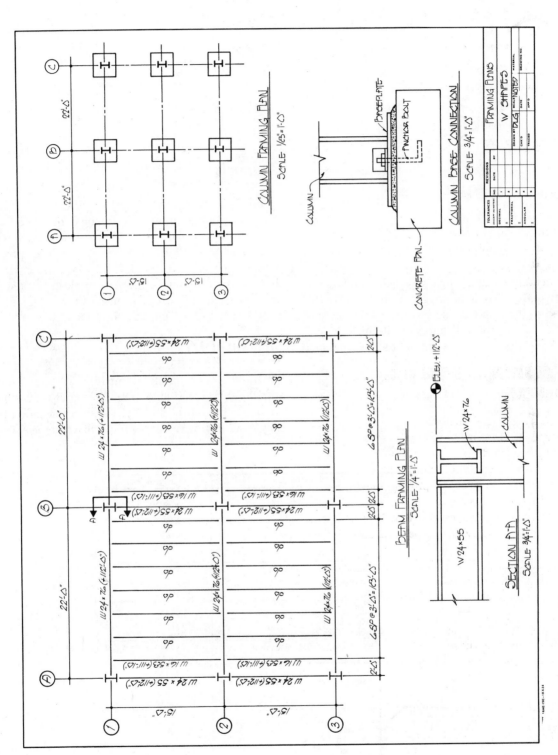

FIGURE 6.1 ■ Framing plans.

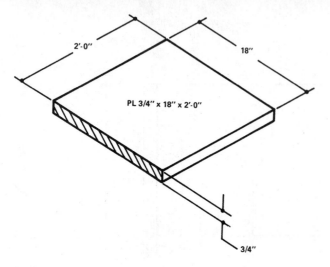

FIGURE 6.2 ■ Steel plate designations.

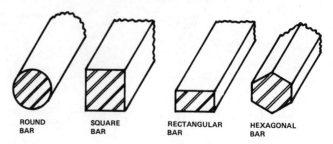

FIGURE 6.3 ■ Structural steel bars.

second example denotes an equal angle in which both legs are 4″ long by 1/4″ thick. The length of any given angle can also be designated by adding the required length at the end of the call-out in feet and inches. If it was desired to have the angles called out as 6″ long, as in the previous example, the designations would read L 6 × 5 × 1/2 × 0′-6″ and L 4 × 4 × 1/4 × 0′-6″.

Angle designations on older drawings read exactly as the previous examples with the exception of the uppercase L. The old symbol for angle was ∠. Because of its similarity to the letter L, drawings lettered freehand will show very little difference.

Channels

Channels are properly designated as C shapes. They have a squared C configuration with sloping flanges and a web of constant thickness (Figure 6.7). Channels are of two types: American Standard and Miscellaneous. Examples of properly called out channels are C 10 × 25, C 12 × 30, and C 15 × 33.9. All three of these examples denote American Standard channels. Some examples of Miscellaneous channel designations are MC 10 × 8.4, MC 12 × 10.6, and MC 18 × 42.7.

Structural Tees

Structural tees are products cut from W, S, and M shapes by splitting the webs (Figure 6.8). Structural-tee designations are a modification of the structural-shape designation from which they were cut, with the addition of an uppercase T. Sample

New Designation	Type of Shape	Old Designation
W 24 x 76 W 14 x 26	W shape	24 WF 76 14 B 26
S 24 x 100	S shape	24 I 100
M 8 x 18.5 M 10 x 9 M 8 x 34.3	M shape	8 M 18.5 10 JR 9.0 8 x 8 M 34.3
C 12 x 20.7	American Standard Channel	12 [20.7
MC 12 x 45 MC 12 x 10.6	Miscellaneous Channel	12 x 4 [45.0 12 JR [10.6
HP 14 x 73	HP shape	14 BP 73
L 6 x 6 x 3/4 L 6 x 4 x 5/8	Equal Leg Angle Unequal Leg Angle	∠6 x 6 x 3/4 ∠6 x 4 x 5/8
WT 12 x 38 WT 7 x 13	Structural Tee cut from W shape	ST 12 WF 38 ST 7 B 13
ST 12 x 50 MT 4 x 9.25 MT 5 x 4.5 MT 4 x 17.15	Structural Tee cut from S shape Structural Tee cut from M shape	ST 12 I 50 ST M 9.25 ST 5 JR 4.5 ST 4 M 17.15
PL 1/2 x 18	Plate	PL 18 x 1/2
Bar 1 ⊡ Bar 1 1/4 ⌀ Bar 2 1/2 x 1/2	Square Bar Round Bar Flat Bar	Bar 1 ⊡ Bar 1 1/4 ⌀ Bar 2 1/2 x 1/2
Pipe 4 Std. Pipe 4 X-Strong Pipe 4 XX-Strong	Pipe	Pipe 4 Std. Pipe 4 X-Strong Pipe 4 XX-Strong
TS 4 x 4 x .375 TS 5 x 3 x .375 TS 3 OD x .250	Structural Tubing: Square Structural Tubing: Rectangular Structural Tubing: Circular	Tube 4 x 4 x .375 Tube 5 x 3 x .375 Tube 3 OD x .250

FIGURE 6.4 ■ Structural steel designations.

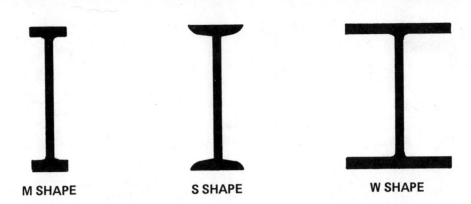

M SHAPE   **S SHAPE**   **W SHAPE**

NOTE
DIMENSIONS FOR DETAILING M, S, AND W SHAPES ARE PROVIDED IN SECTION I OF THE AISC *MANUAL OF STEEL CONSTRUCTION*. EXCERPTS ARE PROVIDED IN THE APPENDIX OF THIS BOOK.

FIGURE 6.5 ■ Structural steel shapes.

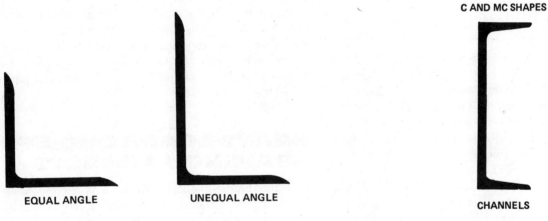

EQUAL ANGLE   **UNEQUAL ANGLE**

C AND MC SHAPES

CHANNELS

NOTE
DIMENSIONS FOR DETAILING L SHAPES ARE PROVIDED IN SECTION I OF THE AISC *MANUAL OF STEEL CONSTRUCTION*. EXCERPTS ARE PROVIDED IN THE APPENDIX OF THIS BOOK.

FIGURE 6.6 ■ L shapes.

NOTE
DIMENSIONS FOR DETAILING C AND MC SHAPES ARE PROVIDED IN SECTION I OF THE AISC *MANUAL OF STEEL CONSTRUCTION*. EXCERPTS ARE PROVIDED IN THE APPENDIX OF THIS BOOK.

FIGURE 6.7 ■ C shapes.

MT   **ST**   **WT**

NOTE
DIMENSIONS FOR DETAILING MT, ST, AND WT SHAPES ARE PROVIDED IN SECTION I OF THE AISC *MANUAL OF STEEL CONSTRUCTION*. EXCERPTS ARE PROVIDED IN THE APPENDIX OF THIS BOOK.

FIGURE 6.8 ■ Structural tees.

SQUARE TUBING **RECTANGULAR TUBING** **ROUND TUBING**

<u>NOTE</u>
DIMENSIONS FOR DETAILING STRUCTURAL TUBING SHAPES ARE PROVIDED IN SECTION I OF THE AISC *MANUAL OF STEEL CONSTRUCTION.* EXCERPTS ARE PROVIDED IN THE APPENDIX OF THIS BOOK.

FIGURE 6.9 ■ Structural tubing.

structural-tee callouts are WT 6 × 95, MT 7 × 8.6, and ST 12 × 60. WT designations indicate that the structural tee was cut from a W shape, MT from an M shape, and ST from an S shape. For a more specific interpretation, the MT 7 × 8.6 was cut from an M 14 × 17.2.

Structural Tubing

Structural tubing is manufactured in square, rectangular, and round cross-sectional configurations. It is often used as a structural column (Figure 6.9). Tubing designations on drawings provide the outside dimensions and the wall thickness dimensions. Sample structural tubing callouts are TS 5 × 5 × 0.375, TS 8 × 4 × 0.375, and TS 3 OD × 0.250. The first example denotes a square structural steel tube that is 5″ square with walls 0.375 (3/8) of an inch thick. The second example denotes a rectangular structural steel tube that has sides 8″ × 4″ and walls that are 0.375 (3/8) of an inch thick. The final example denotes a round structural steel tube 3″ in diameter with walls 0.250 (1/4) of an inch thick.

Structural Pipe

Structural pipe has a round, hollow cross-sectional configuration and is very effective for use as structural columns (Figure 6.10).

Steel pipe is manufactured in three categories of strength: standard, X-strong, and XX-strong. The wall thickness determines the strength and the category. XX-strong pipe has thicker walls and is stronger than X-pipe, which has thicker walls and is stronger than standard pipe.

Examples of structural steel pipe designations are Pipe 3 Std, Pipe 3 X-strong, and Pipe 3 XX-strong. The first example denotes a 3″ diameter pipe of the standard strength category. The second example denotes a 3″ diameter pipe of the X-strong category. The final example denotes a 3″ diameter pipe of the XX-strong category.

HEAVY-LOAD/LONG-SPAN FRAMING PRODUCTS

The structural steel shapes discussed in the preceding paragraphs are the most commonly used structural steel framing products. However, in certain heavy-load or long-span situations, standard rolled steel products do not meet the design requirements. When this is the case, a special built-up framing member can be designed that meets the requirements.

The most common type of built-up framing member is the built-up **girder.** A *built-up girder* is either a standard rolled shape that has been reinforced or a new shape made entirely of steel

STANDARD **X-STRONG** **XX-STRONG**

<u>NOTE</u>
DIMENSIONS FOR DETAILING STRUCTURAL PIPE ARE PROVIDED IN SECTION I OF THE AISC *MANUAL OF STEEL CONSTRUCTION.* EXCERPTS ARE PROVIDED IN THE APPENDIX OF THIS BOOK.

FIGURE 6.10 ■ Structural pipe.

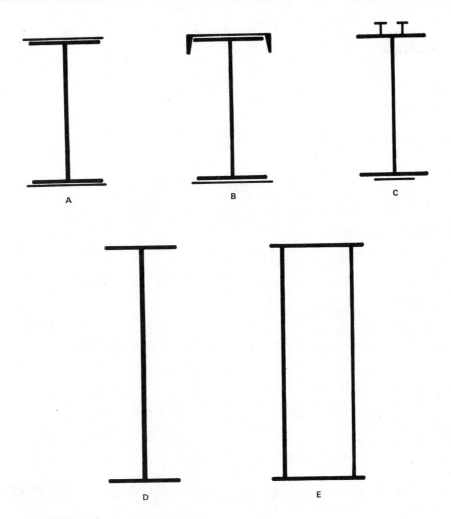

FIGURE 6.11 ■ Built-up girders.

plates. In its publication *Structural Steel Detailing*, the AISC specifies five different types of built-up girders (Figure 6.11).

Figure 6.11A shows the first type of built-up girder in which an I-configured rolled shape is reinforced by the addition of top and bottom plates permanently attached by welding. Figure 6.11B shows a similar built-up girder that substitutes a channel for the top plate. Figure 6.11C shows an example of a built-up girder that is used in composite construction with concrete. This type of girder would support a concrete floor resting on the shear connectors on top of it and would be reinforced on the bottom with a steel plate. Figure 6.11D shows a deep, built-up girder made entirely of plates welded together. Figure 6.11E contains an example of a box girder, which is composed of four welded plates.

STRUCTURAL STEEL FRAMING PLANS

Structural steel framing plans are symbolic representations of columns, beams, built-up girders, and other framing members. They are used primarily for engineering and erection purposes. Structural steel framing plans are drawn by CAD technicians

from information provided by engineers through sketches. Figure 6.12 illustrates an engineer's sketch from which a structural steel drafter might prepare a framing plan. Figure 6.13 shows the framing plan that was prepared from this sketch.

A structural steel framing plan is a plan-view representation showing all **columns, beams, girders, joists**, bridging, and so on as they will appear when the framing for the structure being built is erected. Column centerlines are given number and letter designations and are completely dimensioned. Each structural member represented in the framing plan is given an identifying callout for easy reference. For example, beams may be labeled according to the designation given them in the AISC's *Manual of Steel Construction* or they may be given a mark number or both.

Drafting time in preparing structural steel framing plans is reduced through use of the symbol *do* which means ditto, or the same. When several members have the same designation or mark number and are located together on the framing plan, the *do* symbol may be used in place of the member designation. A note indicating the top of steel elevation is placed next to the member designation or on the drawing under the heading, *Notes*.

Structural steel framing plans are of two types: column framing plans and beam framing plans. Figure 6.14 is an example of

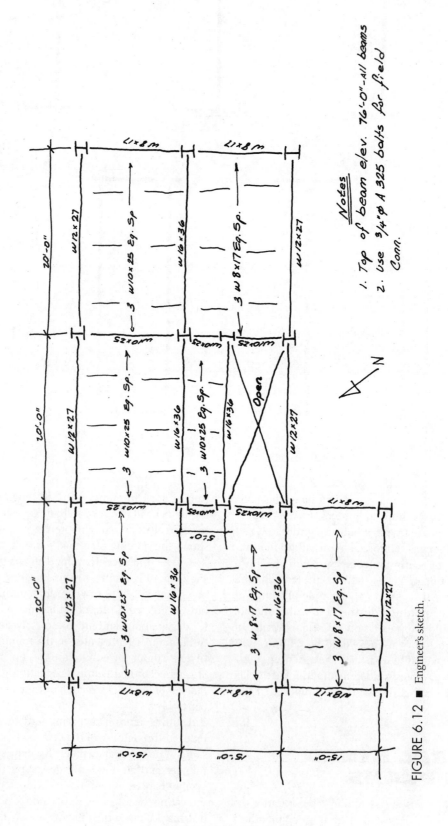

FIGURE 6.12 ■ Engineer's sketch.

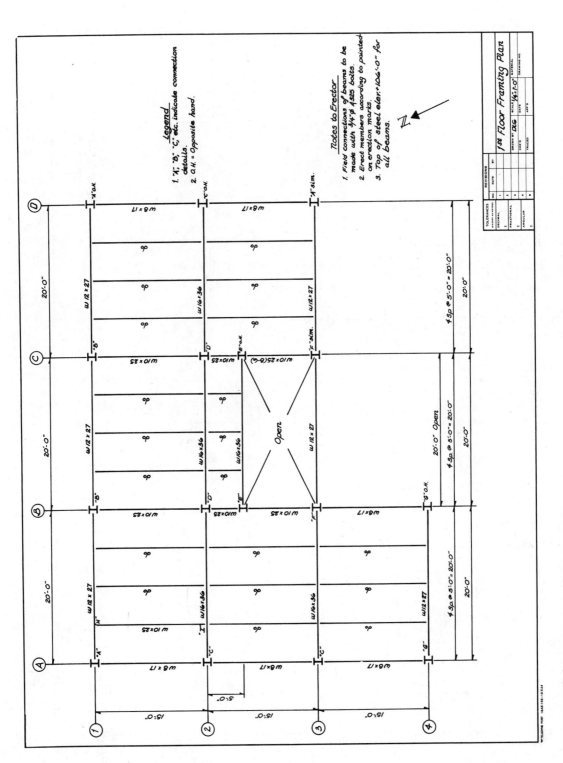

FIGURE 6.13 ■ Column-and-beam framing plan.

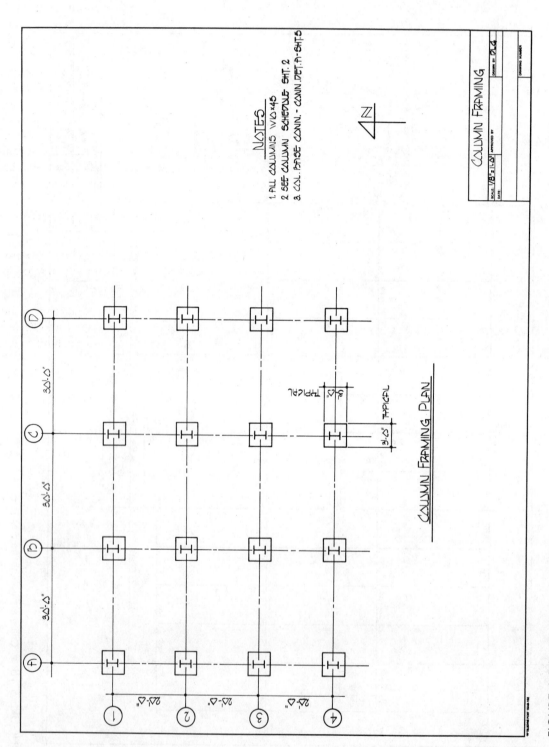

FIGURE 6.14 ■ Column framing plan.

a column framing plan. An example of a beam framing plan is shown in Figure 6.13.

Drawing Structural Steel Framing Plans

Structural steel framing plans consist of column framing plans showing the vertical structural members (columns) and the foundations on which they bear and beam framing plans showing the horizontal structural members (beams, girders, joists). The procedures used by drafters in drawing column framing and beam framing plans differ. The column framing plan is usually prepared first.

Drawing Column Framing Plans. The column framing plan is a plan view of all columns used in a job and the foundations on which they bear. Accompanying the column framing plan is a column schedule (Figure 6.15). The column schedule is drawn on the same sheet as the column framing plan if there is room. Column framing plans and schedules are prepared according to the following procedures:

1. Closely examine the engineer's sketches provided. Select an appropriate scale that will fit the framing plan comfortably on the sheet.

2. Lay out the centerline of column grid lines and draw in the plan view of each column (Figure 6.14).

3. Label vertical centerlines with letter designations and horizontal centerlines with numbers (Figure 6.14).

4. Draw in the foundations that support the columns (Figure 6.14).

5. Complete the foundation play by adding appropriate dimensions and a north arrow (Figure 6.14).

6. Add the column schedule as shown in Figure 6.15. Information required includes floor and roof elevations, top of foundation elevation, column designations, and splice points where applicable (Figure 6.15).

Drawing Beam Framing Plans. Floor and roof framing plans are commonly referred to as *beam framing plans*. The beam framing plan repeats the plan view of the columns and centerline of column designations. In addition, every beam, girder, and/or joist used in the job is shown in the plan (Figure 6.13).

The beam framing plan is also coded to identify each structural connection that requires a connection detail to guide the erection crew in erecting the structural connection. Each individual connection situation is given a letter designation on the framing plan (Figure 6.13). To save drawing time, similar connection situations may share a common connection detail. The abbreviation *Sim* is used on the framing plan as a suffix to the letter designation to indicate two or more connections sharing the same detail. The abbreviation *Opp Hd*, meaning opposite hand, may be applied to connection situations that are exactly opposite so they may also share a common connection detail (Figure 6.13).

Beam framing plans are drawn according to the following procedures:

1. Closely examine the engineer's sketch(es) provided. Select an appropriate scale that will fit the framing plan comfortably on the sheet.

2. Lay out the column grid lines and draw in the columns. Label the centerlines of columns with number and letter designations as was done on the column framing plan (Figure 6.13).

3. Draw in all horizontal members (beams, girders, and joists). Label each member with the proper shape designation (Figure 6.13).

4. Label the top of steel elevation for each beam next to its shape designation or place a top of steel elevation indication on the drawing as a note (Figure 6.13).

5. Complete the beam framing plan by adding appropriate dimensions, notes, connection detail designations, and a north arrow (Figure 6.13).

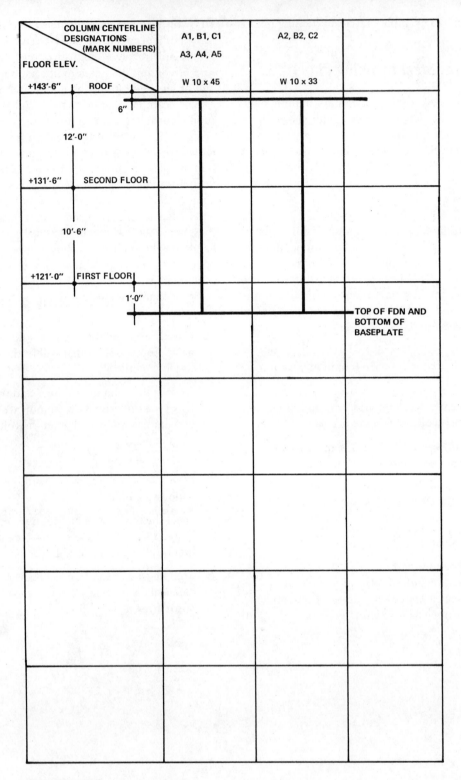

FIGURE 6.15 ■ Column schedule.

SUMMARY

■ Structural CAD technicians prepare engineering drawings and shop drawings. Engineering drawings are sometimes called erection drawings.

■ Framing plans are engineering drawings.

■ Rolled steel products are classified as plates, bars, or shapes.

■ Shapes are the most common structural steel framing products and include W shapes, S shapes, M shapes, angles, channels, structural tees, structural tubing, and pipe.

■ W shape is the new designation for shapes that were designated WF in the past.

■ S shape is the new designation for shapes that were designated I in the past.

■ M shape is the new designation for shapes that were designated M or Jr in the past.

■ Proper callout designations and dimensions for detailing steel shapes are provided in the *Manual of Steel Construction* published by the American Institute of Steel Construction (AISC).

■ Angles are properly designated as L shapes; channels are properly designated as C or MC shapes; and structural tubing is properly designated as TS shapes.

■ Structural tees are cut from W, M, and S shapes and are designated ST, MT, or WT depending on the shape from which they were cut.

■ Structural pipe has three classifications depending on the wall thickness of the pipe: standard, X-strong, and XX-strong.

■ The AISC, in its publication *Structural Steel Detailing*, specifies five types of built-up girders that are used in heavy-load or long-span situations.

■ Structural steel framing plans are of two types: column framing plans and beam framing plans.

■ Information for preparing structural steel framing plans is provided through engineering sketches.

■ CAD technicians prepare structural steel framing plans from information contained in the engineering sketches and the AISC *Manual of Steel Construction*.

REVIEW QUESTIONS

1. What is another name for an engineering drawing?

2. What category of drawings are framing plans?

3. What are the three classifications of rolled steel products?

4. List the most common structural steel framing products.

5. What is the CAD technician's most reliable reference source for determining proper callout designations and dimensions for detailing?

6. List the old designations for the following:
   a. W shape   b. M shape   c. C shape

7. List the three designations for structural tees.

8. List the three classifications of structural pipe.

9. Sketch an example of the five types of built-up girders used in long-span or heavy-load situations.

10. Name the two types of structural steel framing plans.

CAD ACTIVITIES

GENERAL INSTRUCTIONS

The following activities may be completed on any CAD system. Before reading the *specific instructions* for each activity (below), go through each step in the following planning checklist. The checklist applies to any CAD system and will help ensure the optimum use of your time and resources.

1. Analyze the problem carefully. Decide exactly what you are being asked to do.

2. Determine what resources and references you will need in order to complete the problem and collect them.

3. Decide if any particular standards apply to the project and have those standards available.

4. Determine what types of views will be required and how many of each.

5. Determine what the final plotted scale of the drawing will need to be, and select the appropriate paper size for plotting/printing (make sure the appropriate paper size is available).

6. Plan your drawing sequence. In what order will you develop the drawing (i.e., lines, features, dimension lines, leaders, dimensions, notes, etc.)?

7. Review the various CAD commands you will have to use in order to develop the drawing.

8. Examine your CAD system to ensure that everything is in working order, then begin the project.

SPECIFIC INSTRUCTIONS

Activity 6.1—This activity is to be prepared in accordance with the information supplied in the engineer's sketch in Figure 6.16. At an appropriate scale, draw a complete column framing plan from the sketch provided. Refer to the AISC *Manual of Steel Construction* for answers to questions concerning dimensions of structural shapes.

Activity 6.2—Prepare a complete column schedule for the column framing plan in CAD Activity 6.1.

Activity 6.3—Prepare a complete beam framing plan based on the information provided in the engineer's sketch in Figure 6.16. Refer to the AISC *Manual of Steel Construction* to ensure that all beam designations are properly called out.

Activity 6.4—This activity is to be prepared in accordance with the information supplied in the engineer's sketch in Figure 6.17. At an appropriate scale, draw a complete column framing plan from the sketch provided. Refer to the AISC *Manual of Steel Construction* for answers to questions concerning dimensions of structural shapes.

Activity 6.5—Prepare a complete column schedule for the column framing plan in CAD Activity 6.4.

Activity 6.6—Prepare a complete beam framing plan based on the information provided in the engineer's sketch in Figure 6.17. Refer to the AISC *Manual of Steel Construction* to ensure that all beam designations are properly called out.

Activity 6.7—This activity is to be prepared in accordance with the information supplied in the engineer's sketch in Figure 6.18. At an appropriate scale, draw a complete column framing plan from the sketch provided. Refer to the AISC *Manual of Steel Construction* for answers to questions concerning dimensions of structural shapes.

Activity 6.8—Prepare a complete column schedule for the column framing plan in CAD Activity 6.7.

Activity 6.9—Prepare a complete beam framing plan based on the information provided in the engineer's sketch in Figure 6.18. Refer to the AISC *Manual of Steel Construction* to ensure that all beam designations are properly called out.

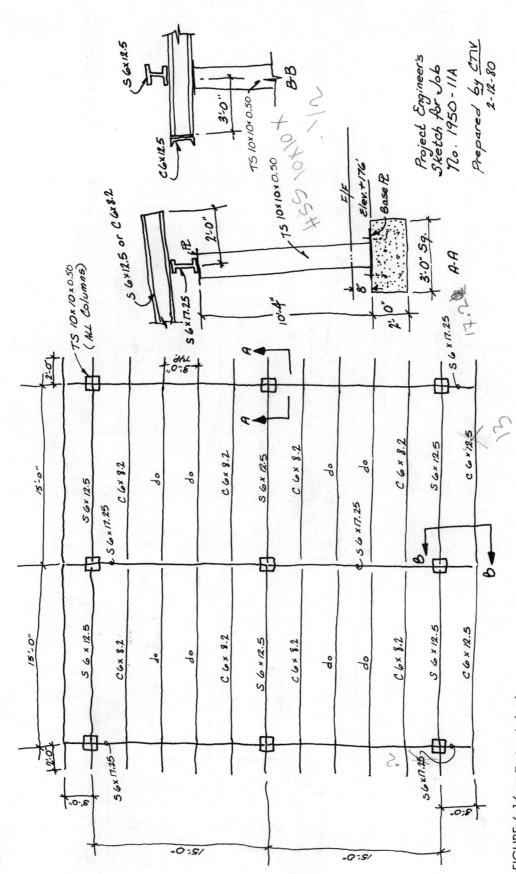

FIGURE 6.16 ■ Engineer's sketch.

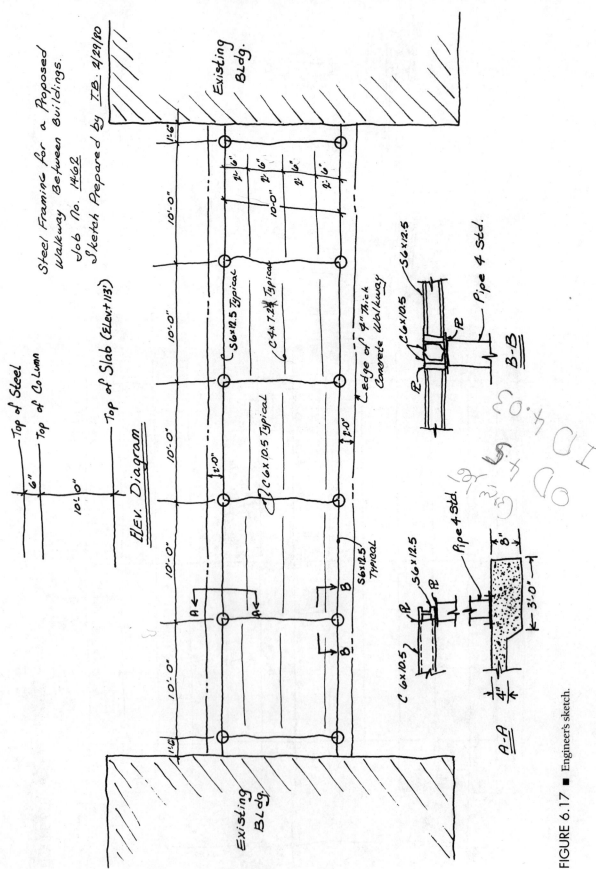

FIGURE 6.17 ■ Engineer's sketch.

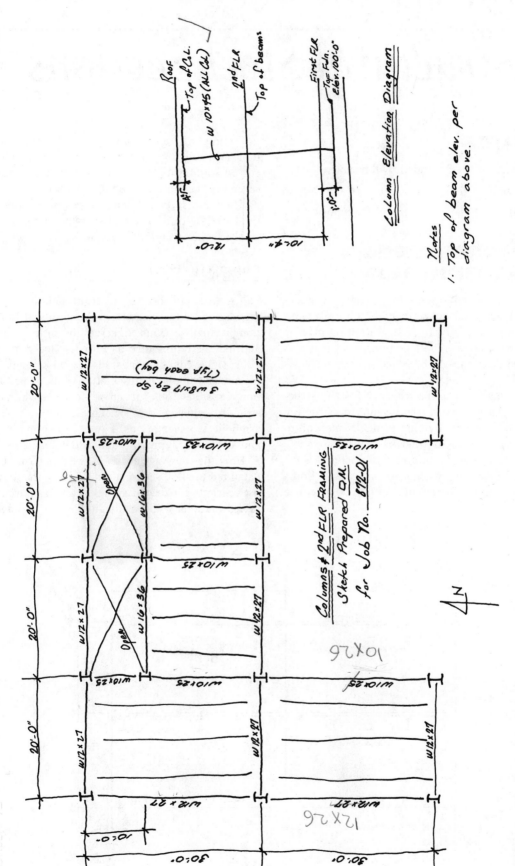

Column Elevation Diagram

Roof
Top of Col.
W 10×45 (All Col.)
2nd FLR
Top of beams
First FLR
Top Fdn.
Elev. 100'-0"

12'-0"
10'-4"

Notes
1. Top of beam elev. per diagram above.

Columns & 2nd FLR Framing
Sketch Prepared D.M.
for Job No. 892-01

N

FIGURE 6.18 ■ Engineer's sketch.

Structural Steel Sections

OBJECTIVES

Upon completion of this unit, the student will be able to:

■ Define structural steel sections.

■ Prepare structural steel full, partial, and offset sections.

STRUCTURAL STEEL SECTIONS DEFINED

Sections are drawings provided to show the reader what materials are used in a job and how they fit together to form a completed structure. Sections in structural steel drafting may be prepared as single-line symbolic representations. They may also be prepared as scaled duplicates of the actual structural shapes being represented (Figure 7.1 and Figure 7.2).

Sections are very important to the reader of a set of structural drawings. They clarify internal relationships in a structure that are not well defined on the framing plans. In addition to showing how a structure fits together, sections also show height information. This height information includes such things as distances between floors, distances between floors and the roof, top of steel information, bottom of baseplate information, and so on.

Sections are *cut* on framing plans but drawn on separate section sheets. Figure 7.3 shows several ways in which sections are cut on framing plans. Figure 7.4 contains an explanation of several of the more commonly used section cutting symbols.

There are several different types of sections used in structural steel drafting. The most commonly used are full sections, partial sections, and offset sections.

Full Sections

Full sections cut through an entire building or structure. Full sections are of two basic types: longitudinal and cross sections. **Longitudinal sections** cut across the entire length of a building and may be represented as actual scaled drawings or as symbolic, single-line drawings. Figure 7.5 contains a scaled drawing and a symbolic representation of the same longitudinal section. **Cross sections** cut across the entire width of a building or structure and may also be represented as scaled drawings or symbolic representations. Figure 7.6 shows a scaled drawing and a symbolic representation of the same cross section.

How the drawings are to be used determines the extent of detail required on sectional drawings. If the drawings are to be used strictly for design and engineering purposes, single-line representations will usually suffice. If the drawings are to be

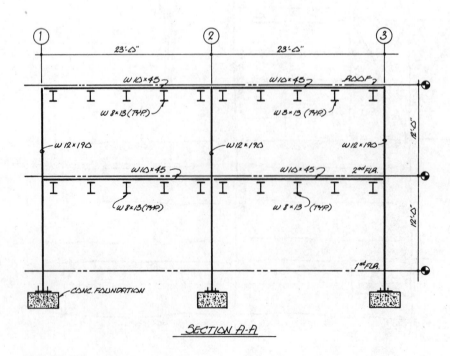

SECTION A-A

FIGURE 7.1 ■ Sample section.

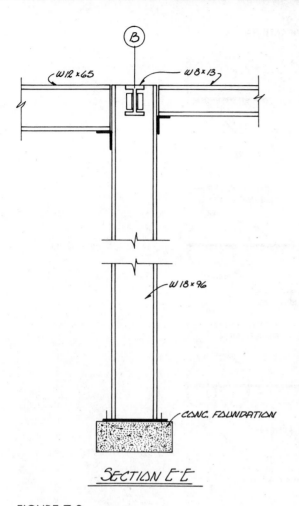

FIGURE 7.2   ■ Sample section.

used for guiding the erection crew or by other drafters in preparing fabrication details, actual scaled drawing representations are required.

All sections cut in structural steel drafting are structural sections. *Structural sections* show only the structural or load-bearing components of the building (foundation, floors, columns, and beams/girders). *Architectural sections,* on the other hand, show all of the interior features of the structure (carpet, flooring, paneling, suspended ceiling, baseboards, etc.). Since none of these architectural features is load bearing or structural, drafting time is shortened by excluding them.

Partial Sections

Isolated situations on a structural steel framing plan may be clarified without requiring full sections by using partial sections. *Partial sections* may be used to clarify internal relationships that are not cleared up by the longitudinal or cross sections. A framing plan may show both full and partial section cuts. Most framing plans have both because few framing plans are so simple that one longitudinal and one cross section will clarify all situations Figure 7.7 has two examples of partial sections as they are used in structural steel drafting.

Offset Sections

All section cutting plane lines do not run straight across the framing plan. Often, interior situations requiring clarification do not fall along a straight extension of the cutting plane line.

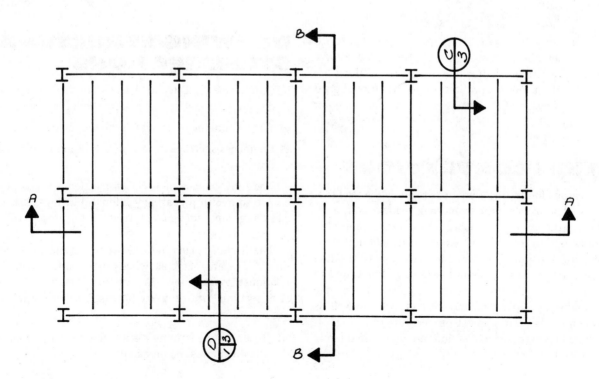

FIGURE 7.3   ■ Sample cut sections.

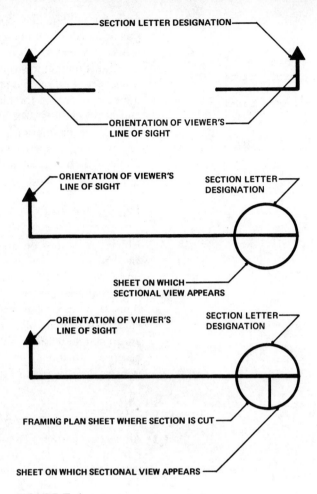

FIGURE 7.4 ■ Section cutting symbols.

When this is the case, an *offset section* may be cut. Figure 7.8 illustrates how an offset section cutting plan appears on a framing plan. The sectional drawings for offset sections are drawn as if the offset interior portion of the structure aligned with a straight extension of the cutting plane line from its point of origin (Figure 7.9).

SECTION CONVENTIONS

One of the reasons for cutting sections is to show the reader of a set of plans what materials the structure is made of. All materials used in a structure have standard symbols that are used in sectional drawings. In concrete and wood construction, there is an extensive list of materials that might be used in a structure. In steel construction, however, the list can be abbreviated to include only concrete and steel. Figure 7.10 gives an example of the sectioning symbols for concrete and steel as well as examples of how they are applied on sectional drawings. It should be noted that section conventions are used infrequently in structural steel drafting.

DRAWING STRUCTURAL STEEL SECTIONS

Structural steel CAD technicians obtain the information needed to prepare sectional drawings from the following two sources:

■ The framing plans on which the sections were cut

■ Engineer's sketches from which the framing plans were drawn

The AISC *Manual of Steel Construction* is also a valuable reference source in preparing sectional drawings. It is helpful in clarifying any ambiguities relative to proper callouts and dimensions.

The framing plans indicate what sections must actually be drawn. The engineer's sketches provide information on height relationships and how the structure fits together. Structural steel sectional drawings are prepared according to the following procedures:

1. Examine the framing plan(s) to determine what sectional drawings must be prepared.

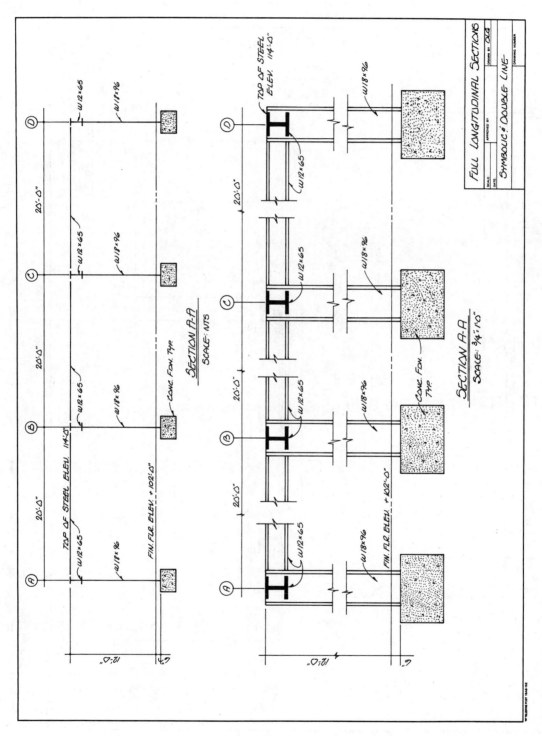

FIGURE 7.5 ■ Sample sections.

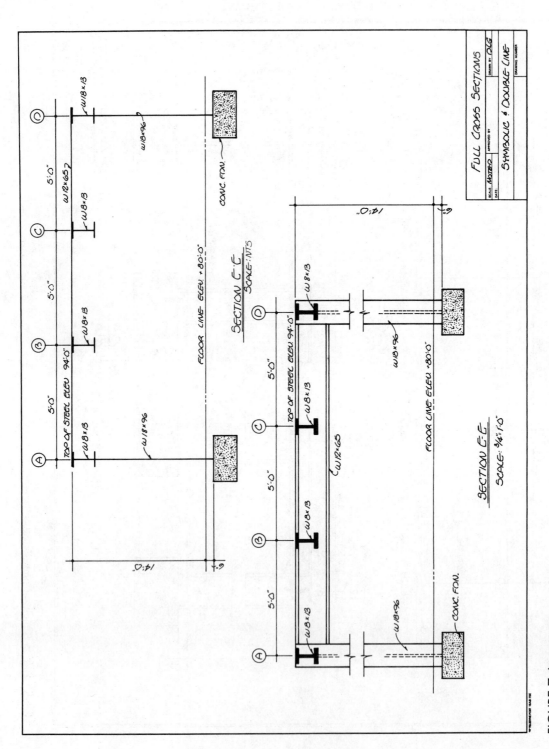

FIGURE 7.6 ■ Sample sections.

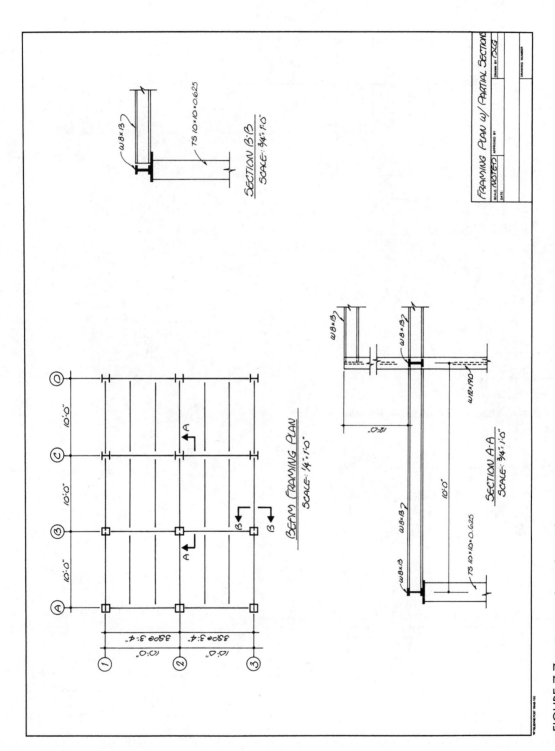

FIGURE 7.7 ■ Framing plan with partial sections.

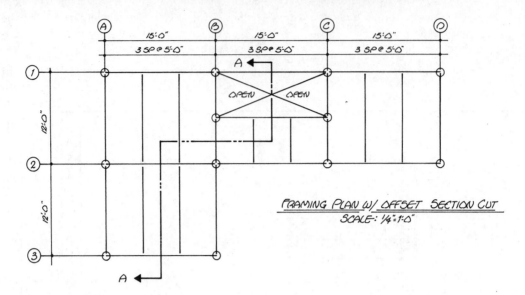

FIGURE 7.8 ■ Offset section cut.

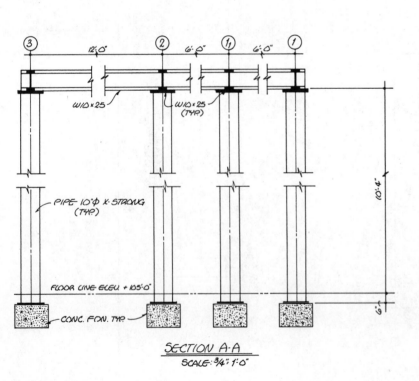

FIGURE 7.9 ■ Sample section.

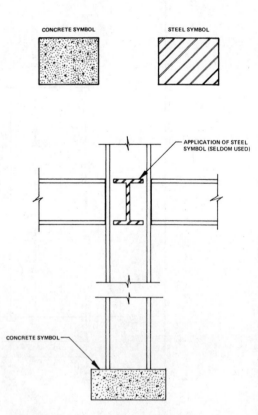

FIGURE 7.10 ■ Section and symbols.

2. Examine each sectioned situation individually and determine the following information:
   ■ What materials are used in the section?
   ■ How do the structural components fit together?
   ■ What are the important height distances that should appear on the sectional drawing? These will include such distances as: bottom of baseplate elevation, top of foundation elevation, floor elevations, roof elevations, column heights, and top of steel elevations.

3. Determine whether scaled drawings or symbolic representations are required and draw the sectional drawings at any appropriate scale. Symbolic representations may be drawn at smaller scales or not to scale.

■ Sections in structural steel drafting may be drawn as scaled duplicates of the actual structural shapes or as single-line symbolic representations.

■ All sections drawn in structural steel drafting are structural sections and only show load-bearing features such as the foundation, floor, columns, and beams/girders.

■ Partial sections are used to clarify isolated situations not cleared up by full sections.

■ Offset sections are used to clarify detailed interior situations in a structure that do not fall along a straight extension of the cutting plane line from its point of origin.

SUMMARY

■ Sections are drawings that show the materials used in a job and how they fit together to form a completed structure.

■ Sections clarify internal relationships in a job that are not clear or do not appear on the framing plan.

■ Sections are important sources of height information for readers of structural steel drawings.

■ The most commonly used types of sections in structural steel drafting are full, partial, and offset sections.

■ Full sections cut through an entire building or structure along an imaginary cutting plane.

■ Full sections are of two types: longitudinal and cross section.

■ Longitudinal sections cut across the entire length while cross sections cut across the entire width of a structure.

REVIEW QUESTIONS

1. Define *sections* as used in structural steel drafting.

2. List the three most common types of sections used in structural steel drafting.

3. Define *full section*.

4. List the two types of full sections.

5. Distinguish between the two types of sections in Question 4.

6. All sections in structural steel drafting are classified as what type of section? What does this mean?

7. Sketch an example of a partial section.

8. Explain what an offset section is and how one is used in structural steel drafting.

CAD ACTIVITIES

GENERAL INSTRUCTIONS

The following activities may be completed on any CAD system. Before reading the *specific instructions* for each activity (below), go through each step in the following planning checklist. The checklist applies to any CAD system and will help ensure the optimum use of your time and resources.

1. Analyze the problem carefully. Decide exactly what you are being asked to do.

2. Determine what resources and references you will need in order to complete the problem and collect them.

3. Decide if any particular standards apply to the project and have those standards available.

4. Determine what types of views will be required and how many of each.

5. Determine what the final plotted scale of the drawing will need to be, and select the appropriate paper size for plotting/printing (make sure the appropriate paper size is available).

6. Plan your drawing sequence. In what order will you develop the drawing (i.e., lines, features, dimension lines, leaders, dimensions, notes, etc.)?

7. Review the various CAD commands you will have to use in order to develop the drawing.

8. Examine your CAD system to ensure that everything is in working order, then begin the project.

SPECIFIC INSTRUCTIONS

The student may wish to refer to the AISC *Manual of Steel Construction* for dimensional information. See Figure 7.11 for the information required to complete these activities.

Activity 7.1—Prepare a single-line representation of the view indicated by Section AA. Refer to Figure 7.1 and Figure 7.5 for examples.

Activity 7.2—Prepare a double-line drawing of the representation completed in Activity 7.1. Refer to Figure 7.5 for an example.

Activity 7.3—Prepare a single-line representation of the view indicated by Section BB. Refer to Figure 7.6 for an example.

Activity 7.4—Prepare a double-line drawing of the representation completed in Activity 7.3. Refer to Figure 7.5 for an example.

Activity 7.5—Prepare a single-line representation of the offset section indicated by Section CC. Refer to Figure 7.1, Figure 7.5, and Figure 7.6 for examples.

Activity 7.6—Prepare a double-line drawing of the representation completed in Activity 7.5. Refer to Figure 7.9 for an example.

Activity 7.7—Prepare a double-line drawing of the view indicated by Section DD. Refer to Figure 7.2 for an example.

Activity 7.8—Prepare a double-line drawing of the view indicated by Section EE. Refer to Figure 7.2 for an example.

Activity 7.9—Prepare a double-line drawing of the view indicated by Section FF. Refer to Figure 7.2 for an example.

Activity 7.10—Prepare a double-line drawing of the view indicated by Section GG. Refer to Figure 7.2 for an example.

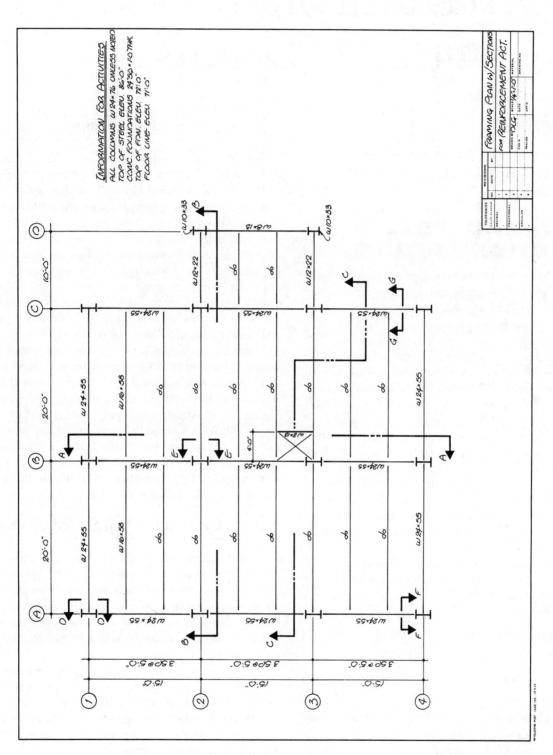

FIGURE 7.11 ■ Framing plan for drafting activities.

8 UNIT

Structural Steel Connection Details

OBJECTIVES

Upon completion of this unit, the student will be able to:

■ Prepare complete structural steel baseplate, framed, and seated connections.

STRUCTURAL STEEL CONNECTION DETAILS

Connection details in structural drafting are used to show exactly how all structural members are to be connected during erection. They are very similar to sections except they are much more detailed and they isolate on the area in which the connection is located. For example, a column baseplate connection shows only that portion of the column immediately around the baseplate and the foundation on which it rests. Beam-to-column and beam-to-beam connections also isolate on the connection. Figure 8.1 shows several examples of structural steel connection details.

Connection details are used in a number of different ways and are an important part of a set of structural steel working drawings. CAD technicians use the details, along with sections and framing plans, in preparing fabrication details of structural members. Erection crews use connection details to guide them in putting a structure together. Designing connection details is the responsibility of an engineer or a designer. Drawing them is the responsibility of the drafter.

Structural steel connections may be either bolted or welded. Riveting, once widely used, is no longer considered a major steel-connecting process. Welded and/or bolted connections are used for the following:

■ Connecting column baseplates to a foundation

■ Making column splices

■ Connecting beams to columns

■ Connecting beams to beams

Column baseplate connections are usually bolted. Beam-to-column and beam-to-beam connections may be either bolted or welded, and they may be either framed or seated or a combination of both.

Baseplate Connection Details

Structural steel columns are fitted with a steel baseplate designed to fit over **anchor bolts** cast into a concrete foundation. The detail showing exactly how this connection is made and specifying all necessary information about the connection is called a *baseplate connection detail*.

A baseplate connection detail must be provided for every different connection situation. This means that baseplate connection situations in which all vital information (baseplate size, anchor bolt size, and connector specifications) is the same may share a common connection detail. If any of this information or any other information relating to the connection differs, a separate connection detail must be provided.

A common practice is to call out the various connection details that are to be drawn on the framing plan. Letter designations such as A, B, C, and so on are used to do this. Baseplate connection details are called out on the column framing plan. Figure 8.2 illustrates commonly used baseplate connection details for W-, S-, and M-shaped columns. Figure 8.3 shows baseplate connections for tubular and pipe columns. The sizes and grades of plates, angles, bolts, and so on in these details were specified by an engineer. The details themselves were drawn by a CAD technician.

Framed Connections and Seated Connections

A common connection method for joining beams to columns and beams to beams is framed connection. A *framed connection* involves connecting one member to another member at the webs. This is accomplished by attaching angles to the webs of one member and connecting these angles to the web of the other member. All connections in a framed connection may be either welded or bolted.

It is common practice to connect angles or any other connectors that are to be attached in the shop by welding and then to make field connections by bolting. Figure 8.4, Figure 8.5, and Figure 8.6 show examples of framed and seated connections. Note that a connection detail is actually a plan view of the connection with sections cut through it to show elevations of the connection.

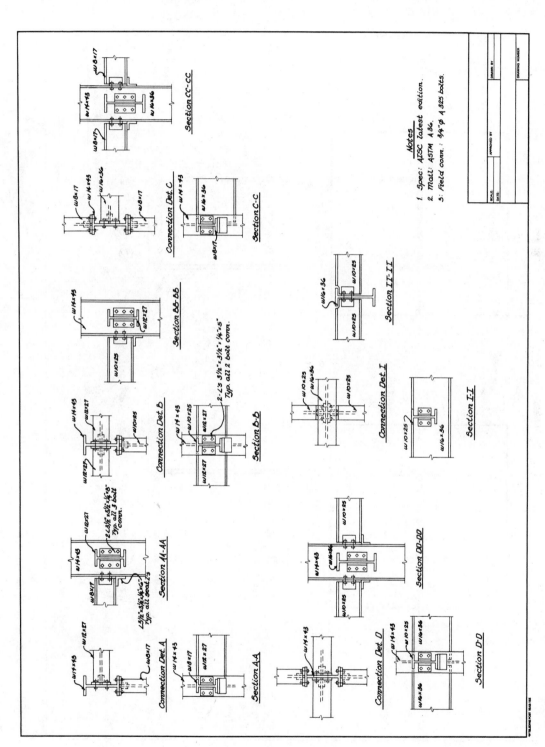

FIGURE 8.1 ■ Connection details.

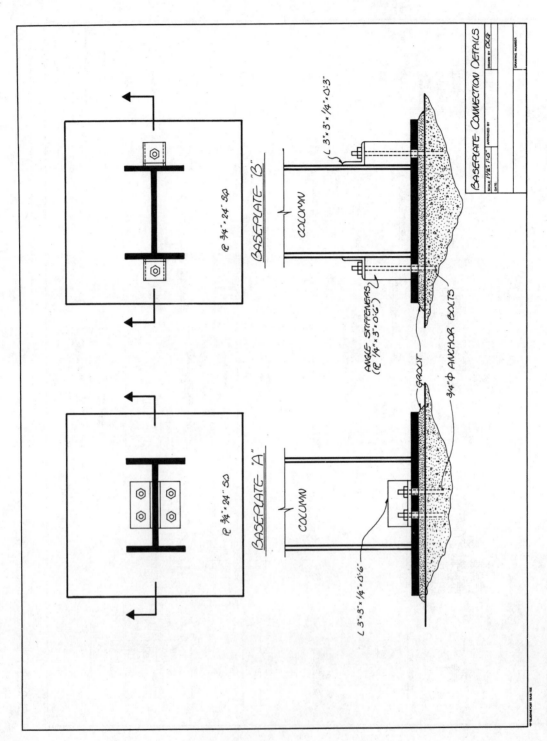

FIGURE 8.2 ■ Baseplate connection details.

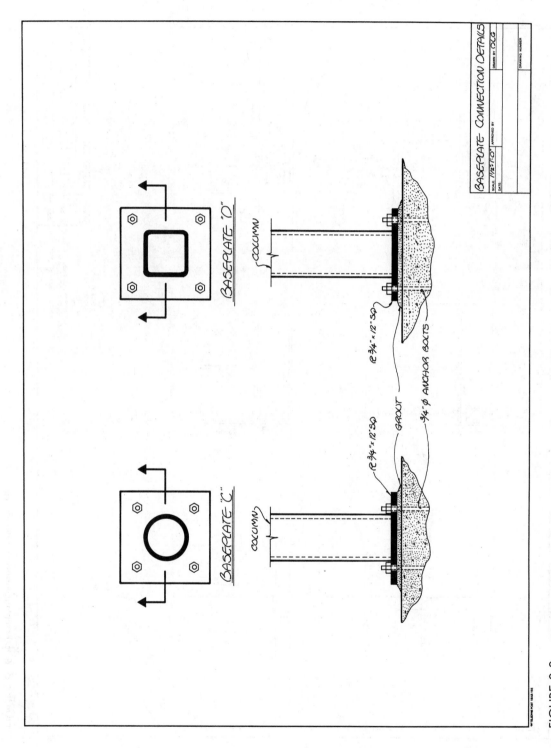

FIGURE 8.3 ■ Baseplate connection details.

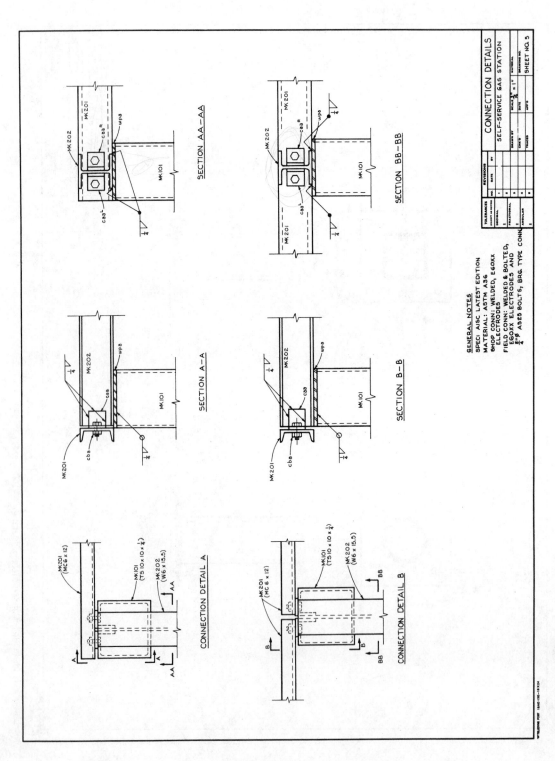

FIGURE 8.4 ■ Connection details.

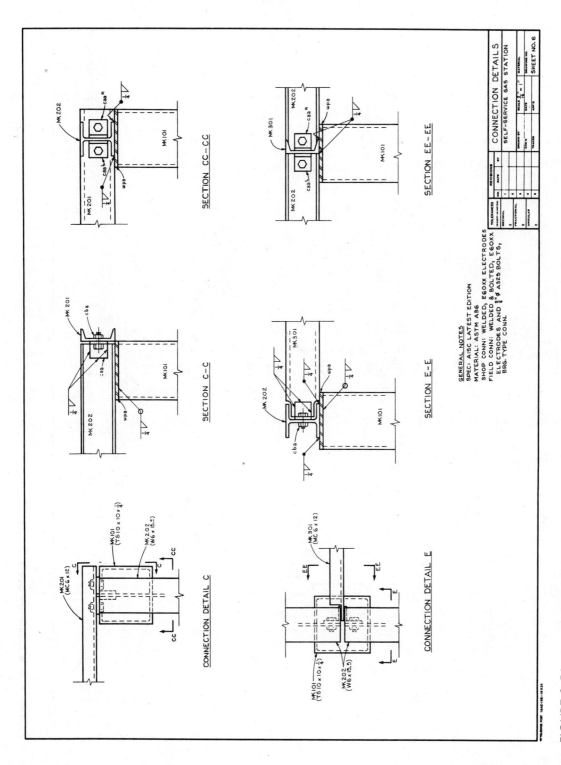

FIGURE 8.5 ■ Connection details.

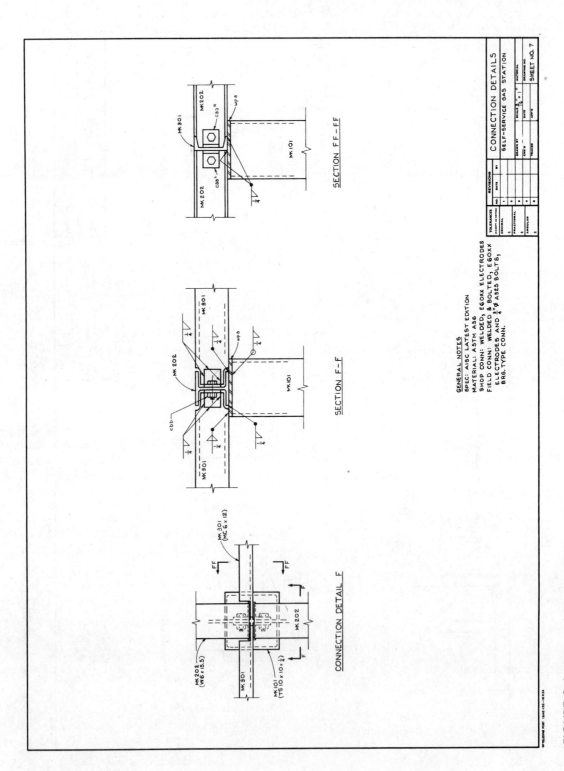

FIGURE 8.6 ■ Connection details.

Figure 8.4 contains two connection details, A and B, which were properly prepared. Connection Detail A shows a W6 × 15.5 resting on a plate that has been welded to the top of a 10″ square tubular column. A framed connection has been provided for connecting an MC 6 × 12 to the web of the W6 × 15.5. This connection was accomplished by preparing two angles with one hole apiece to accept a bolt. These angles were then welded to opposite sides of the web of the W6 × 15.5 beam in the shop.

Upon erection of the structure, a bolt was passed through aligning holes in the angles and the channel to be connected. Connection Detail B in Figure 8.4 is very similar to Connection Detail A. All of the remaining connection details in Figure 8.5 and Figure 8.6 were accomplished in much the same manner as Connection Detail A.

Seated connections occur when a beam rests on top of a column or another beam or when a beam must be attached to the flange of a column. Seated connections are often used together with framed connections. By doing this, erection is made easier. The erectors are given a surface on which to place the beam while connections are being made.

All of the connection details contained in Figure 8.4, Figure 8.5, and Figure 8.6 combine seated and framed connections. In Connection Detail A (Figure 8.4), the W6 × 15.5 beam resting on the TS 10 × 10 × 1/4 represents a seated connection. The beam actually sits on the column and is connected to it by welding. The connection of the beam and the channel is a framed connection.

A more common type of seated connection involves a beam framing into the flange of a W-, S-, or M-shaped column. In this case, an angle is attached to the column during fabrication and the beam is seated on the angle during erection (Figure 8.7).

Connection details for beam-to-beam and beam-to-column connections are called out by letter designations on the beam framing plan (Figure 6.13). All connector specifications are provided by an engineer or a designer.

Drawing Structural Steel Connection Details

Designing structural connections is the job of an engineer or an experienced designer. Drawing connection details is the job of the CAD technicians. In order to draw connection details, the drafter must have the following items:

■ All pertinent framing plans

■ All pertinent sections

■ Engineering specifications for all connections

When these things are in hand, the drafter proceeds as follows:

1. Examine the column framing plan(s) to determine what and how many baseplate details are required. Examine the beam framing plan(s) to determine what and how many connection details are required.

2. Using the framing plan(s) and the sections together, construct plan views for each connection detail required.

3. Cut necessary sections on the connection detail plan views and draw the required elevation views.

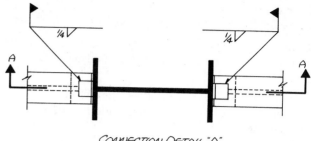

CONNECTION DETAIL "A"

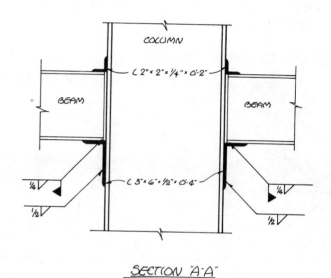

SECTION "A·A"

FIGURE 8.7 ■ Connection detail.

4. Using the engineering specifications as a guide, label all connectors and connecting material. Add any additional notes that may be needed for clarification.

SUMMARY

■ Structural steel connection details are used to show exactly how all structural members in a job are connected during erection.

■ Connection details are similar to sections except they are more detailed and isolate on the connection area only.

■ Connection details are used by CAD technicians in preparing fabrication details of structural members and by the erection crew in putting the structure together.

■ Designing structural connections is the job of the engineer or designer. Drawing them is the job of the CAD technician.

■ Structural connections may be either bolted or welded. Riveting is no longer considered a major steel-connecting process.

■ Typical structural steel connection situations that must be detailed are column baseplate connections, beam-to-column connections, and beam-to-beam connections.

■ Baseplate connection details are called out on the column framing plan and must be provided for every different column connection situation.

■ Beam-to-column connection details may be either framed or seated. Beam-to-beam details are usually framed.

■ A framed connection joins two structural members at the webs. A seated connection provides a bearing surface for one member to sit on.

■ A connection detail contains a plan view of the connection with appropriate sections cut to provide elevations of the connection.

■ In order to draw connection details for a job, the structural drafter must have the framing plan(s), the sections, and engineering data on connector specifications.

REVIEW QUESTIONS

1. Why are connection details included in a set of structural steel working drawings?

2. How do connection details differ from sections?

3. How and by whom are connection details used?

4. In preparing connection details, what is the division of labor between engineering and drafting?

5. Name the two most common connecting processes used in structural steel connections.

6. Name three structural steel connection situations that must be detailed.

7. Explain how and where connection details are called out.

8. Explain the difference between a framed and a seated connection by constructing a simple sketch of each.

9. Provide a simple sketch that illustrates all of the components of a beam-to-column connection detail.

10. What three things are needed by the structural drafter in preparing connection details for a job?

CAD ACTIVITIES

GENERAL INSTRUCTIONS

The following activities may be completed on any CAD system. Before reading the *specific instructions* for each activity (below), go through each step in the following planning checklist. The checklist applies to any CAD system and will help ensure the optimum use of your time and resources.

1. Analyze the problem carefully. Decide exactly what you are being asked to do.

2. Determine what resources and references you will need in order to complete the problem and collect them.

3. Decide if any particular standards apply to the project and have those standards available.

4. Determine what types of views will be required and how many of each.

5. Determine what the final plotted scale of the drawing will need to be, and select the appropriate paper size for plotting/printing (make sure the appropriate paper size is available).

6. Plan your drawing sequence. In what order will you develop the drawing (i.e., lines, features, dimension lines, leaders, dimensions, notes, etc.)?

7. Review the various CAD commands you will have to use in order to develop the drawing.

8. Examine your CAD system to ensure that everything is in working order, then begin the project.

SPECIFIC INSTRUCTIONS

Column-and-beam framing information for the activities is contained in Figure 8.8. Sectional information and engineer-ing data for the activities is contained in Figure 8.9. Several of the activities listed may be drawn on one sheet.

Activity 8.1—At an appropriate scale, prepare baseplate connection detail "A."

Activity 8.2—At an appropriate scale, prepare baseplate connection detail "B."

Activity 8.3—At an appropriate scale, prepare baseplate connection detail "C."

Activity 8.4—At an appropriate scale, prepare connection detail "A" (beam to column).

Activity 8.5—At an appropriate scale, prepare connection detail "B" (beam to column).

Activity 8.6—At an appropriate scale, prepare connection detail "C" (beam to column).

Activity 8.7—At an appropriate scale, prepare connection detail "D" (beam to column).

Activity 8.8—At an appropriate scale, prepare connection detail "E" (beam to column).

Activity 8.9—At an appropriate scale, prepare connection detail "F" (beam to column).

Activity 8.10—At an appropriate scale, prepare connection detail "G" (beam to column).

Activity 8.11—At an appropriate scale, prepare connection detail "H" (beam to column).

Activity 8.12—At an appropriate scale, prepare connection detail "I" (beam to beam).

Activity 8.13—At an appropriate scale, prepare connection detail "J" (beam to beam).

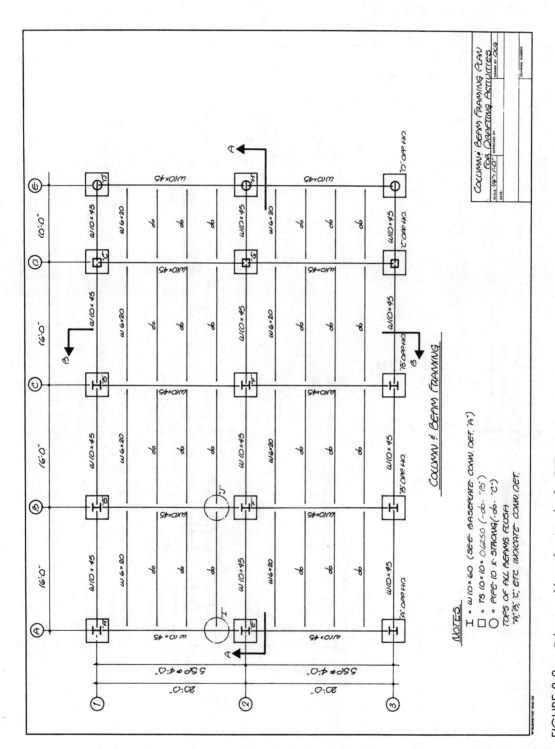

FIGURE 8.8 ■ Column-and-beam framing plan for CAD activities.

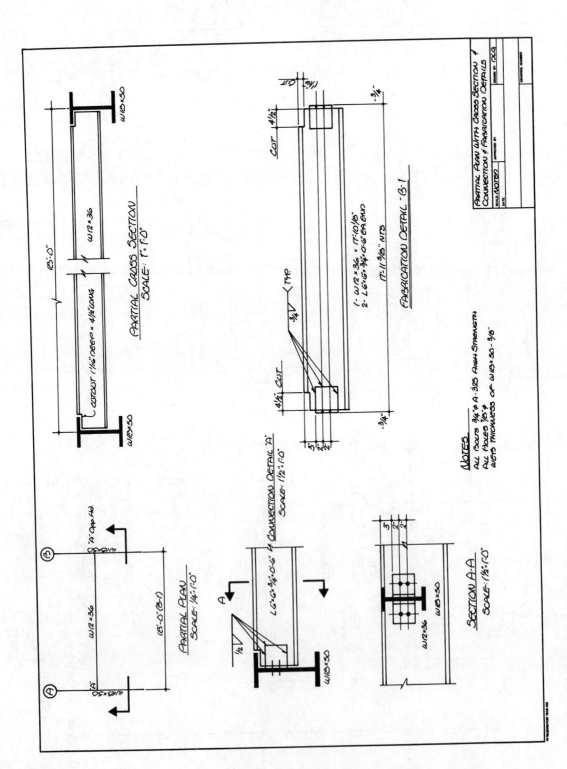

FIGURE 8.9 ■ Sections for CAD activities.

Structural Steel Fabrication Details

OBJECTIVES

Upon completion of this unit, the student will be able to:

- Define structural steel shop drawings.
- Define structural steel fabrication details.
- Construct fabrication details for structural steel columns and beams.

STRUCTURAL STEEL SHOP DRAWINGS DEFINED

In Unit 1, it was learned that a complete set of structural working drawings consists of engineering drawings for engineering purposes and shop drawings for fabrication purposes. *Engineering drawings* in structural steel drafting include framing plans, sections, and connection details. These were covered in Unit 6, Unit 7, and Unit 8. *Shop drawings* include fabrication details and bills of material. This unit deals with fabrication details. Bills of material are covered in Unit 10.

Shop drawings are comprehensive, precisely detailed drawings accompanied by bills of material. Shop drawings are prepared for use by shop workers in fabricating structural steel members for a job. The primary component of shop drawings is the fabrication detail. Fabrication details are prepared by structural steel drafters. The information required in constructing the fabrication details for a job is contained in the framing plan(s), sections, and connection details.

STRUCTURAL STEEL FABRICATION DETAILS DEFINED

Fabrication details are orthographic drawings of structural steel columns and beams. They contain all of the information required by shop workers in fabricating the structural members used in any given job. Each different column and beam used in the framing of a job must have an individual fabrication detail. This fabrication detail shows length information, locations of plates and angles, positions of holes that are to be drilled, and any other information relating to the fabrication process (Figure 9.1).

CONSTRUCTING STRUCTURAL STEEL FABRICATION DETAILS

In order to construct the fabrication details for a job, the drafter must have the framing plan(s) for the job, the sections, and the connection details. The framing plan(s) provides such information as:

- How many columns must be detailed?
- What is the structural shape of each column?
- How many of a certain column designation are required?

The framing plan provides this same information for beams.

Sections provide height information valuable in calculating the height of columns. Connection details provide the information necessary for converting centerlines of column distances into actual beam lengths. The details also provide information needed for locating holes, plates, angles, and so on. Connection specifications also appear on the connection details.

Figure 9.2 shows a removed portion of a framing plan, a section through the removed portion, and a connection detail. The fabrication detail shown was developed from information provided on the framing plan, section, and connection detail. Examine Figure 9.1 closely to determine how the actual member length of the beam and hole locations were arrived at.

Column Fabrication Details

Structural steel fabrication details contain the following information:

- At least two views (plan and elevation) of the column
- Complete locational dimensions for holes, plates, and angles
- The baseplate
- Connection specifications
- Miscellaneous notes for the fabricator (Figure 9.3)

Column fabrication details are drawn according to the following procedures:

1. The column length is determined. Column length calculations are illustrated in Figure 9.4 and Figure 9.5.

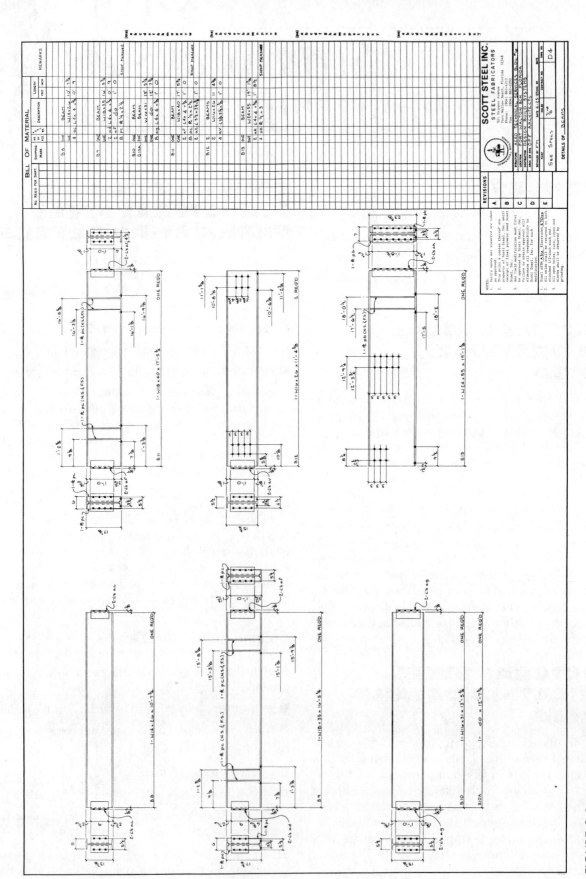

FIGURE 9.1 ■ Fabrication detail drawing.

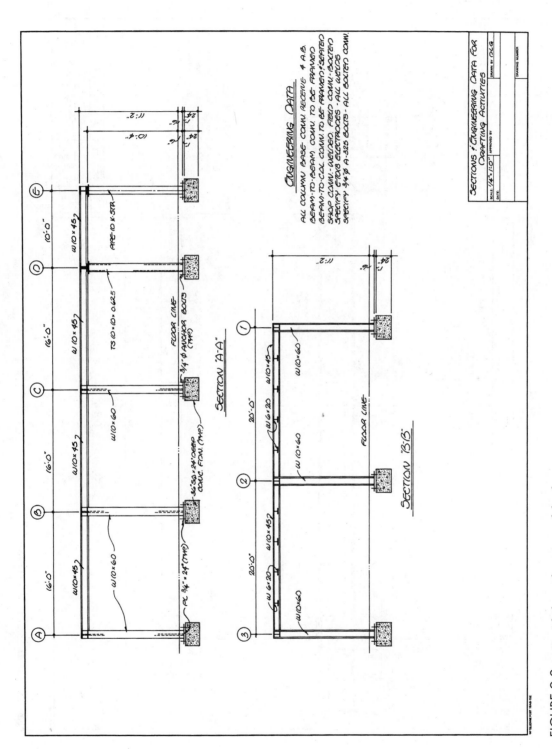

FIGURE 9.2 ■ Partial plan with section and details.

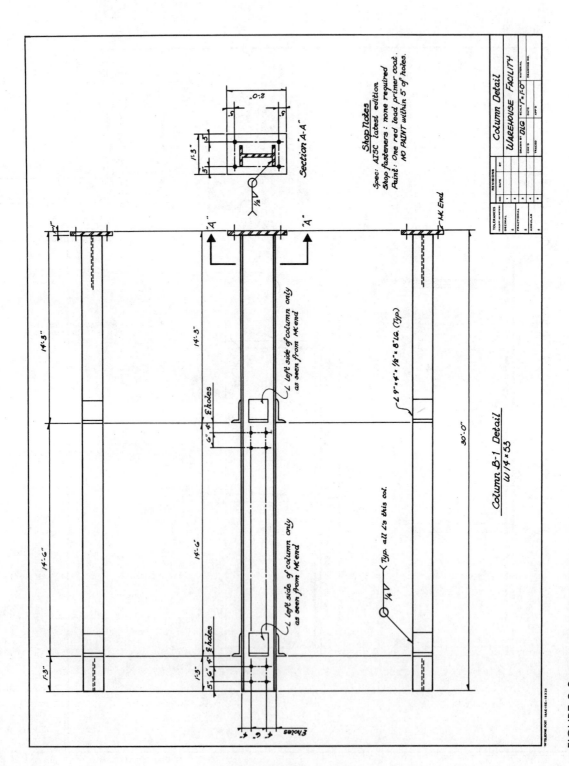

FIGURE 9.3 ■ Sample column detail.

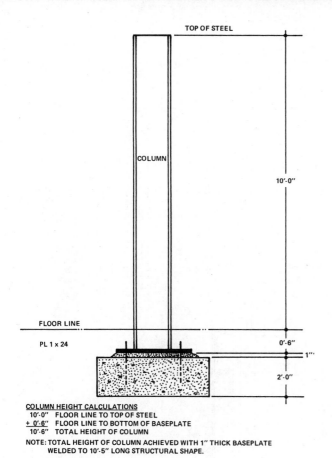

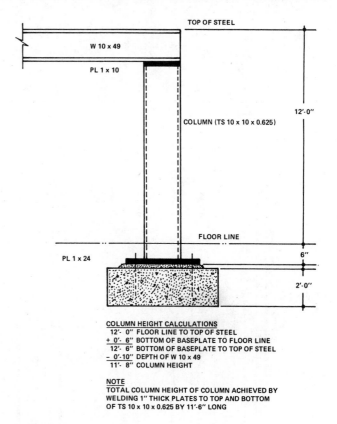

COLUMN HEIGHT CALCULATIONS
10'-0"   FLOOR LINE TO TOP OF STEEL
+ 0'-6"   FLOOR LINE TO BOTTOM OF BASEPLATE
10'-6"   TOTAL HEIGHT OF COLUMN

NOTE: TOTAL HEIGHT OF COLUMN ACHIEVED WITH 1" THICK BASEPLATE
WELDED TO 10'-5" LONG STRUCTURAL SHAPE.

FIGURE 9.4 ■ Column height calculations.

COLUMN HEIGHT CALCULATIONS
12'- 0"   FLOOR LINE TO TOP OF STEEL
+ 0'- 6"   BOTTOM OF BASEPLATE TO FLOOR LINE
12'- 6"   BOTTOM OF BASEPLATE TO TOP OF STEEL
− 0'-10"   DEPTH OF W 10 x 49
11'- 8"   COLUMN HEIGHT

NOTE
TOTAL COLUMN HEIGHT OF COLUMN ACHIEVED BY
WELDING 1" THICK PLATES TO TOP AND BOTTOM
OF TS 10 x 10 x 0.625 BY 11'-6" LONG

FIGURE 9.5 ■ Column height calculations.

2. The required orthographic views of the column are drawn to the proper length, and the overall length dimension is placed on the drawing. It is common practice in structural steel drafting to draw width and depth dimensions to scale but the length not to scale (NTS). (Figure 9.3).

3. Locations of holes, plates, angles, and so on are determined from the connection details and placed on the drawing with accompanying dimensions (Figure 9.3).

4. A section is cut through the column facing the baseplate. The sectional view is drawn to show how the baseplate is attached and its configuration (Figure 9.3).

5. Any connection specifications or other information that should be noted is taken from the connection details and entered on the drawing (Figure 9.3).

Beam Fabrication Details

Structural steel beam fabrication details contain the following information:

■ An elevation of the beam with end views or sections when required for clarity

■ Complete locational dimensions for holes, plates, and angles

■ Length dimensions

■ Connection specifications

■ Cutouts

■ Miscellaneous notes for the fabricator (Figure 9.6)

Beam fabrication details are drawn according to the following procedures:

1. The beam length is determined. Beam length calculations are illustrated in Figure 9.7 and Figure 9.8.

2. The required orthographic views of the beam are drawn to the proper length, and the overall dimension is placed on the drawing. It is common practice in structural steel drafting to show depth and width dimensions to scale but the length not to scale (NTS) (Figure 9.6).

3. Locations of holes, plates, angles, and so on are determined from the connection details and placed on the drawing with accompanying dimensions (Figure 9.6).

4. Any connection specifications or other information that should be noted is taken from the connection details and entered on the drawing (Figure 9.6).

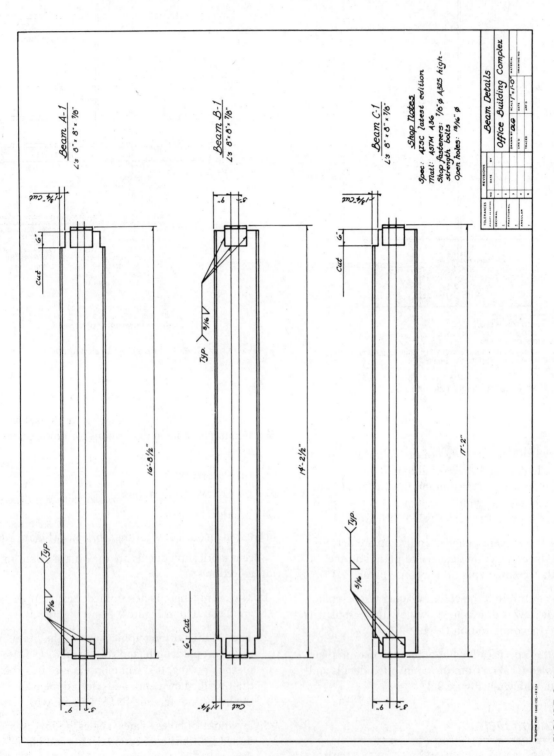

FIGURE 9.6 ■ Sample beam details.

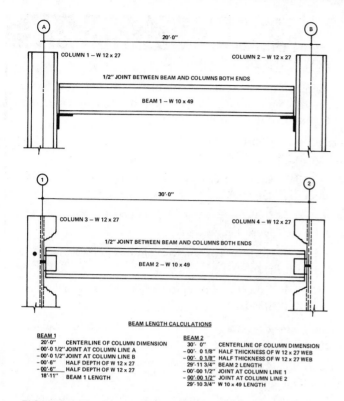

FIGURE 9.7 ■ Beam length calculations.

BEAM LENGTH CALCULATIONS

BEAM 1
20'-0"   CENTERLINE OF COLUMN DIMENSION
– 00'-0 1/2"   JOINT AT COLUMN LINE A
– 00'-0 1/2"   JOINT AT COLUMN LINE B
– 00'-6"   HALF DEPTH OF W 12 x 27
– 00'-6"   HALF DEPTH OF W 12 x 27
18'-11"   BEAM 1 LENGTH

BEAM 2
30'- 0"   CENTERLINE OF COLUMN DIMENSION
– 00'- 0 1/8"   HALF THICKNESS OF W 12 x 27 WEB
– 00'- 0 1/8"   HALF THICKNESS OF W 12 x 27 WEB
29'-11 3/4"   BEAM 2 LENGTH
– 00'-00 1/2"   JOINT AT COLUMN LINE 1
– 00'-00 1/2"   JOINT AT COLUMN LINE 2
29'-10 3/4"   W 10 x 49 LENGTH

FIGURE 9.8 ■ Beam length calculations.

BEAM LENGTH CALCULATIONS

BEAM 1
30'- 0"   CENTERLINE OF COLUMN DIMENSION
– 00'-00 1/2"   HALF THICKNESS WEB OF W 14 x 228
– 00'-00 1/2"   HALF THICKNESS WEB OF W 14 x 228
29'-11"   BEAM 1 LENGTH
– 00'-00 1/2"   JOINT AT COLUMN LINE A
– 00'-00 1/2"   JOINT AT COLUMN LINE B
29'-10"   W 12 x 53 LENGTH
(CUTOUT DOES NOT AFFECT LENGTH)

BEAM 2
24'- 0"   CENTERLINE OF COLUMN DIMENSION
– 00'- 0 1/8"   HALF THICKNESS WEB OF W 12 x 40
– 00'- 0 1/8"   HALF THICKNESS WEB OF W 12 x 40
23'-11 3/4"   BEAM 2 LENGTH
– 00'-00 1/2"   JOINT AT COLUMN LINE 1
– 00'-00 1/2"   JOINT AT COLUMN LINE 2
23'-10 3/4"   W 6 x 12 LENGTH

NOTE
WEB THICKNESSES ARE AVAILABLE IN THE APPENDIX OR
AISC *MANUAL OF STEEL CONSTRUCTION.*

SUMMARY

■ A complete set of structural steel working drawings consists of engineering drawings for engineering purposes and shop drawings for fabrication purposes.

■ Shop drawings include fabrication details and bills of material.

■ The primary component of shop drawings is the fabrication detail.

■ Fabrication details are orthographic drawings of structural steel columns and beams showing all information necessary for fabrication.

■ Fabrication details are prepared by structural drafters from information found on the framing plan(s), sections, and connection details.

■ Column fabrication details contain: orthographic views of the column; locational dimensions for plates, angles, and holes; length dimensions; the baseplate configuration; connection specifications; and miscellaneous notes for the fabricator.

■ Beam fabrication details contain: orthographic views of the beam; sections or end views when required for clarity; length dimensions; locational dimensions for holes, angles and plates; cutouts; connection specifications; and miscellaneous notes for the fabricator.

REVIEW QUESTIONS

1. Define the term *shop drawing* as used in structural steel drafting.

2. Define the term *fabrication detail* as used in structural steel drafting.

3. What is the primary component of shop drawings?

4. What must the structural steel CAD technician have in order to prepare the fabrication details for a job?

5. What information is contained in a column fabrication detail?

6. What information is contained in a beam fabrication detail?

7. Construct a simple sketch that illustrates how to calculate column heights for I-shaped columns.

8. Construct a simple sketch that illustrates how to calculate the beam length for a beam that frames into the webs of two I-shaped columns.

9. Construct a simple sketch that illustrates how to calculate the beam length for a beam that frames into the flanges of two I-shaped columns.

10. Construct a simple sketch that illustrates how to calculate the beam length for a beam that frames into the web of two other beams.

CAD ACTIVITIES

GENERAL INSTRUCTIONS

The following activities may be completed on any CAD system. Before reading the *specific instructions* for each activity (below), go through each step in the following planning checklist. The checklist applies to any CAD system and will help ensure the optimum use of your time and resources.

1. Analyze the problem carefully. Decide exactly what you are being asked to do.

2. Determine what resources and references you will need in order to complete the problem and collect them.

3. Decide if any particular standards apply to the project and have those standards available.

4. Determine what types of views will be required and how many of each.

5. Determine what the final plotted scale of the drawing will need to be, and select the appropriate paper size for plotting/printing (make sure the appropriate paper size is available).

6. Plan your drawing sequence. In what order will you develop the drawing (i.e., lines, features, dimension lines, leaders, dimensions, notes, etc.)?

7. Review the various CAD commands you will have to use in order to develop the drawing.

8. Examine your CAD system to ensure that everything is in working order, then begin the project.

SPECIFIC INSTRUCTIONS

All information necessary for completing the activities is contained in Figure 9.9. In all activities, the students must refer to Figure 9.10 through Figure 9.13.

Activity 9.1—Construct a complete column fabrication detail for column A2, Figure 9.9.

Activity 9.2—Construct a complete column fabrication detail for column B1, Figure 9.9.

Activity 9.3—Construct a complete column fabrication detail for column C1, Figure 9.9.

Activity 9.4—Construct a complete column fabrication detail for column E2, Figure 9.9.

Activity 9.5—Construct a complete beam fabrication detail for beam B-1, Figure 9.9.

Activity 9.6—Construct a complete beam fabrication detail for beam B-2, Figure 9.9.

Activity 9.7—Construct a complete beam fabrication detail for beam B-3, Figure 9.9.

Activity 9.8—Construct a complete beam fabrication detail for beam B-4, Figure 9.9.

Activity 9.9—Construct a complete beam fabrication detail for beam C-1, Figure 9.9.

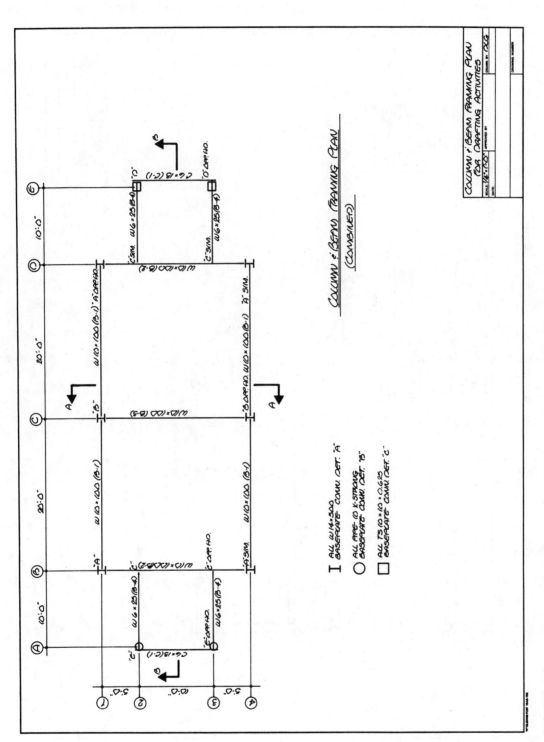

FIGURE 9.9 ■ Column-and-beam framing plan for CAD activities.

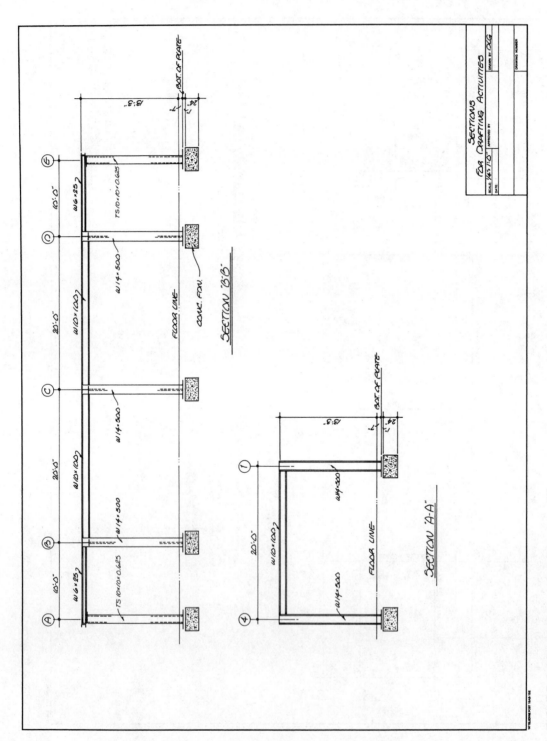

FIGURE 9.10 ■ Sections for CAD activities.

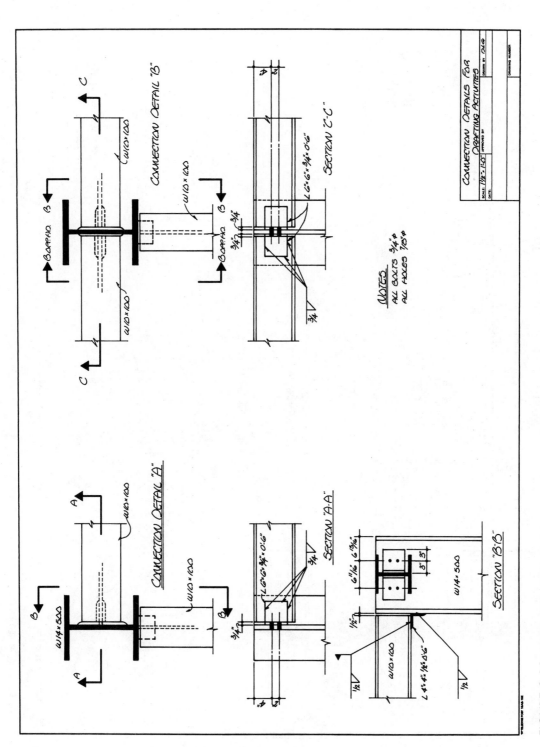

FIGURE 9.11 ■ Details for CAD activities.

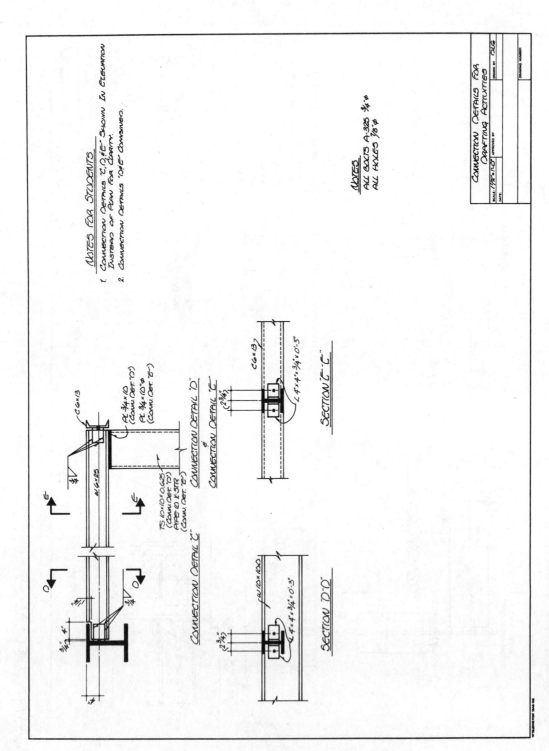

FIGURE 9.12 ■ Connection details for CAD activities.

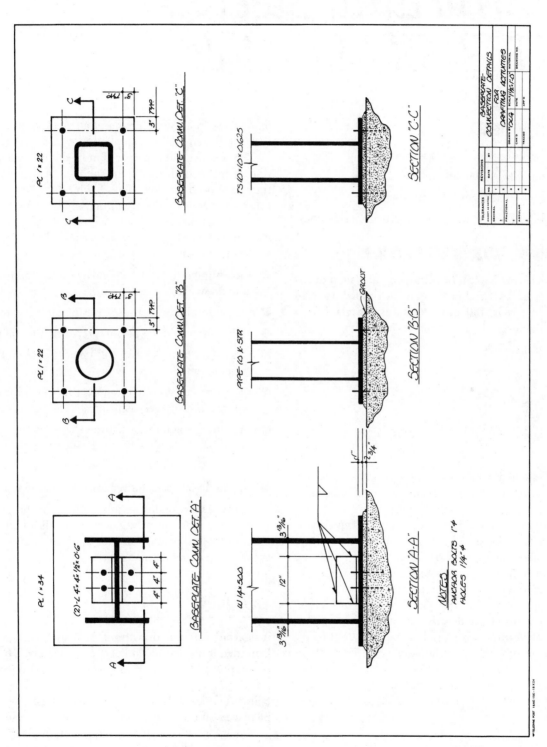

FIGURE 9.13 ■ Baseplate connection details.

Structural Steel Bills of Materials

OBJECTIVES

Upon completion of this unit, the student will be able to:

■ Define the terms *advance bill* and *shop bill*.

■ Distinguish between advance bills and shop bills.

■ Prepare advance bills and shop bills for structural steel jobs.

BILLS OF MATERIALS

In addition to preparing engineering drawings and shop drawings, structural steel CAD technicians are also called upon to prepare bills of materials. **Bill of materials** is a general term including other more specific terms such as the following:

■ Advance bill of materials

■ Advance bill

■ Materials list

■ Preliminary bill

■ Shop bill of materials

■ Shop bill

There are actually two types of bills of materials with which the CAD technician must be concerned. All of the preceding terms are different names for these two types.

The first type involves drafting and the purchasing department. It is used to order all of the steel shapes and pieces required to fabricate an entire job. It is called alternately *advance bill of materials*, *material lists*, *preliminary bill*, and *advance bill*, with the latter term being the most commonly used.

The second type involves drafting and the shop or fabrication workers. It is used to identify all of the steel pieces required in the fabrication of each individual structural member for a job and accompanies the shop drawings. This type of bill of materials is alternately called the *shop bill of materials* or more commonly the *shop bill*. A shop bill may be placed right on the shop drawing or may be attached to it as a separate sheet.

Advance Bills

Most structural steel fabrication companies do not stockpile large quantities of steel shapes and materials. Rather, the amounts of steel needed for a job are ordered on a job-by-job basis from rolling mills and steel warehousing firms. An *advance*

bill is a comprehensive listing of all steel shapes and materials that will be required to complete a given job (Figure 10.1).

The advance bill is prepared immediately following completion of initial drafting on the fabrication details, often before they have been checked. This is why the advance bill is sometimes referred to as the preliminary bill.

The exact makeup of advanced bill forms varies from company to company. However, all advance bill forms contain the following information at a minimum:

■ A job number that corresponds with the job number on the shop drawings

■ A sheet number that is used for the bill of materials sheet

■ A space in which the person preparing the bill is identified

■ A list of numbers so that every piece listed on the bill has an item number

■ A space beside each item number in which the number required of each piece is listed

■ A space in which the structural shape for each piece is identified

■ A space in which the length of each piece is listed

■ An additional space for remarks pertaining to each piece

Figure 10.2 contains a sample bill form that might be used in industry. This form has been generalized from numerous other forms designed for specific companies. However, it serves the purpose of familiarizing the CAD student with advance bill forms.

Shop Bills

A *shop bill* is used to draw from the material ordered that which is needed in the fabrication of specific members in a job. Methods of preparation vary from company to company. A common practice is to prepare a shop bill to accompany each sheet of fabrication details in a set of shop drawings. The shop bill may be placed on the drawing or attached to it as a separate form.

The information contained in a shop bill is the same as that contained in the advance bill in terms of the form itself. Item numbers, the number of each piece required, item lengths, and so on are also all listed on the shop bill. The difference between the two is twofold. First, the advance bill covers an entire job while the shop bill is geared toward individual members. Second, the advance bill lists all of the columns for a job together,

ADVANCE BILL

JOB NUMBER __2632__   PREPARED BY __OLG__   SHEET NUMBER __1__ OF __1__

ITEM NUMBER	NUMBER REQUIRED	SHAPE	DESCRIPTION	LENGTH	REMARKS
			COLUMNS		
1	1	W	12 × 190	15'-6"	B4
2	2	W	12 × 72	15'-6"	B5, B6
3	4	W	8 × 40	15'-6"	A4, A5, A6, A7
4	2	W	8 × 24	15'-6"	C4, C7
5	2	PIPE	10 X-STRONG	12'-6½"	C5, C6
6	2	TS	10×10×0.625	12'-6½"	C1, C3
			BEAMS		
7	3	W	16 × 40	22-1½"	
8	6	W	16 × 40	23'-0"	
9	2	W	10 × 45	16'-11"	
10	5	S	7 × 20	10'-6"	
11	2	S	7 × 20	9'-5½"	
			PLATES		
12	2	PL	3/4 × 24	18'-0"	CUT FOR BASEPLATES
13	1	PL	3/4 × 22	8'-0"	CUT FOR BASEPLATES
			ANGLES		
14	1	L	6 × 6 × 5/8	46'-0"	CUT FOR CONNECTIONS
15	1	L	4 × 6 × 5/8	32'-0"	CUT FOR CONNECTIONS

FIGURE 10.1 ■ Sample of a handwritten advance bill.

BILL OF MATERIALS FORM

JONES STEEL FABRICATION COMPANY, INC.
201 COUNT RD
STEELTOWN, FLORIDA

JOB NUMBER _____ PREPARED BY _____ BILL OF MATERIALS SHEET _____ OF _____

ITEM NUMBER	NUMBER REQUIRED	SHAPE	DESCRIPTION	LENGTH	REMARKS

FIGURE 10.2 ■ Sample bill of materials form.

all of the beams for a job together, all of the plates together, and so on. The shop bill lists all of the materials required for each specific member in a job together.

Like the advance bill form, shop bill forms vary from company to company. Within a specific company, however, the two forms are often the same or very similar. Figure 10.3 shows a completed shop bill that was prepared on a form that was generalized from several used in industry.

PREPARING BILLS OF MATERIALS

A number of different people are involved in the preparation of a bill of materials. If it is an advance bill, a CAD technician prepares it, a checker checks it, and then it is passed along to the purchasing department for their input. If it is a shop bill, a CAD technician prepares it, a checker checks it, and it is sent to fabrication along with the shop drawings.

The forms for the advance bill and the shop bill may be the same or very similar. However, the procedures for preparing the two different types of bills of materials vary due to their differing uses.

Preparing Advance Bills

A CAD technician assigned to prepare the advance bill for a job needs prints of all of the fabrication details for the job and an advance bill form. With these things in hand, the drafter proceeds as follows:

1. The fabrication details are arranged into columns and beams and listed on the form. The columns are listed first

and the beams second. Next, all miscellaneous pieces such as angles, plates, bars, and so on are grouped together and listed (Figure 10.2).

2. All pertinent information such as the number required of each piece, shape, description, length, and remarks when applicable are entered on the form (Figure 10.2).

3. The entire bill is given a thorough examination by the drafter preparing it. This is done to ensure that all necessary information has been included and is correct.

Preparing Shop Bills

A CAD technician assigned to prepare the shop bill(s) for a job or portion of a job needs prints of all fabrication details to be billed and a shop bill form. With these things in hand, the CAD technician proceeds as follows:

1. Each individual member to be billed is isolated and taken separately. The main structural shape and all pertinent data about it are listed first. This is followed by a listing of all other items such as plates, angles, bars, bolts, and so on required in the fabrication of the member (Figure 10.3).

2. Step 1 is repeated for every structural member that has been detailed (Figure 10.3).

3. The entire bill is given a thorough examination by the drafter preparing it. This is done to ensure that all necessary information has been included and is correct.

SHOP BILL

JOB NUMBER 76E3   PREPARED BY DLG   SHEET NUMBER 1 OF 1

ITEM NUMBER	NUMBER REQUIRED	SHAPE	DESCRIPTION	LENGTH	REMARKS
			2 BEAMS	1-B-1 & 1 B-2	
1	1	W	10 × 49	20'-10¾"	
2	1	W	10 × 49	19'-10¾"	
3	8	L	6 × 6 × 1/2	6"	
4	16	BOLTS	3/4"φ	–	A325 HIGH STRENGTH
			1 BEAM	B-3	
5	1	W	16 × 40	12'-10½"	
6	4	L	6 × 5 × 1/2	10"	
7	12	BOLTS	3/4φ	–	A325 HIGH STRENGTH
			2 CHANNELS	C-1 & C-2	
8	1	C	10 × 15.3	16'-0"	
9	1	C	10 × 15.3	15'-8¾"	
			1 CHANNEL	C-3	
10	1	C	15 × 40	19'-5"	
11	2	L	7 × 7 × 1/2	9"	
12	1	PLATE	15 × 1/2	3¾"	
13	6	BOLTS	7/8"φ	–	A307

FIGURE 10.3 ■ Sample of a handwritten shop bill.

SUMMARY

■ The term *bill of materials* is a general term that includes a number of terms that are alternately used in structural steel drafting departments.

■ Common alternative terms for bill of materials are *materials list, preliminary bill, advance bill, shop bill of materials,* and *shop bill.*

■ All of the various terms for bill of materials are simply other names for the two types of bills that are used in structural steel drafting. These two types are most commonly referred to as *advance bill* and *shop bill.*

■ An advance bill is a comprehensive listing of all steel shapes and materials that are needed to fabricate an entire job. In lay terms, it might be considered a grocery list for a structural steel job.

■ A shop bill is a listing of the steel shapes and materials that are needed to fabricate one or several individual structural members in a job. In lay terms, the shop bill can be viewed as a recipe for an individual item included in the grocery list.

■ Advance bill and shop bill forms may be exactly alike or very similar. The makeup of the forms varies from company to company. However, all forms contain the following information at a minimum: a job number, a sheet number, item numbers, shape descriptions, lengths, and remarks.

■ Advance bills are prepared immediately following completion of the fabrication details, often before they have been checked.

■ Advance bills go from drafting to the purchasing department for further preparation.

■ Shop bills are prepared after the fabrication details have been checked and approved.

■ Shop bills accompany the shop drawings to fabrication.

REVIEW QUESTIONS

1. The term *bill of materials* is a general term that includes several other terms. List four of these terms.

2. There are two types of bills of materials used in structural steel drafting departments. List the most common names for the two.

3. Define the term *advance bill.*

4. Define the term *shop bill.*

5. Distinguish between advance bill and shop bill.

6. Sketch an example of a bill form that includes all of the information minimally required for an advance bill or a shop bill.

CAD ACTIVITIES

GENERAL INSTRUCTIONS

The following activities may be completed on any CAD system. Before reading the *specific instructions* for each activity (below), go through each step in the following planning checklist. The checklist applies to any CAD system and will help ensure the optimum use of your time and resources.

1. Analyze the problem carefully. Decide exactly what you are being asked to do.

2. Determine what resources and references you will need in order to complete the problem and collect them.

3. Decide if any particular standards apply to the project and have those standards available.

4. Determine what types of views will be required and how many of each.

5. Determine what the final plotted scale of the drawing will need to be and select the appropriate paper size for plotting/printing (make sure the appropriate paper size is available).

6. Plan your drawing sequence. In what order will you develop the drawing (i.e., lines, features, dimension lines, leaders, dimensions, notes, etc.)?

7. Review the various CAD commands you will have to use in order to develop the drawing.

8. Examine your CAD system to ensure that everything is in working order, then begin the project.

SPECIFIC INSTRUCTIONS

Activity 10.1—Using the dimensions set forth in Figure 10.4 and the format shown in Figure 10.1, prepare a blank advance bill form.

Activity 10.2—Using the dimensions set forth in Figure 10.4 and the format shown in Figure 10.3, prepare a blank shop bill form.

Activity 10.3—Using the blank advance bill form prepared in Activity 10.1, prepare a complete advance bill for the fabrication details contained in Figure 10.5 and Figure 10.6.

Activity 10.4—Activity 10.4 through Activity 10.12 are to be completed on the shop bill form prepared in Activity 10.2. Enter the necessary information for the column in Figure 10.5 onto the shop bill form.

Activity 10.5—Prepare the shop bill for Beam 1, Figure 10.6.

Activity 10.6—Prepare the shop bill for Beam 2, Figure 10.6.

Activity 10.7—Prepare the shop bill for Beam 3, Figure 10.6.

Activity 10.8—Prepare the shop bill for Beam 4, Figure 10.6.

Activity 10.9—Prepare the shop bill for Beam 5, Figure 10.6.

Activity 10.10—Prepare the shop bill for Beam 6, Figure 10.6.

Activity 10.11—Prepare the shop bill for Beam 7, Figure 10.6.

Activity 10.12—Prepare the shop bill for Beam 8, Figure 10.6.

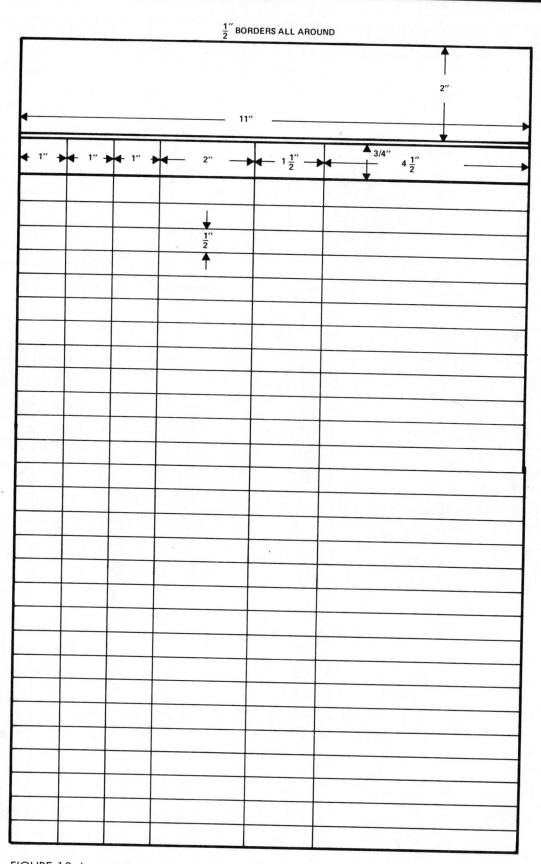

FIGURE 10.4 ■ Dimensions for bill of materials forms.

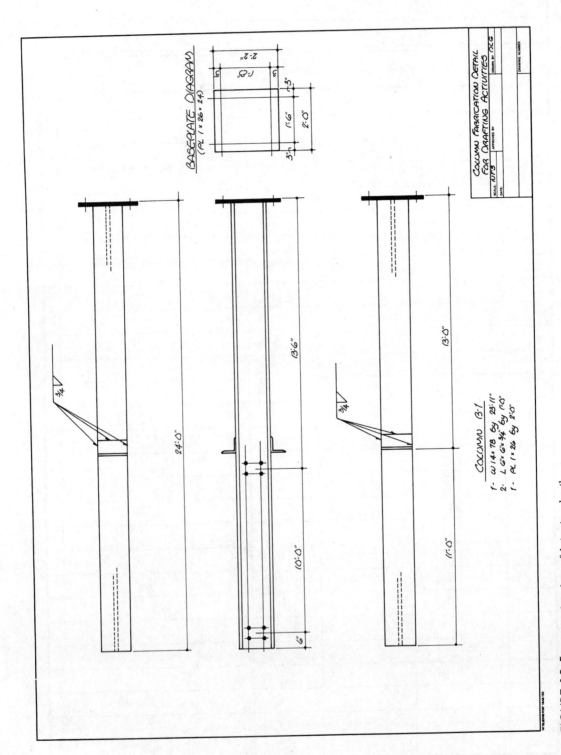

FIGURE 10.5 ■ Sample column fabrication detail.

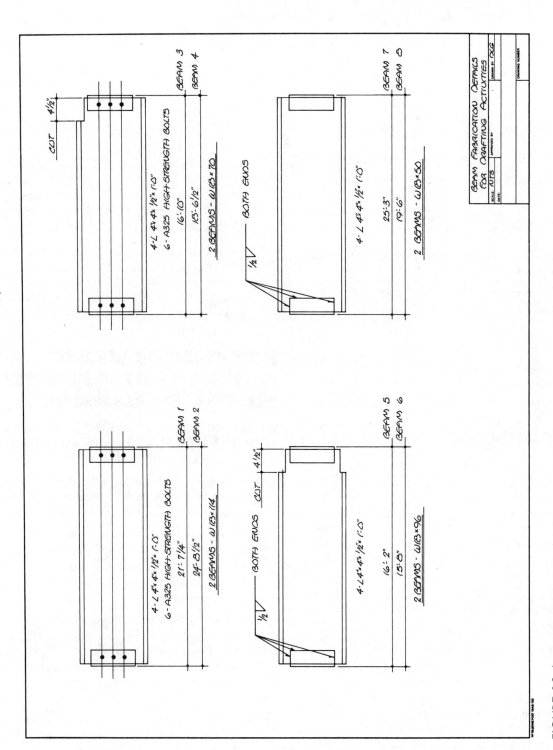

FIGURE 10.6 ■ Sample beam fabrication detail.

Pre-Engineered Metal Buildings

OBJECTIVES

Upon completion of this unit, the student will be able to:

- Describe the products used to construct pre-engineered metal buildings.
- Develop anchor bolt plans for pre-engineered metal buildings.
- Develop rigid-frame cross sections for pre-engineered metal buildings.
- Develop roof framing and sheeting plans for pre-engineered metal buildings.
- Develop sidewall framing and sheeting plans for pre-engineered metal buildings.
- Develop endwall framing and sheeting plans for pre-engineered metal buildings.
- Develop connection details for pre-engineered metal buildings.

APPLICATIONS

A special subset of the field of structural steel design, drafting, and constructions is the pre-engineered metal building. Pre-engineered metal buildings can be used for a variety of applications including warehouses, manufacturing facilities, churches, office facilities, schools, gymnasiums, airplane hangars, retail facilities, mini-storage facilities, and numerous other uses. Figure 11.1, Figure 11.2, and Figure 11.3 are examples of applications of metal buildings.

The principal differences between the construction of a regular structural steel building and a pre-engineered metal building are as follows:

1. With a regular structural steel building, engineers begin with an architect's plan and develop a structural steel framework that will accommodate the building in question. With pre-engineered metal buildings, the engineering calculations for the various building products have already been done. Design personnel need only select the appropriate materials in the proper sizes for the situation in question.

2. With regular structural steel buildings, the steel components are typically (although not always) larger and of a heavier grade. Pre-engineered buildings typically use smaller, lighter steel components for the building frame.

3. With regular steel buildings, there is often much cutting, fitting, and welding to do on site during the course of constructing the building. With pre-engineered metal buildings, the various metal components are not just pre-engineered, they are presized, prefabricated, and numbered. The author often describes pre-engineered metal-building projects as LEGOS® for adults.

4. Regular structural steel buildings typically require more equipment, more tools, and more expertise to erect than pre-engineered metal buildings. Regular structural steel buildings usually take more time to erect than pre-engineered metal buildings.

PRODUCTS USED IN PRE-ENGINEERED METAL BUILDINGS

Pre-engineered metal buildings are constructed of a variety of standard metal products. The following components are commonly used in pre-engineered metal buildings:

Eave strut—A cold-formed C-section fabricated for the proper roof pitch. Eave struts ensure that the metal building is weathertight at the eaves of the roof (Figure 11.4).

Purlin and girt—Cold-rolled Z-sections are the principal structural members attached to the building's frame (Figure 11.5). Girts are used in the walls of the building. Purlins are used in the roof. Roof and wall sheeting is attached to the girts and purlins.

Base angle and sheeting angle—Cold-rolled angles. Base angles are attached to the concrete floor of the building and provide a continuous base to which the bottom of the sheeting panels can be attached (Figure 11.6). Sheeting angles are similar but are used for attaching the sheeting panels at the rake of the building.

Roof and wall sheeting panels—There are a number of different types of roof and wall sheeting panels available for use in constructing pre-engineered metal buildings. Some are used primarily for walls, some just for roofs, and some for both; and others are architectural or decorative in nature. Some of the more commonly used shapes of sheeting panels are shown in Figure 11.7 through Figure 11.10.

FIGURE 11.1 ■ Pre-engineered metal building used in a manufacturing application.

FIGURE 11.2 ■ Pre-engineered metal building used in a warehouse application.

FIGURE 11.3 ■ Pre-engineered metal building used in an office application.

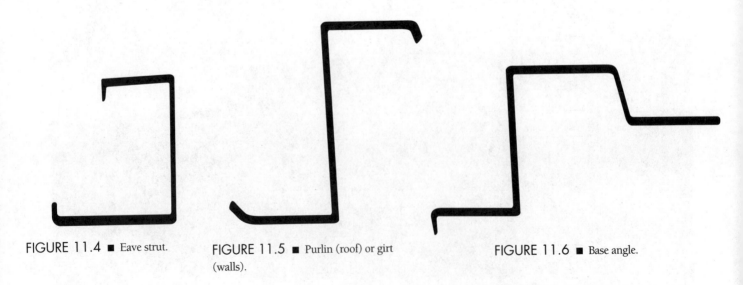

FIGURE 11.4 ■ Eave strut.

FIGURE 11.5 ■ Purlin (roof) or girt (walls).

FIGURE 11.6 ■ Base angle.

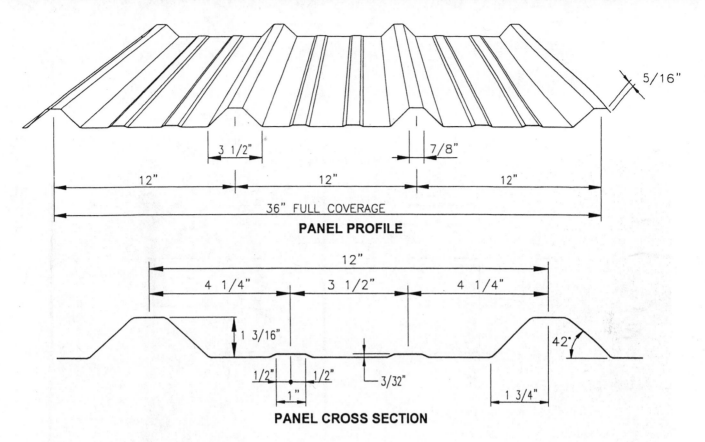

FIGURE 11.7 ■ Standard sheeting panel.

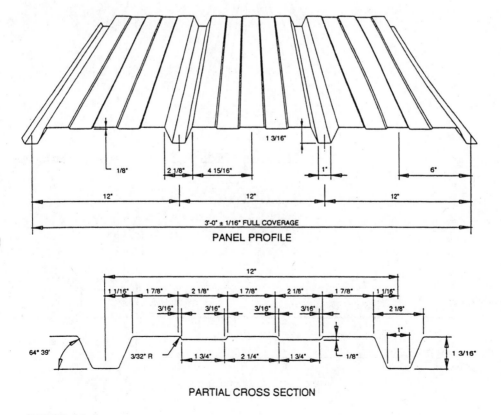

FIGURE 11.8 ■ Standard sheeting panel.

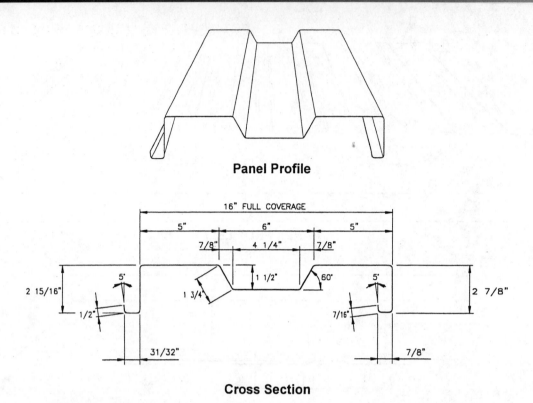

FIGURE 11.9 ■ Standard sheeting panel.

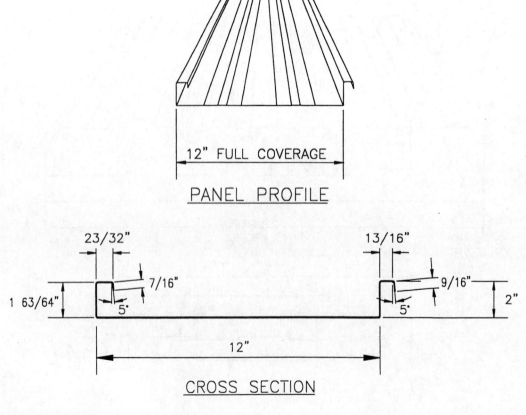

FIGURE 11.10 ■ Standard sheeting panel.

Columns and beams—The rigid-frame system for a pre-engineered metal building consists of columns and beams, as it does with any structural steel building. They types of steel shapes used for columns and beams are the same as those shown earlier in Unit 4, Figure 4.1. Figure 11.11 is a cross section showing a widely used rigid-frame system for pre-engineered metal buildings.

ANCHOR BOLT PLAN

An *anchor bolt plan* is a symbolic representation of the layout of the anchor bolts in the building's foundation/floor system. It is a plan-view drawing that shows where the anchor bolts are to be located in the floor system. The bases of the various columns used in the rigid-frame system for the building will be attached to the anchor bolts. Consequently, getting the locations of the anchor bolts correct is critical. Figure 11.12 is

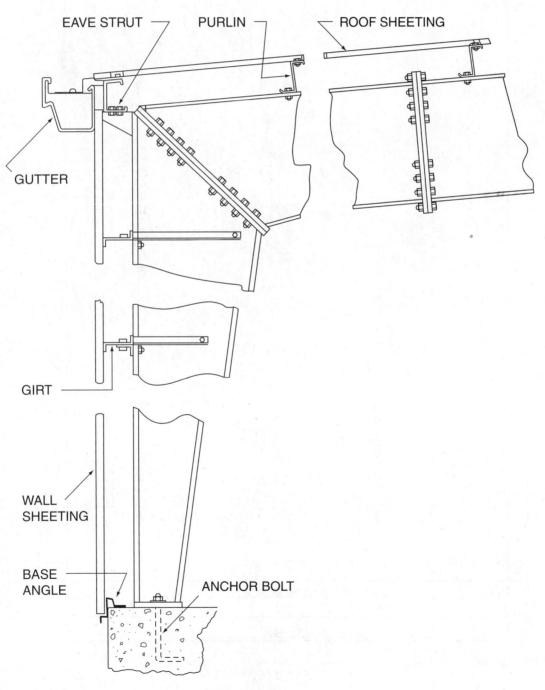

FIGURE 11.11 ■ Typical rigid-frame section.

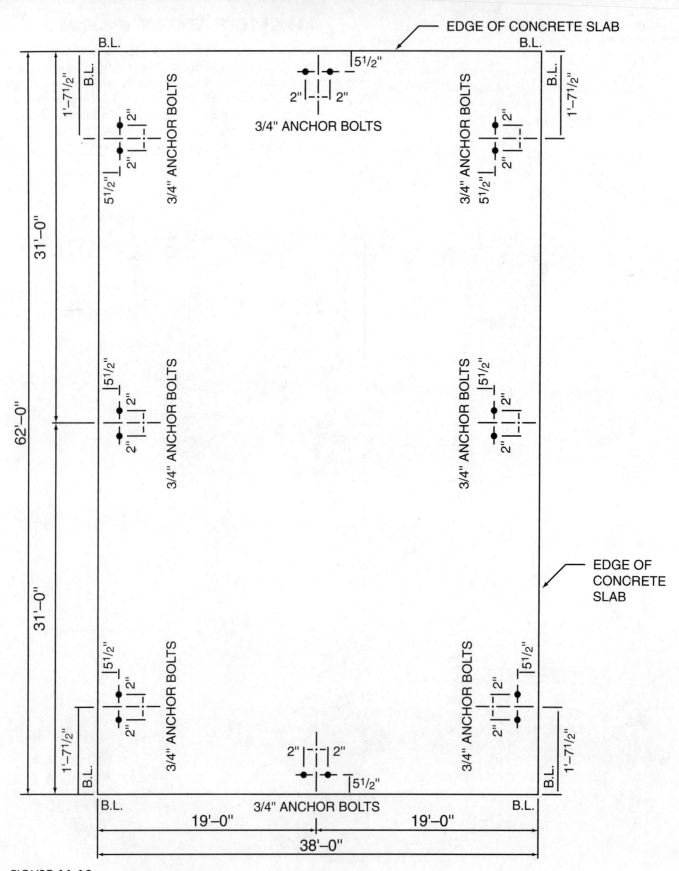

FIGURE 11.12 ■ Anchor bolt plan.

an example of an anchor bolt plan for a pre-engineered metal building. Figure 11.13 is a sectional view cut through the foundation/floor system to show the vertical placement of anchor bolts and to give various notes relating to the foundation and anchor bolts.

Drawing Anchor Bolt Plans

Using the architectural plans for the building in question and any engineering sketches provided, prepare the anchor bolt plan in the following manner:

1. Lay out the external dimensions (length and width) of the concrete floor in plan view.

2. Lay out the centerline grid for each group of anchor bolts showing the dimensions from the edges of the concrete floor system. Add dimensions that show the placement of each anchor bolt from the centerlines (Figure 11.12).

3. Add notes and a sectional view of the foundation/flooring system to show vertical placement of the anchor bolts (Figure 11.13).

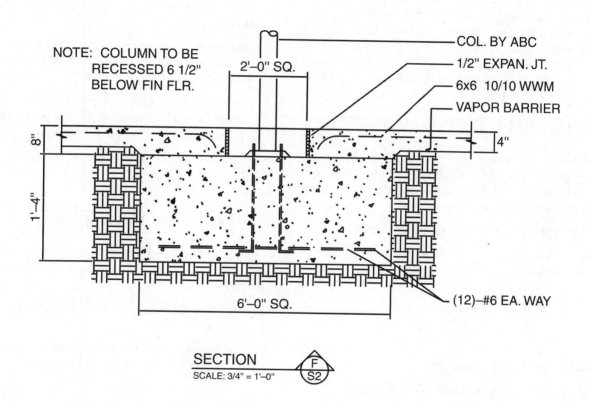

NOTE: COLUMN TO BE RECESSED 6 1/2" BELOW FIN FLR.

2'-0" SQ.

COL. BY ABC
1/2" EXPAN. JT.
6x6 10/10 WWM
VAPOR BARRIER

8"

4"

1'-4"

6'-0" SQ.

(12)–#6 EA. WAY

SECTION
SCALE: 3/4" = 1'-0"

F
S2

FOUNDATION NOTES:

1. CONCRETE SHALL BE 3000 PSI IN 28 DAYS
2. CONCRETE SLAB, APRONS AND DOOR STOOPS SHALL BE REINFORCED WITH 6x6 10/10 WWM.
3. 6 MIL POLY VAPOR BARRIER SHALL BE PLACED BETWEEN FLOOR SLAB AND COMPACTED SUB-GRADE.
4. ALL REINFORCEMENT BARS SHALL BE GRADE 60.
5. VERIFY ANCHOR BOLT SIZE AND LOCATIONS WITH BUILDING MANUFACTURER'S DRAWINGS PRIOR TO CONCRETE PLACEMENT.
6. ANCHOR BOLTS SHALL HAVE A 3" PROJECTION AND EXTEND TO LOWER 1/3 OF FOOTING DEPTH.
7. ANCHOR BOLTS STRENGTH SHALL BE A36 GRADE.

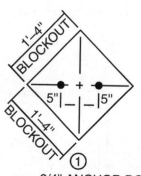

1'-4" BLOCKOUT

BLOCKOUT 1'-4"

5" — | — 5"

1

3/4" ANCHOR BOLTS
NOTE: TOP OF FOOTING AT ANCHOR BOLT
1 SHALL BE 8" BELOW FINISH FLOOR.

FIGURE 11.13 ■ Foundation section accompanying the anchor bolt plan.

RIGID-FRAME CROSS SECTIONS

The rigid frame for a pre-engineered metal building consists of the columns that are attached to the floor system by anchor bolts and the beams that are attached to the columns. A cross section through the building shows the configuration used for the building in question. Figure 11.14 shows a typical rigid-frame cross section without dimensions. The dimensions are selected by drafting personnel from a chart such as the one shown in Figure 11.15 based on the desired width and height of the building. These dimensions for various widths and heights of buildings using 20, 30, or 40 psf steel are an example of *pre-engineering*. Rigid-frame cross sections come in a variety of designs. Figure 11.16 and Figure 11.17 show some of the more widely used designs. Each individual configuration must have its own set of pre-engineered calculations, such as those shown in Figure 11.15.

Drawing the Rigid-Frame Cross Section

Using the instructions or sketches provided by the project engineer (these instructions will include the height, width, and weight of steel for the building in question), prepare the rigid-frame cross sections in the following manner:

1. Select the appropriate cross-sectional configuration.

2. Select the chart of pre-engineered dimensions for the cross-sectional configuration in question, for the height and width of the building, and for the weight of steel.

3. Lay out the rigid-frame cross section and add all necessary dimensions.

ROOF FRAMING AND SHEETING PLANS

The roof framing plan shows the arrangement and spacing of the purlins that are attached to the roof beams. The sheeting plan shows the spacing and arrangement of the sheeting panels for the roof. Figure 11.18 is an example of a roof framing plan for a pre-engineered metal building. Figure 11.19 is an example of a roof sheeting plan for a metal building.

Drawing the Roof Framing Plan

Using the anchor bolt plan, the rigid-frame sections, and any instructions or sketches provided by the engineer as guides, develop the roof framing plan in the following manner:

1. Lay out the column grid for the length and width of the building.

2. Determine the spacing for purlins that will be attached to the roof beams (typically, an engineer will provide the spacing).

3. Lay out the purlins according to the specified spacing pattern (typically, an engineer will determine the distance that purlins extend beyond the columns).

4. Add the necessary dimensions.

Drawing the Roof Sheeting Plan

Using the roof framing plan and any instructions provided by the engineer as guides, develop the roof sheeting plan in the following manner:

1. Determine the length of the sheeting panels (include the overhang).

2. Determine the spacing for the sheeting panels (typically, an engineer will provide the spacing).

3. Lay out the sheeting panels according to the specified spacing pattern.

4. Add all necessary dimensions.

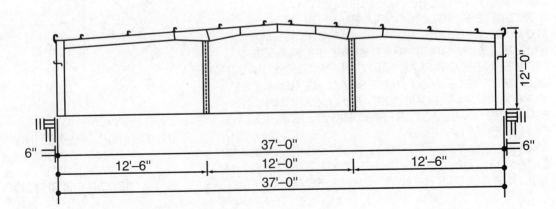

FIGURE 11.14 ■ ■ Typical rigid-frame cross section.

TABLE OF ENGINEERING CALCULATIONS

		20 PSE LL					30 PSE LL		
WD	**HT**	**B**	**C**	**E**	**WD**	**HT**	**B**	**C**	**E**
20	10	8'–0"	8"	14"	20	10	8'–0"	8"	17"
20	12	10'–0"	8"	14"	20	12	10'–0"	8"	17"
20	14	12'–0"	8"	14"	20	14	12'–0"	8"	17"
20	16	14'–0"	8"	14"	20	16	14'–0"	8"	17"
20	20	18'–0"	8"	14"	20	20	18'–0"	8"	17"
30	10	8'–0"	8"	14"	30	10	8'–0"	8"	17"
30	12	10'–0"	8"	14"	30	12	10'–0"	8"	17"
30	14	12'–0"	8"	14"	30	14	12'–0"	8"	17"
30	16	14'–0"	8"	14"	30	16	14'–0"	8"	17"
30	20	18'–0"	8"	14"	30	20	18'–0"	8"	17"
40	12	10'–0"	9"	15"	40	12	10'–0"	9"	20"
40	14	12'–0"	9"	15"	40	14	12'–0"	9"	20"
40	16	14'–0"	9"	15"	40	16	14'–0"	9"	20"
40	20	18'–0"	9"	15"	40	20	18'–0"	9"	20"

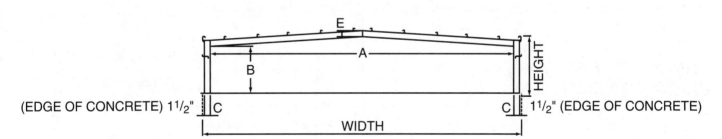

FIGURE 11.15 ■ Dimensions are pre-engineered for different heights and widths.

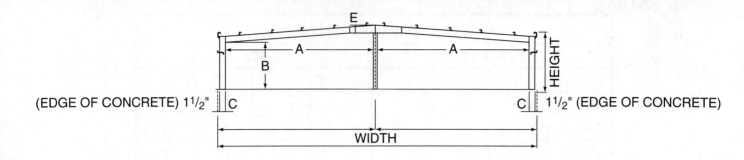

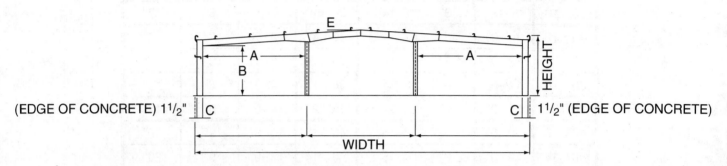

FIGURE 11.16 ■ Typical configurations for rigid-frame cross sections.

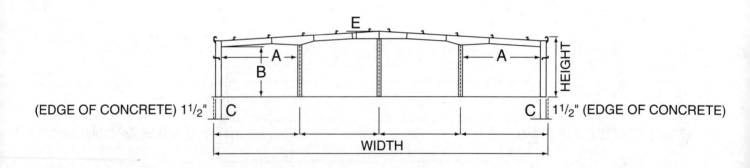

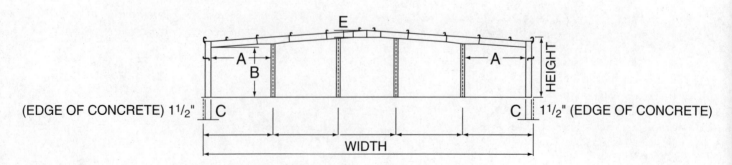

FIGURE 11.17 ■ Typical configurations for rigid-frame cross sections.

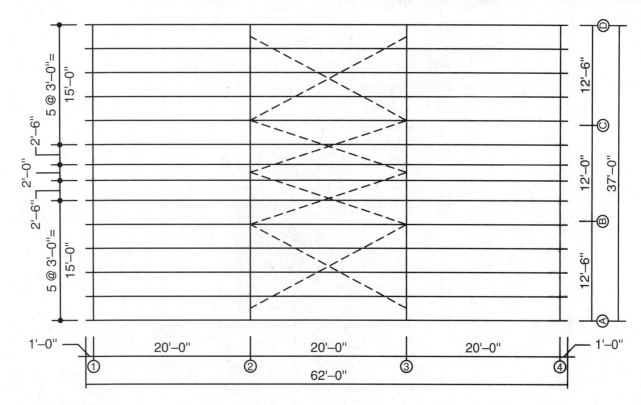

FIGURE 11.18 ■ Roof framing plan.

SIDEWALL AND ENDWALL FRAMING AND SHEETING PLANS

Sidewall framing plans show the location of the girts, eave struts, and base angles to which the side sheeting panels will be attached. Side sheeting plans show the general arrangement of the side sheeting panels for the side of the building in question. Endwall framing and sheeting plans serve the same purpose and show the same information but for the ends of the building.

Drawing the Sidewall and Endwall Framing and Sheeting Plans

Figure 11.20 is an example of a sidewall framing plan. Figure 11.21 is the corresponding sidewall sheeting plan. Very little detail is required in either of these plans. The sidewall framing plan uses single lines to show the spacing of the columns that make up the rigid frame for the building as they sit on the concrete slab. It also shows the location (vertically) of girts and eave struts. It is understood that base angles are located at the top edge of the concrete slab.

The sidewall sheeting plan requires even less detail. It simply shows the number of panels and their arrangement. Length and width dimensions of the individual panels are precalculated and do not need to be shown on the sheeting plan. Notice in Figure 11.21 that each panel has a *mark number*. Panels that are

the same width and height have the same mark number. In the case of Figure 11.21, all of the panels with mark number 2 are of one width and length. Those with mark number 1 have a different width than those with mark number 2, but both panels numbered with a 1 are the same width and length. The framing plan tells erectors to begin on either end with a mark number 1 panel, attach 23 mark number 2 panels, and finish off at the other end with a mark number 1 panel.

The same rules and practices apply when developing the endwall framing and sheeting plans (Figure 11.22 and Figure 11.23). Notice in Figure 11.23 that, beginning at either the left or right side of the plan, each sheeting panel gets successively longer in the height direction until reaching the two center panels (mark number 12). Each panel is clearly marked to guide the erectors in attaching it to the building frame.

CONNECTION DETAILS

The connection details required in a set of plans for a pre-engineered metal building can vary depending on the nature and characteristics of the building. However, regardless of the size of the building, the configuration of the rigid frame, and the types of sheeting panels selected, at least three connection details are always needed:

■ Girt connection detail

■ Purlin connection detail

■ Eave strut connection detail

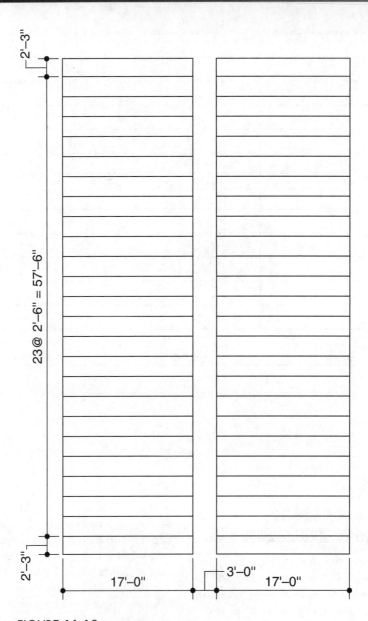

FIGURE 11.19 ■ Roof sheeting plan.

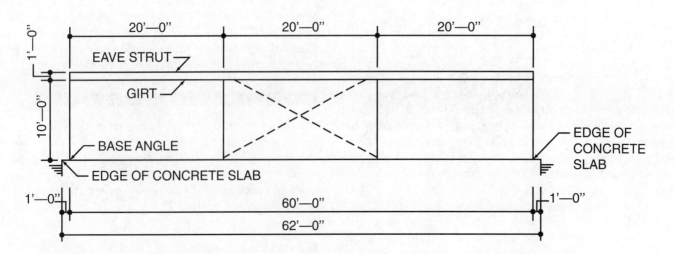

FIGURE 11.20 ■ Sidewall framing plan.

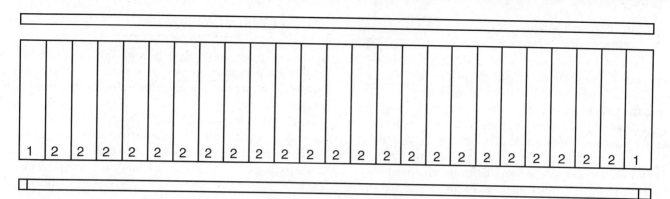

FIGURE 11.21 ■ Sidewall sheeting plan.

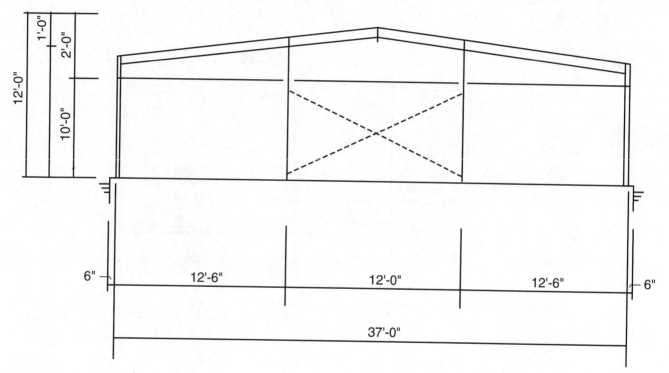

FIGURE 11.22 ■ Endwall framing plan.

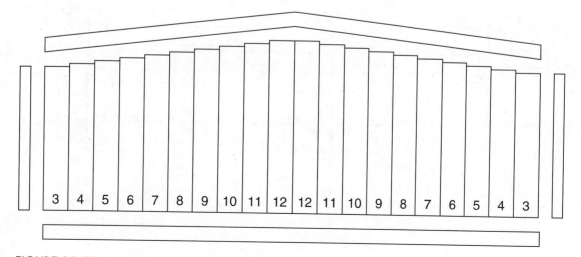

FIGURE 11.23 ■ Endwall sheeting plan.

Girt Connection Details

Girt connection details show how the girts are connected to the columns in the rigid frame. A typical girt-to-column connection involves bolting girt clips to the web of the column and bolting the girts to the girt clips (see Figure 11.24). Girt connection details should show the connection, give bolt spacing dimensions, and specify the size of bolts to be used. Engineering and design personnel determine the size and spacing of bolts for the connections and provide this information to drafting personnel either through sketches or pre-engineered tables (although experienced drafting personnel will usually know since connections for pre-engineered metal buildings are typically standardized within companies).

Purlin Connection Details

Purlin connection details show how purlins are connected to the rafters in the rigid frame. A typical purlin-to-rafter connection involves bolting the purlin to the rafter, overlapping purlins, and bolting the overlapped purlins together (see Figure 11.25). Purlin connection details should show the purlin connected to the rafter, the amount of overlap between purlins, and the bolt spacings. Engineering and design personnel determine the size and spacing of bolts and the amount of overlap required and provide this information to drafting personnel either through sketches or pre-engineered tables.

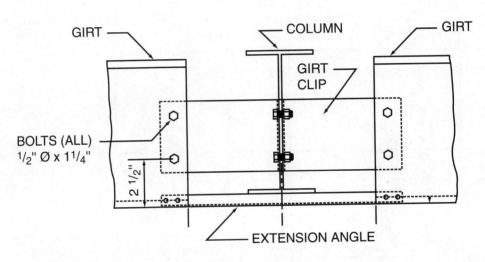

FIGURE 11.24 ■ Typical girt connection detail (to column).

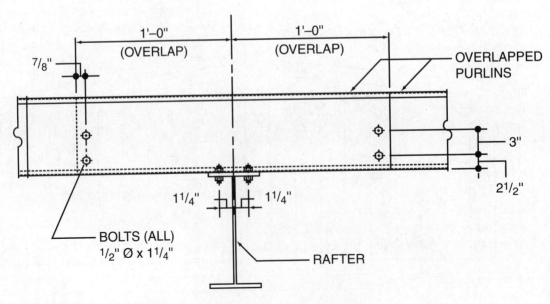

FIGURE 11.25 ■ Typical purlin connection detail (to rafter) showing overlap.

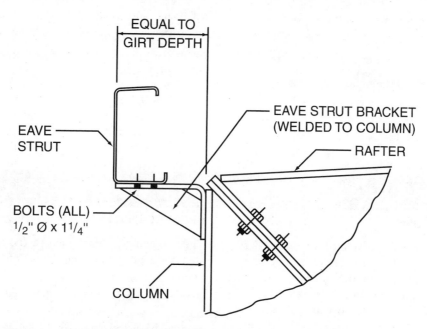

FIGURE 11.26 ■ Typical eave strut connection detail.

Eave Strut Connection Details

Eave strut connection details show how eave struts are connected to the rigid frame. This can be accomplished in various ways depending on the type of rigid frame used. An eave strut connection detail should show the eave strut connected to the column in the rigid frame or to an eave strut bracket that is connected to the column by a shop weld when the rigid frame is fabricated (see Figure 11.26). Engineering and design personnel determine how the actual connection will be made and convey this information to drafting personnel either through sketches or by using standardized connection details that have been pre-engineered for all possible situations.

SUMMARY

■ Pre-engineered metal buildings represent a subset of structural steel design, drafting, and construction. They can be used for a variety of applications including warehouses, manufacturing facilities, churches, office facilities, schools, gymnasiums, airplane hangars, retail facilities, and ministorage facilities.

■ The products most commonly used in the construction of pre-engineered metal buildings are as follows: eave struts, purlins, girts, base angles, sheeting angles, roof and wall sheeting panels, columns, and beams.

■ The anchor bolt plan is a symbolic representation of the layout of the anchor bolts in the building's foundation/floor system.

■ The rigid frame of a pre-engineered metal building consists of the steel columns that are attached to the anchor bolts and the beams that are attached to the columns.

■ The roof framing plan shows the arrangement and spacing of the purlins that are attached to the roof beams. The sheeting plan shows the spacing and arrangement of the sheeting panels for the roof.

■ Sidewall framing plans show the location of the girts, eave struts, and base angles to which the sheeting panels will be attached. Sidewall sheeting plans show the general arrangement of the sidewall sheeting panels for the side of the building in question. Endwall framing and sheeting plans serve the same purpose and show the same information for the ends of the building.

■ The connection details required in a set of plans for a pre-engineered metal building can vary depending on the nature and characteristics of the building. However, regardless of these things, at least three connection details are always needed: girt, purlin, and eave strut connection details.

REVIEW QUESTIONS

1. What is a pre-engineered metal building?

2. Explain the differences between the construction of a regular structural steel building and a pre-engineered metal building.

3. Describe the primary products used to construct pre-engineered metal buildings.

4. What is the purpose of the anchor bolt plan for a pre-engineered metal building?

5. What is the rigid frame for a pre-engineered metal building?

6. What purpose do the roof framing and sheeting plans serve?

7. What purpose do the endwall and sidewall framing plans serve?

8. What purpose do the endwall and sidewall sheeting plans serve?

9. What types of connection details are typically included in the plans for a pre-engineered metal building?

CAD ACTIVITIES

GENERAL INSTRUCTIONS

The following activities may be completed on any CAD system. Before reading the *specific instructions* for each activity (below), go through each step in the following planning checklist. The checklist applies to any CAD system and will help ensure the optimum use of your time and resources.

1. Analyze the problem carefully. Decide exactly what you are being asked to do.

2. Determine what resources and references you will need in order to complete the problem and collect them.

3. Decide if any particular standards apply to the project and have those standards available.

4. Determine what types of views will be required and how many of each.

5. Determine what the final plotted scale of the drawing will need to be, and select the appropriate paper size for plotting/printing (make sure the appropriate paper size is available).

6. Plan your drawing sequence. In what order will you develop the drawing (i.e., lines, features, dimension lines, leaders, dimensions, notes, etc.)?

7. Review the various CAD commands you will have to use in order to develop the drawing.

8. Examine your CAD system to ensure that everything is in working order, then begin the project.

SPECIFIC INSTRUCTIONS

The following reinforcement activities represent individual steps in the development of a set of plans for erecting a pre-engineered metal building:

Activity 11.1—Using Figure 11.12 and Figure 11.13 as guides, develop an anchor bolt plan for a pre-engineered metal building that will be placed on a concrete slab that is 30 feet wide and 40 feet long.

Activity 11.2—Select the top configuration shown in Figure 11.16. Develop a rigid-frame cross section for your pre-engineered metal building.

Activity 11.3—Develop the roof framing plan for your pre-engineered metal building.

Activity 11.4—Develop the roof sheeting plan for your pre-engineered metal building.

Activity 11.5—Develop a sidewall framing plan for your pre-engineered metal building.

Activity 11.6—Develop a sidewall sheeting plan for your pre-engineered metal building.

Activity 11.7—Develop an endwall framing plan for your pre-engineered metal building.

Activity 11.8—Develop an endwall sheeting plan for your pre-engineered metal building.

Activity 11.9—Develop girt, purlin, and eave strut connection details for your pre-engineered metal building.

SECTION III

Structural Precast Concrete Drafting

Precast Concrete Framing Plans

OBJECTIVES

Upon completion of this unit, the student will be able to:

■ Construct precast concrete column framing plans, beam framing plans, floor/roof framing plans, and wall framing plans from raw data available.

PRECAST CONCRETE FRAMING PLANS

Structural precast concrete drafters prepare two types of drawings: engineering drawings and shop drawings. Framing plans are engineering drawings. Taken together, the engineering drawings and the shop drawings make up a set of structural working drawings.

The four basic types of framing plans in precast concrete drafting are column, beam, floor/roof, and wall. Figure 12.1 through Figure 12.6 show examples of these types of framing plans. The examples are a complete set of framing plans for a warehouse facility. They were prepared from guidelines set forth in architectural plans.

Framing plans are the heart of a set of structural steel precast concrete working drawings and serve the same purposes as the floor plan in a set of architectural plans. Most companies require separate column, beam, floor, wall, and roof framing plans. Information required on framing plans includes the following:

■ A plan-view layout of the structural members

■ Mark numbers for each structural member

■ Product schedule, general notes, and legend

■ North arrow

■ Complete dimensions

■ Centerline of column designations (column framing plan only)

Each of these items is illustrated in Figure 12.1 through Figure 12.6. Several of these require further explanation before proceeding to the various steps in preparing precast concrete framing plans.

Mark Numbers

Precast concrete structures are composed of numerous individual structural members. These individual members are fabricated in a manufacturing plant, shipped to the jobsite, and erected. In order to manage the process, a system of mark numbers is used by most companies. Assigning mark numbers to each individual structural ember is much like numbering the pieces of a jigsaw puzzle so that it can be easily assembled.

In order to distinguish structural members within a job, it is common practice to assign mark numbers from different series. For example, columns are usually numbered within the 100 series. This means that the numbers assigned to columns fall between 100 and 199. Beams are numbered in the 200 series, while floor, wall, and roof members are numbered from 1 to 99.

The rules for assigning mark numbers are:

■ All members that have the same length, width, thickness, shape, reinforcing, and stranding receive the same mark number.

■ Any member that differs in even one of these ways receives a different mark number.

■ Members that are alike in all of these ways but still differ in some other way receive a mark number suffix (i.e., mark numbers 100, 100A, 100B, 100C, etc.).

Examine the mark numbers in Figure 12.1 through Figure 12.6 and try to determine the reasons for the differences in numbers. These reasons are not always clear from the framing plan. Sometimes, the drafter is required to examine sections and details to find an answer. Sections and details are covered in later units.

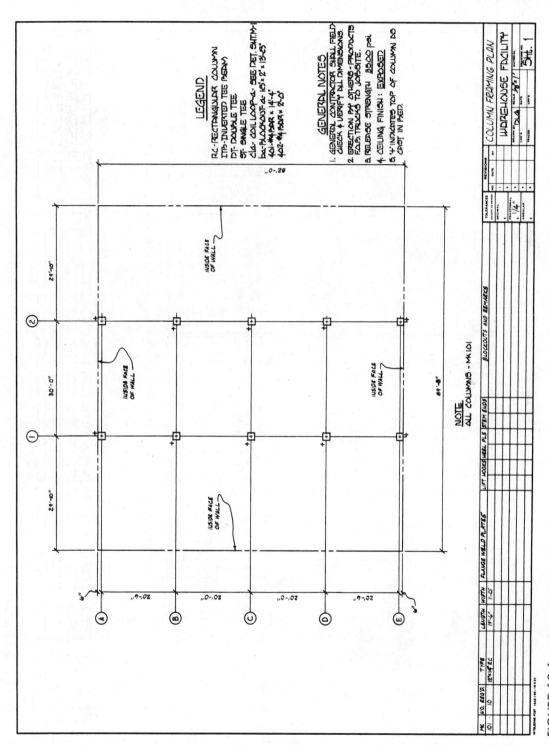

FIGURE 12.1 ■ Column framing plan.

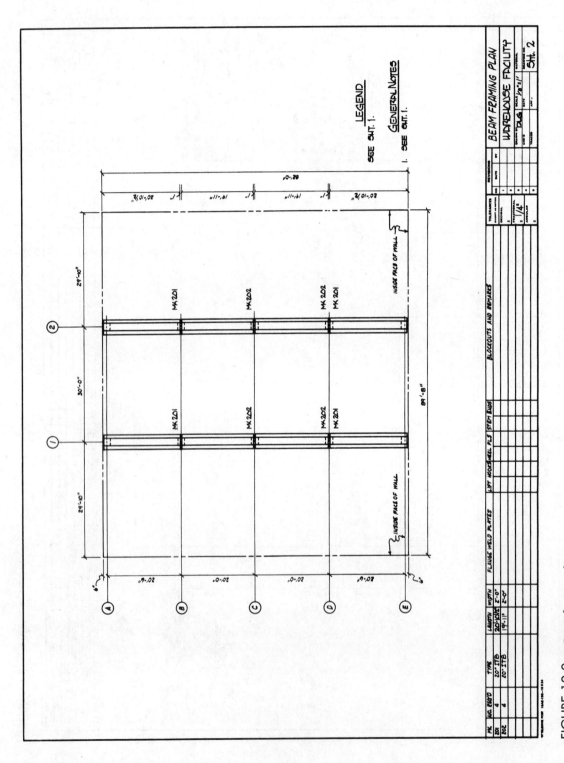

FIGURE 12.2 ■ Beam framing plan.

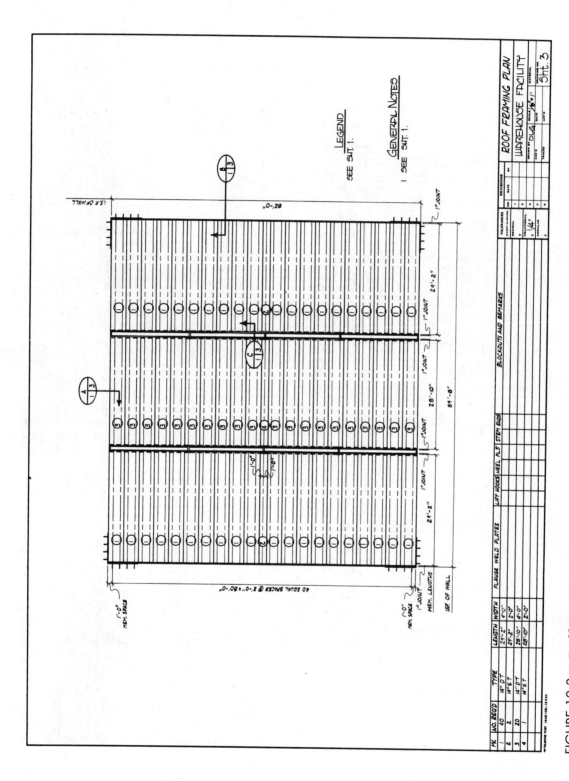

FIGURE 12.3 ■ Roof framing plan.

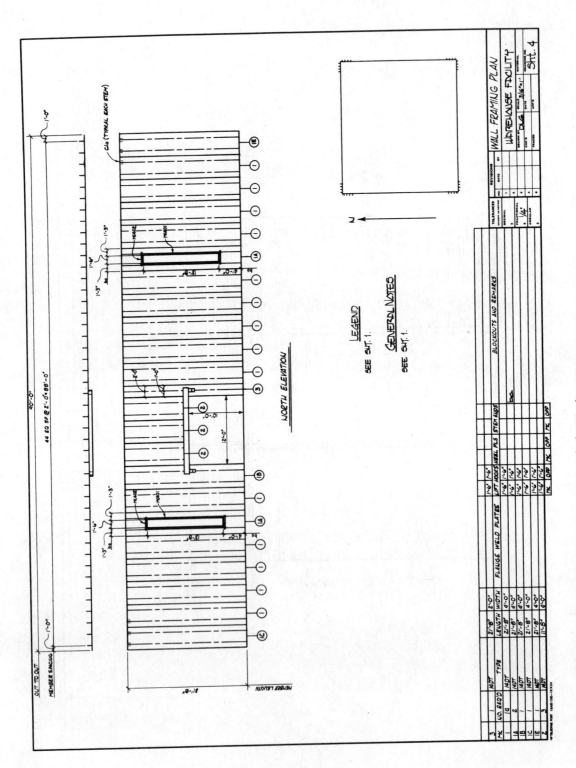

FIGURE 12.4 ■ Double-tee wall panel framing plan (north elevation).

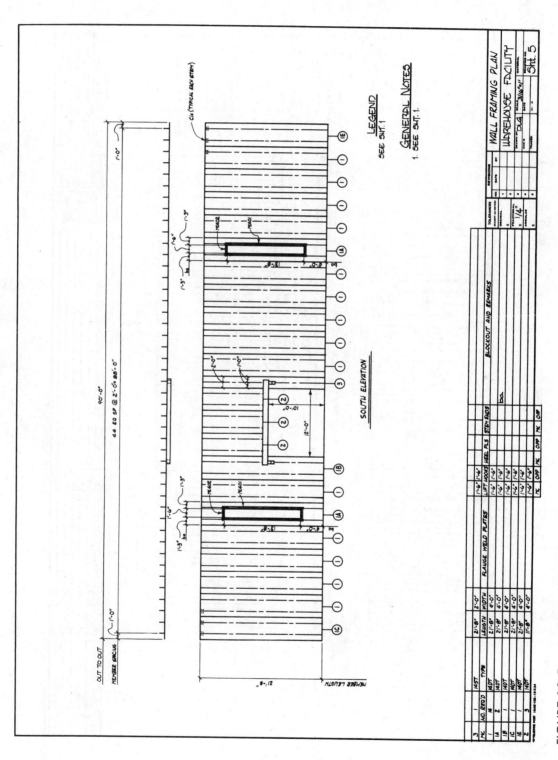

FIGURE 12.5 ■ Double-tee wall panel framing plan (south elevation).

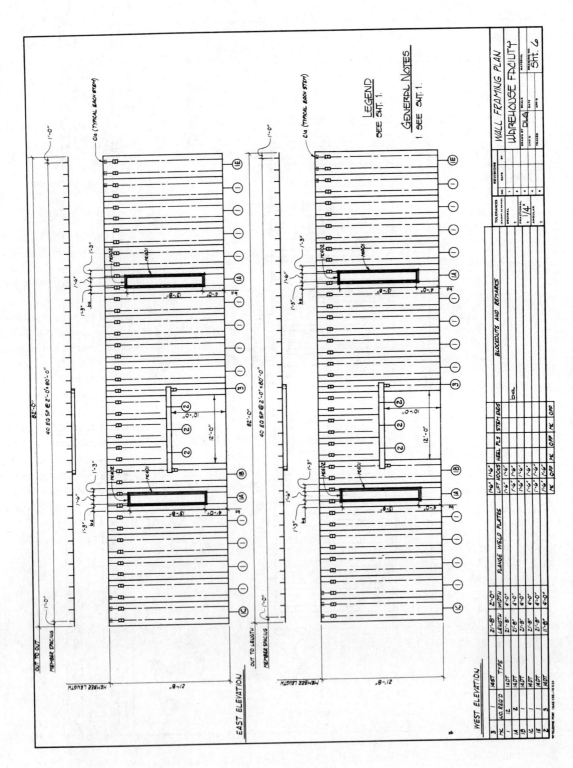

FIGURE 12.6 ■ Double-tee wall panel framing plan (east and west elevations).

Product Schedules

Product schedules provide a summary of all structural products contained in a framing plan. These schedules serve as an easy reference for information such as the following:

- Types of products used

- Length and width of all products

- How many of each product are used on a given framing plan

- Various other items of miscellaneous information

Figure 12.7 shows a product schedule that might be found on a framing plan, usually near the bottom of the page. Individual companies differ slightly in the makeup of their product schedules, but all companies use them. The sample schedule in Figure 12.7 can be used for all drafting activities in this unit.

General Notes

Notes are an important part of framing plans. Notes provide information to further clarify the drawings. Various types of notes are used on framing plans, but the most common are *general notes*.

These are notes that apply to the entire job as opposed to some notes that apply only to a specific part of the job. Figure 12.8 has a sample of the general notes that might be found on a framing plan. To save time, it is common practice to place the general notes for a job on a cover sheet or on the first sheet of the job only (Figure 12.1). On successive sheets, only the notes that specifically apply to each sheet are listed. The reader's attention is directed to the first sheet or the cover sheet for the job-wide general notes (Figure 12.9). Miscellaneous notes to the fabricator, contractor, subcontractor, and architect are also placed on the drawings when required.

Legend

An important part of the framing plans and all sheets in a set of precast concrete working drawings is the legend. The *legend* is a key for readers that explains symbols and abbreviations used on the drawings.

Legends are handled much like the general notes. A master legend for the entire job is compiled and placed on the cover sheet or first sheet of the job rather than on each individual sheet (Figure 12.1). This is common practice; however, some companies do still place a legend on every sheet that contains symbols or abbreviations. Individual company differences relative to standard operating procedures are usually explained in a company's drafting practices manual.

Figure 12.10 lists a number of common abbreviations and symbols used in precast concrete drafting. Any time a symbol or abbreviation is used, it must be explained in a legend. Legends are particularly important in precast concrete drafting because abbreviations and symbols often vary from company to company. Figure 12.11 shows a sample legend that might be found in a set of precast concrete working drawings.

One of the most common entries in a precast concrete legend is the blockout. A *blockout* is an opening left in a precast member for a window, door, skylight, conduit, ductwork, piping, or for numerous other reasons. Blockouts are formed by placing a wooden, styrofoam, or cardboard object the size and shape of the desired opening in the fabrication bed before pouring in the concrete. Once the concrete is dry, the blockout material is removed leaving the desired opening.

Figure 12.4 and Figure 12.5 show examples of blockouts in precast concrete members. Blockouts are labeled in the legend by letter designation, width in inches, depth in inches, and length in feet and inches (Figure 12.11).

DRAWING FRAMING PLANS

Precast concrete CAD technicians obtain the information they need to prepare framing plans from architectural plans, engineers' sketches, or contractors' sketches. The framing plans may either be prepared by one CAD technician or by a team of CAD technicians. If one person is assigned all of the framing plans, he proceeds in the following order: (1) column framing plan, (2) beam framing plans, (3) floor framing plan (if applicable), (4) wall framing plans, and (5) roof framing plan.

If a team of CAD technicians prepares the framing plans simultaneously, they must coordinate their efforts very closely. The procedures for preparing the various types of framing plans vary slightly and should be confronted individually.

Figure 12.12 shows an architect's rough drawing of the structural plans for a small department store. The job calls for precast columns, beams, walls, and roof members. Figure 12.13 through Figure 12.16 illustrate precast concrete framing plans that were prepared according to the architect's rough drawing. These plans were drawn according to the procedures that follow.

Column Framing Plan

1. From the raw data supplied (architect's drawings), determine the centerline of column dimensions, select an appropriate scale, and lay out construction lines for the centerlines of the columns.

2. Determine the depth and width of the columns and draw the plan view of each column locating it on the proper centerline (Figure 12.13).

3. Add centerline of column number and letter designations and dimensions as shown in Figure 12.13.

4. Assign mark numbers to each column based on information available. Mark number changes or suffixes are subject to occur as the drawings progress.

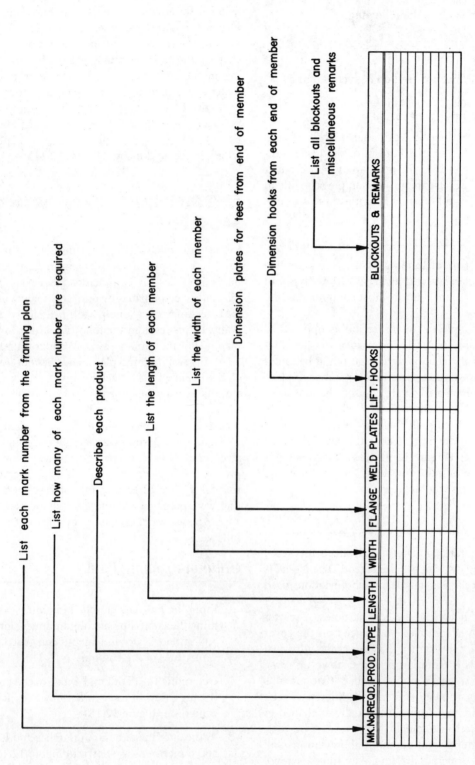

FIGURE 12.7 ■ Sample product schedule.

GENERAL NOTES

1. GENERAL CONTRACTOR SHALL FIELD CHECK AND VERIFY ALL DIMENSIONS AND CONDITIONS AT JOBSITE.
2. ERECTION INCLUDES PLACING OUR PRODUCTS AND MAKING OUR MEMBER CONNECTIONS ONLY.
3. ERECTION BY OTHERS, PRODUCTS F.O.B. TRUCKS JOBSITE.
4. NO GROUTING, POINTING, CAULKING, OR FIELD-POURED CONCRETE.
5. RELEASE STRENGTH: _____ psi (UNLESS OTHERWISE NOTED).
6. CEILING FINISH.
7. BLOCKOUTS SMALLER THAN 10″ × 10″ (EXCEPT AS NOTED ON SHOP DRAWING) TO BE FIELD-CUT BY PROPER TRADES.
8. WALL PANEL FINISHES.

FIGURE 12.8 ■ Sample general notes that apply to an entire job.

General Notes

1. These notes apply to sht. 6 only.

2. See sht. 1 for job-wide general notes.

3. Shipped loose items reqd. this sht:

   32 nbpa *14* nbpa *18* nbpc

FIGURE 12.9 ■ General notes that apply to a specific framing plan.

PRODUCT ABBREVIATIONS

BM	—	Beam
CF	—	Core floor
COL	—	Column
DT	—	Double tee
FS	—	Flat slab
ITB	—	Inverted tee beam
LB	—	L beam or ledger beam
LT	—	Lin tee
RB	—	Rectangular beam
SqB	—	Square beam
SqCol	—	Square column
S	—	Single tee
WP	—	Wall panel

MISCELLANEOUS ABBREVIATIONS AND SYMBOLS

ab — anchor bolt (aba = anchor bolt a)

b — blockout (ba = blockout a)

bp — baseplate (bpa = baseplate a)
     bearing pad (bpa = bearing pad a)

cb — connection bolt (cba = conn. bolt a)

cp — connection plate (cpa = conn. plate a)

hp — heel plate (hpa = heel plate a)

tr — threaded rod (tra = threaded rod a)

wa — weld angle (waa = weld angle a)

wp — weld plate (wpa = weld plate a)

#3 Bar — 3/8″ Ø Reinforcing bar (#301, 302, 303, etc.)

#4 Bar — 4/8″ Ø Reinforcing bar (#401, 402, 403, etc.)

#5 Bar — 5/8″ Ø Reinforcing bar (#501, 502, 503, etc.)

#6 Bar — 6/8″ Ø Reinforcing bar (#601, 602, 603, etc.)

#7 Bar — 7/8″ Ø Reinforcing bar (#701, 702, 703, etc.)

#8 Bar — 8/8″ Ø Reinforcing bar (#801, 802, 803, etc.)

FIGURE 12.10 ■ Common abbreviations and symbols in precast concrete drafting.

Legend

ba —Blockout -a- 5″ × 6″ × 1′-2″

bb —   ″   -b- 6″ × 3″ × 1′-0″

bc —   ″   -c- 10″ × 2″ × 0′-9″

bd —   ″   -d- 3″ × 3″ × 0′-3″

be —   ″   -e- 6″ × 9″ × 2′-1″

#301—#3 Bar × 6′-0″ lg. straight

#302—#3 Bar × 1′-2″ lg. Bt. See Detail

#401—#4 Bar × 4′-2″ ″ ″ ″ ″

#501—#5 Bar × 18′-0″ lg. straight

#502—#5 Bar × 1′-9″ ″ ″ ″

aba— Anchor bolt a-¹/₂″ Ø × 1′-6″ lg.

bpa— Baseplate a- See Detail

cpa— Connection Plate a- See Detail

wpa— Weld Plate a- See Detail

FIGURE 12.11 ■ Sample legend.

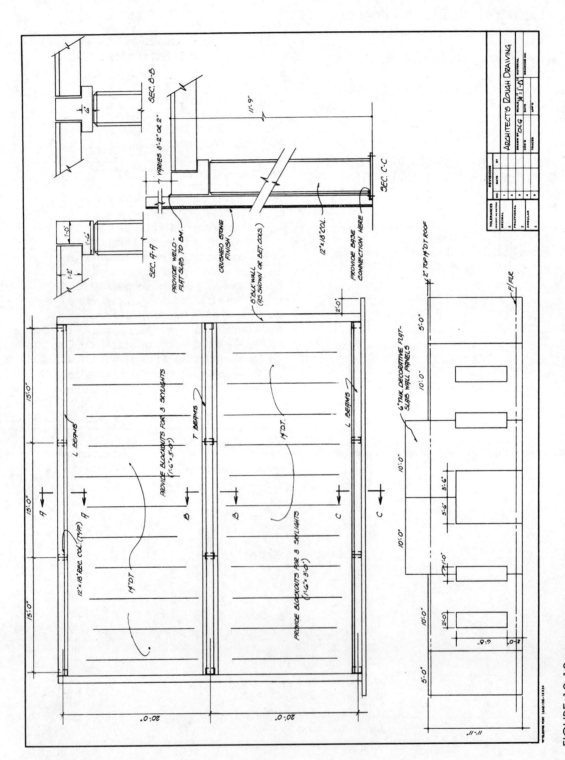

FIGURE 12.12 ■ Architect's rough drawing of the structural plans for a small department store.

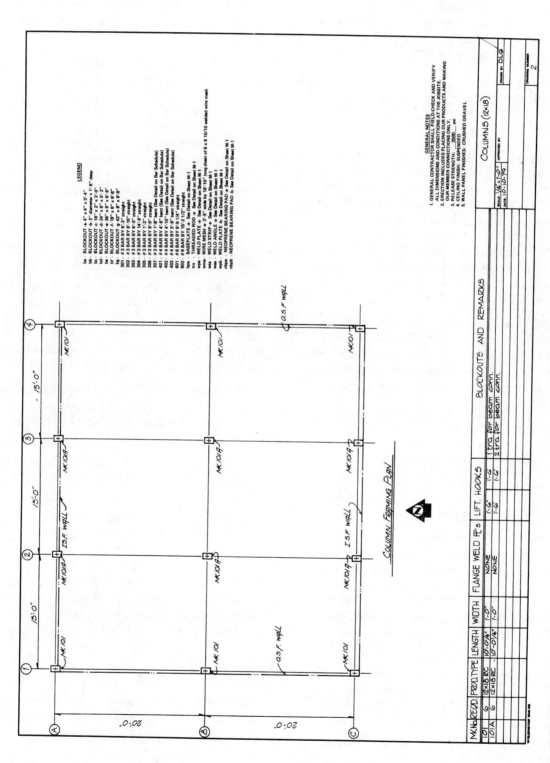

FIGURE 12.13 ■ Column framing plan prepared from architect's rough drawing.

5. Add the products schedule, fill in the title block, add general notes for the entire job, and add a north arrow. North arrows are presumed to be straight up on the sheet unless otherwise specified (Figure 12.13).

6. On a pad of paper, begin compiling a master legend for the entire job. Once the job, including all engineering and shop drawings, is complete, the legend is entered on the first page or on a cover sheet (Figure 12.13).

Beam Framing Plan

1. Determine the types and sizes of beams to be used.

2. Basing your calculations on the centerline of column dimensions, determine the beam lengths. It is common practice to leave a 1″ joint between abutting beams (Figure 12.14). A sample beam length calculation would proceed as follows (for mark number 202, Figure 12.14):
   a. Record the centerline of column dimension for the columns that support mark number 202: 15′-0″.
   b. From this dimension, subtract one-half of the 1″ joint from each end of mark number 202 or a total of 1″. This leaves a total beam length of 14′-11″.

3. Place a new sheet over the completed column framing plan and trace each column using hidden lines. Draw the beams resting on top of the columns and show the appropriate joint between abutting beams. Widths of beams and the joints between them may be exaggerated for clarity (Figure 12.14).

4. Add the appropriate dimensions as shown in Figure 12.14. Centerline of column dimensions are optional on the beam framing plan and are often omitted.

5. Assign mark numbers to each beam based on information available. Mark number changes and suffixes are subject to occur as the drawings progress.

6. Complete the products schedule and title block; add any notes that apply specifically to the beam framing plan (see the fabricator's note in Figure 12.14); refer readers to the first sheet for the legend and add the north arrow (Figure 12.14). Note that the blockout section of the products schedule is not filled in until later when the beam-to-column connection details have been drawn.

7. Examine each place where a beam rests on a column and assign letter designations for each different situation. Situations that are very close to being the same receive the same letter with the abbreviation *Sim*, meaning similar. Situations that are a mirror image of another situation receive the same letter with the abbreviation *Opp Hd*, meaning opposite hand.

Floor/Roof Framing Plan

1. Determine what product is to be used for framing the roof or floor and then look up the width of the framing product.

2. You may use the beam framing plan as the starting point to save time. The bearing surfaces of the beams should be drawn with hidden lines.

3. Calculate the required amount of framing. In Figure 12.12, the amount of framing can be determined to be 46′-0″. This amount is arrived at by first adding together the three 15′-0″ centerline of column dimensions for a total of 45′-0″. An additional 6″ of framing is required to cover from the centerline of the outside columns to the edge of the walls, requiring an additional 1′-0″ to be added to the initial 45′-0″. This makes a total of 46′-0″ of framing required. Since 14″ double tees are 4′-0″ wide, the roof may be framed with 11 full double tees and 1 cut in half and used as a 2′-0″ wide single tee (Figure 12.15).

4. Based on the calculations from Step 3, draw the roof members and add necessary dimensions (Figure 12.15).

5. Examine the raw data closely to determine if openings (blockouts) are required in the roof. In Figure 12.12, the architect has requested 3, 1′-6″ × 3′-0″ openings in each of the two bays of framing for skylights. The requested blockouts are placed in the roof members, dimensioned, and assigned a letter designation (Figure 12.15).

6. Cut sections to clarify internal relationships as shown in Figure 12.15. The actual section breakouts are drawn after the framing plans have been completed.

7. Assign a mark number to each roof member based on information available. Mark number changes or suffixes are subject to occur as the drawings progress.

8. Complete the products schedule and title block; add any notes that apply specifically to the roof framing plan; refer readers to the first sheet for the legend and add the north arrow (Figure 12.15). Note that the information for the flange weld plates section of the products schedule is used for floor or roof framing plans when double or single tees are the framing product. However, the information that is entered in this section is supplied by the engineer and not computed by the CAD technician. See Figure 12.17.

Wall Panel Framing Plan

1. From the raw data, determine the size and type of product used to frame the walls. In Figure 12.12, 6″ thick, flat-slab wall panels with a decorative coating of crushed gravel are specified.

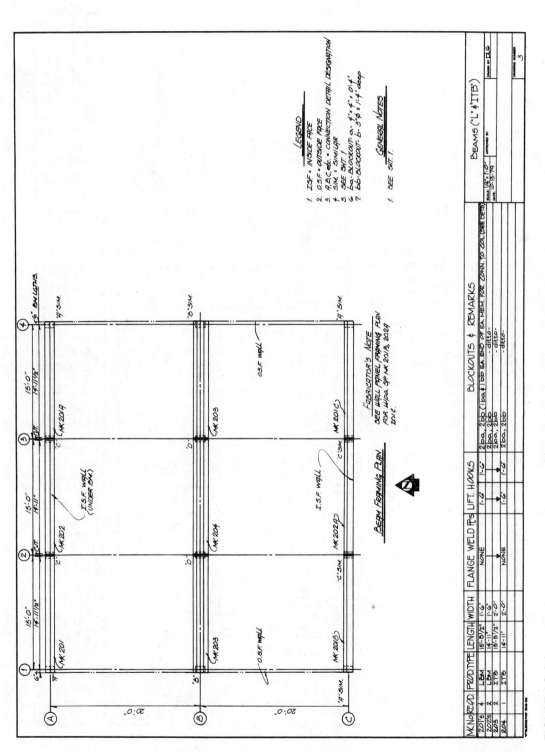

FIGURE 12.14 ∎ Beam framing plan prepared from architect's rough drawing.

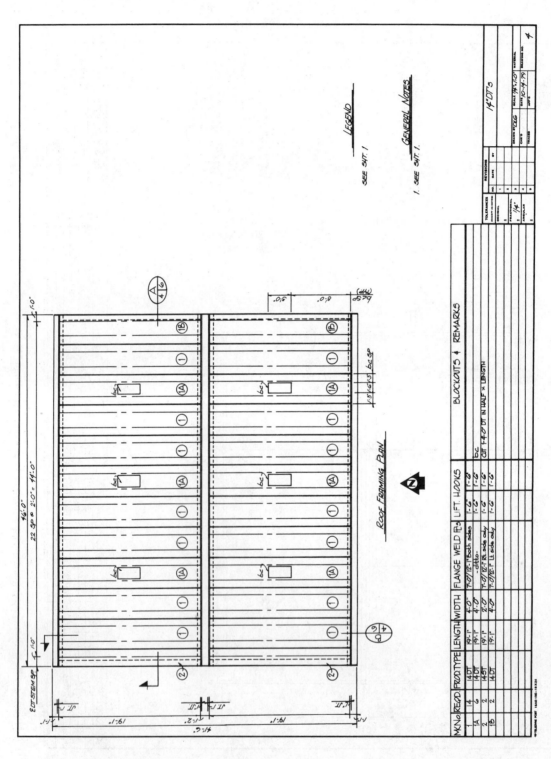

FIGURE 12.15 ■ Roof framing plan prepared from architect's rough drawing.

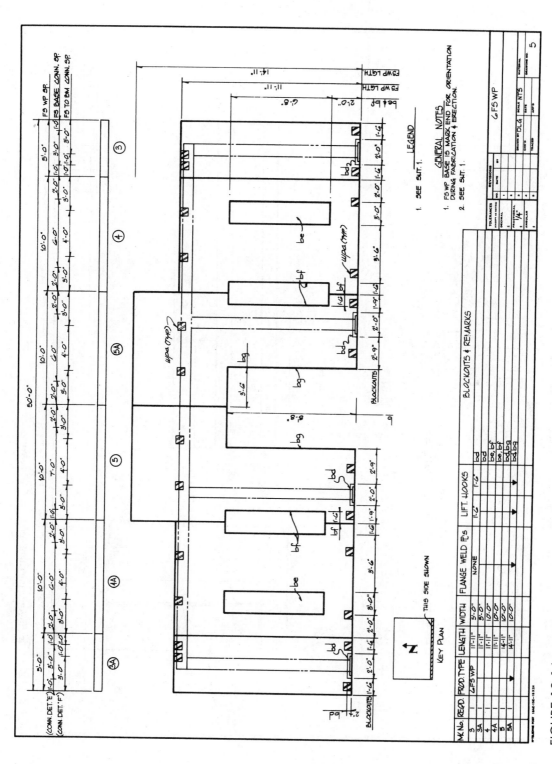

FIGURE 12.16 ■ Wall framing plan prepared from architect's rough drawing.

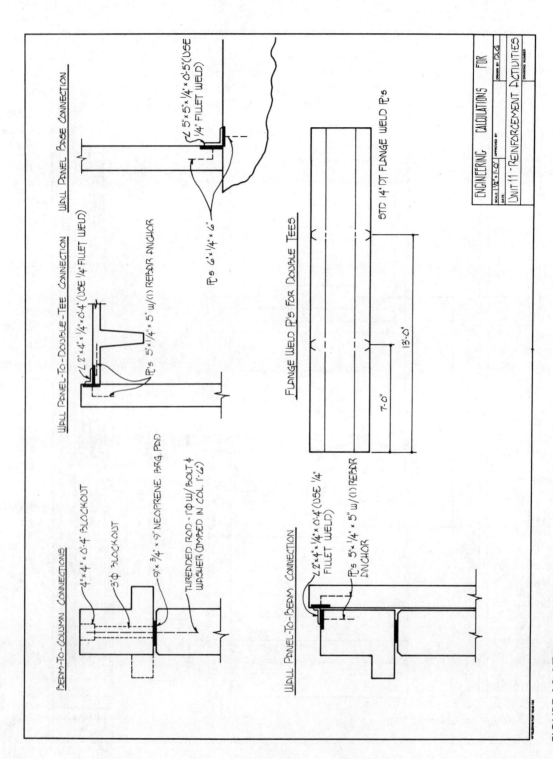

FIGURE 12.17 ■ Engineering calculations for CAD activities.

2. Determine the total amount of framing, the widths of framing members, and the heights of framing members. From Figure 12.12, it can be determined that the total amount of framing required is 50'-0". To frame this, the architect has requested 4 full, 10'-0" wide panels and 2 panels cut to 5'-0" wide.

3. Using the information determined in Step 2, lay out the wall panel framing. Wall panel framing plans consist of a plan view and an elevation of each wall panel drawn as if the viewer is on the inside of the building looking out. The columns and beams are superimposed on the wall panel elevations with phantom lines (Figure 12.16).

4. Locate all required openings for doors, windows, and so on in the wall panel elevations, assign each opening a blockout designation, and add dimensions (Figure 12.16).

5. Identify any connections that are required involving the wall panels, obtain the engineer's design for the connections, and draw the connectors in the appropriate wall panels as indicated in the calculations. Figure 12.17 has sample engineer's connection calculations.

6. Add necessary dimensions as shown in Figure 12.16. Assign mark numbers for each wall panel based on information available. Mark numbers and suffixes are subject to occur as the drawings progress.

7. Complete the products schedule and title block; add any notes that apply specifically to the wall panel framing plan; add a key plan and code it to show which wall is being drawn (Figure 12.16); refer readers to the first sheet for the legend and add the north arrow next to the key play.

SUMMARY

∎ Precast concrete CAD technicians prepare two types of drawings: engineering drawings and shop drawings. Framing plans are engineering drawings.

∎ There are four basic types of framing plans to precast concrete drafting: column, beam, floor/roof, and wall.

∎ Information required on framing plans includes: a plan-view layout of all structural members; mark numbers for each member; product schedule, general notes, and legend; north arrow; complete dimensions; and centerline of column designations (required on column framing plans; optional on beam framing plans; and not required on floor, roof, or wall panel framing plans).

∎ All members in a precast concrete job must be assigned mark numbers.

∎ Floor, roof, and wall members receive mark numbers between 1 and 99, columns receive mark numbers between 100 and 199, and beams receive mark numbers between 200 and 299.

∎ All precast members in a job that have the same length, width, thickness, shape, reinforcing, and stranding receive the same mark number.

∎ Any member that meets the preceeding criteria but still differs in some way receives a mark number suffix.

∎ Any member that does not have the same length, width, thickness, shape, reinforcing, or stranding receives a different mark number.

∎ Product schedules are provided on framing plans as summaries of the products used in the plan.

∎ General notes pertain to an entire job and are usually placed on the first sheet of the job or on a cover sheet.

∎ Only notes that apply specifically to a particular sheet and are not contained in the general notes are placed on successive sheets of a job.

∎ A legend is a guide to readers of a set of plans that explains abbreviations and symbols.

∎ A master legend for a job is usually compiled and placed on the first sheet or a cover sheet rather than on each sheet in a job.

∎ Blockouts in precast concrete members are given a letter designation and placed in the legend by width in inches, depth in inches, and length in feet and inches.

∎ Precast concrete drafters obtain the information they need to prepare framing plans from architects' drawings, engineers' sketches, or contractors' sketches.

REVIEW QUESTIONS

1. Name the two types of drawings that are prepared by precast concrete CAD technicians.

2. List the four basic types of precast concrete framing plans.

3. List the information that is required on a framing plan.

4. What is the purpose of a product schedule on a framing plan?

5. Why are general notes usually placed on the first sheet of a set of plans rather than repeated on each individual sheet?

6. Explain the purpose of a legend on precast concrete drawings.

7. Provide a legend entry for an opening in a precast concrete member that is to be 3'-0" long, 6" deep, and 1'-0" wide.

8. From what group of numbers would a precast concrete beam be assigned a mark number?
   a. 1–99
   b. 100–199
   c. 200–299
   d. 300–399

9. From what group of numbers would a precast concrete floor, wall, or roof member be assigned a mark number?
   a. 1–99
   b. 100–199
   c. 200–299
   d. 300–399

10. From what group of numbers would a precast concrete column be assigned a mark number?
    a. 1–99
    b. 100–199
    c. 200–299
    d. 300–399

CAD ACTIVITIES

GENERAL INSTRUCTIONS

The following activities may be completed on any CAD system. Before reading the *specific instructions* for each activity (below), go through each step in the following planning checklist. The checklist applies to any CAD system and will help ensure the optimum use of your time and resources.

1. Analyze the problem carefully. Decide exactly what you are being asked to do.

2. Determine what resources and references you will need in order to complete the problem and collect them.

3. Decide if any particular standards apply to the project and have those standards available.

4. Determine what types of views will be required and how many of each.

5. Determine what the final plotted scale of the drawing will need to be, and select the appropriate paper size for plotting/printing (make sure the appropriate paper size is available).

6. Plan your drawing sequence. In what order will you develop the drawing (i.e., lines, features, dimension lines, leaders, dimensions, notes, etc.)?

7. Review the various CAD commands you will have to use in order to develop the drawing.

8. Examine your CAD system to ensure that everything is in working order, then begin the project.

SPECIFIC INSTRUCTIONS

The raw data for the following activities are provided in the architect's drawing contained in Figure 12.18 and the engineer's calculations contained in Figure 12.17.

Activity 12.1—Prepare a column framing plan.

Activity 12.2—Prepare a beam framing plan.

Activity 12.3—Prepare a wall panel framing plan for the north wall.

Activity 12.4—Prepare a wall panel framing plan for the south wall.

Activity 12.5—Prepare a wall panel framing plan for the east wall.

Activity 12.6—Prepare a wall panel framing plan for the west wall.

Activity 12.7—Prepare a roof framing plan.

The raw data for the following activities are provided in the architect's drawing contained in Figure 12.19 and the sample roof framing plan provided in Figure 12.20.

Activity 12.8—The architect's drawing in Figure 12.19 calls for a precast concrete roof of either prestressed flat slabs or cored flat slabs. The sample roof framing plan in Figure 12.20 is an example of how the three-unit apartment complex might be framed in 4″ thick, flat-slab members. Using Figure 12.20 as an example, construct another roof framing plan for the three-unit apartment complex in Figure 12.19 using 8″ thick × 4′ wide cored flat slabs.

Activity 12.9—Using the architect's drawing of a three-unit apartment complex in Figure 12.19 as a starting point, change the overall length of the building to 66′-0″ and the overall width to 42′-0″. Then construct a roof framing plan for the building using 4″ thick × 10′ wide prestressed flat slabs. Use Figure 12.20 as an example.

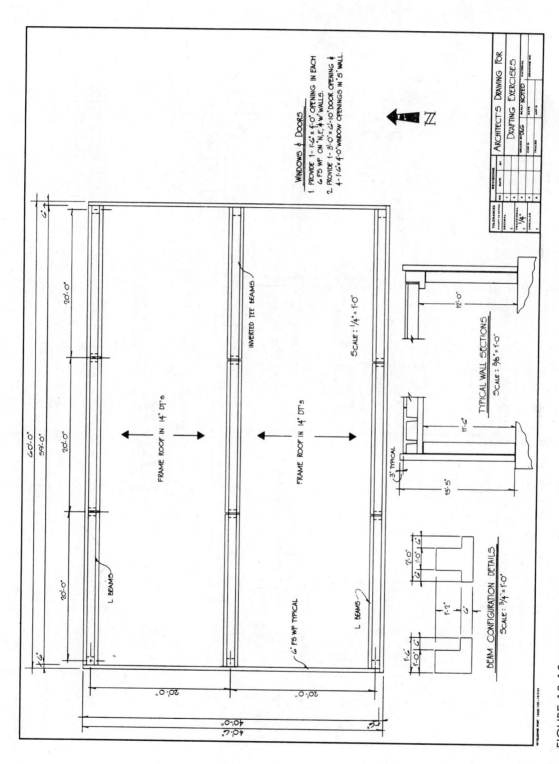

FIGURE 12.18 ■ Raw data for CAD activities.

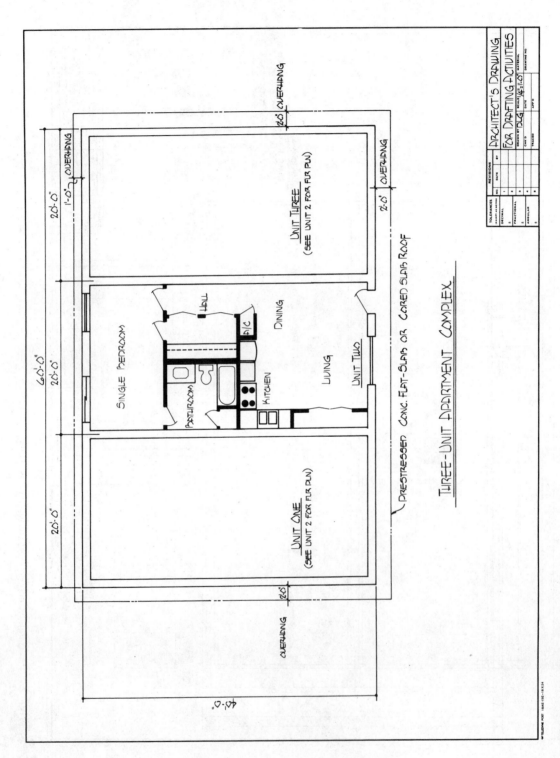

FIGURE 12.19 ■ Architect's drawing for CAD activities.

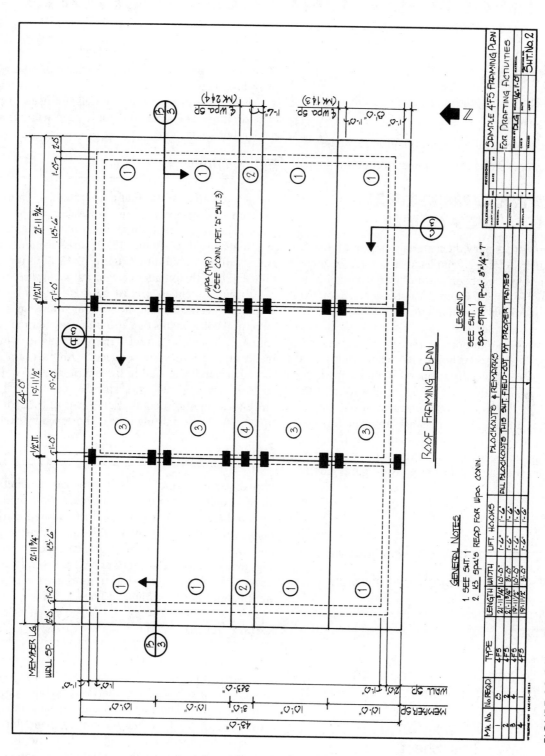

FIGURE 12.20 ■ Sample roof framing plan for CAD activities.

Precast Concrete Sections

OBJECTIVES

Upon completion of this unit, the student will be able to:

- Define precast sections.
- Prepare precast concrete full, partial, and offset sections.
- Illustrate examples of structural section conventions.

PRECAST CONCRETE SECTIONS DEFINED

Sections are drawings that show the materials used in a job and how they fit together to form a completed structure. They are cut on framing plans with a section cutting symbol along imaginary planes through the plan. Figure 13.1 shows an annotated example of a commonly used section cutting symbol. Refer to Figure 7.4 for more examples.

Sections are a very important part in a set of precast concrete working drawings. Sections clarify internal relationships in a structure that either do not appear on the framing plans or appear as hidden lines. In addition to showing how a structure fits together, sections also show height information. This information includes such items as the distance from the finished floor to the ceiling or other important height information. Figure 13.2 shows two full sections that were cut on the framing plan found in Figure 12.15. Examine these two figures closely.

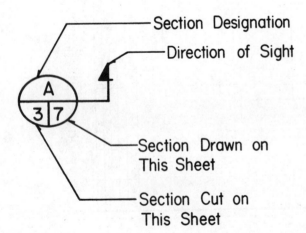

FIGURE 13.1 ■ Annotated section cutting symbol.

Section Designation
Direction of Sight
Section Drawn on This Sheet
Section Cut on This Sheet

Make note of the close relationship that exists between framing plans and sections.

Several different types of sections are used in precast concrete drafting. The most commonly used sections are full, partial, and offset sections.

Full Sections

Full sections are sections that cut through an entire building or structure. Full sections are of two basic types: longitudinal and cross sections. Longitudinal sections cut across the entire length of a building (Figure 13.3). Cross sections cut across the entire width of a building (Figure 13.4). Both types of full sections fall into the broader category, *structural sections*.

All sections cut in precast concrete drafting are structural sections. This means that the sectional drawings show only the structural components of a structure (foundation, floors, walls, columns, beams, and roof), the materials of which they are made, and how they are connected. Architectural sections, on the other hand, show all of the interior details of the structure such as carpet, paneling, suspended ceiling, and baseboards. Since these interior details are not part of the structural make-up of a structure, drafting time is shortened by excluding them.

Partial Sections

Partial sections are often used in precast concrete drafting to clarify isolated portions of a structure. Full sections are very involved and time-consuming. If the desired clarification can be provided with less than a full section, a partial section is cut and drawn. A complete set of precast concrete working drawings may contain full sections as well as several partial sections. Several examples of partial sections as used in precast concrete drafting are shown in Figure 13.5.

Offset Sections

When cutting a section through a framing plan or some portion of a structure, the cutting plane line does not always run straight. To clarify detailed interior situations that may not fall along a straight extension of the cutting plane line, the line may offset (Figure 13.6). When the cutting plane line is offset, the sectional drawing is drawn as if the offset interior portion of the structure aligned with the straight extension of the cutting plane line from its point of origin (Figure 13.7).

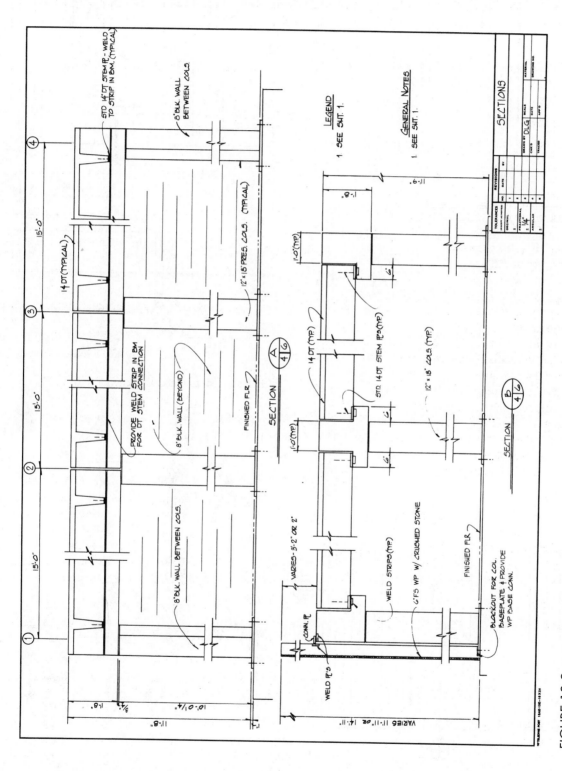

FIGURE 13.2 ■ Precast concrete sections.

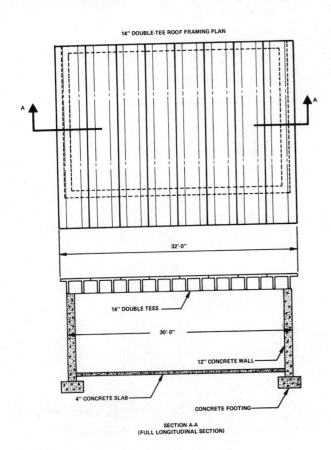

FIGURE 13.3 ■ Full longitudinal section.

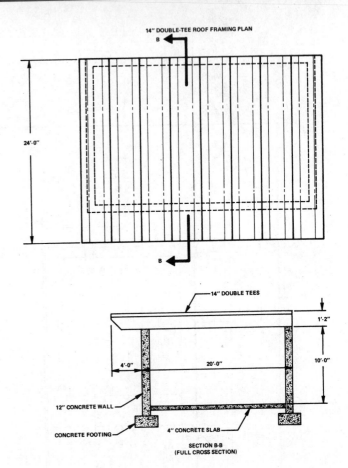

FIGURE 13.4 ■ Full cross section.

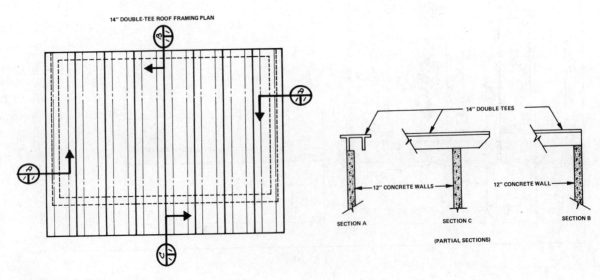

FIGURE 13.5 ■ Partial sections.

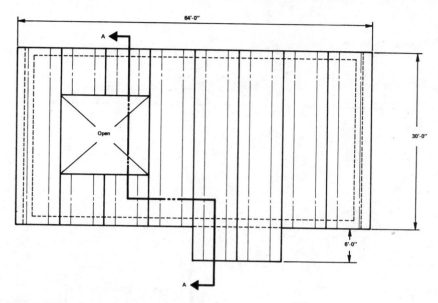

FIGURE 13.6 ■ Framing plan with offset section cut.

SECTION CONVENTIONS

One of the primary purposes of sections is to show the reader of a set of plans what materials the structure is made of. All materials used in building have standard symbols that may be used in sectional drawings, or an individual company may use its own materials symbols and explain them in a legend. Examples of building materials symbols commonly used in precast concrete drafting are shown in Figure 13.8. Although these symbols are used in general, many companies cut down on drafting time by minimizing their use. It is common practice for a company that specializes in the fabrication of precast concrete products to make sectional drawings without using the standard symbols for steel and concrete. This can be done without confusion because only the structural components of a structure are included in the sections and these will be constructed of concrete with steel connectors. The use of sectioning symbols is definitely required only where a structure is composed of several different structural materials.

DRAWING PRECAST CONCRETE SECTIONS

Precast concrete CAD technicians obtain the information they need to prepare sectional drawings from two sources:

■ The framing plans where the sections were cut

■ The raw data from which the framing plans were prepared

The framing plans indicate what sections actually must be drawn. The raw data from which the framing plans were prepared provides information on how the structural members fit together, how they are to be connected, and height information. In Figure 13.9 and Figure 13.10, common precast concrete sectional drawings are illustrated. These sectional drawings were prepared according to the following procedures:

1. From the framing plan(s), determine what sectional drawings must be prepared.

2. Determine whether the sectional drawings are to be full sections, partial sections, or offset sections.

3. Examine each sectioned situation individually and answer the following questions for each:
   ■ What materials are used in the section?
   ■ How do the structural components fit together?
   ■ What are the important height distances in the section?

4. Based on the information obtained in Step 3, select an appropriate scale and draw the sectional drawing. Sections may be drawn to a number of different scales, depending on the degree of detail desired, the complexity of the section, and the amount of space available for drawing.

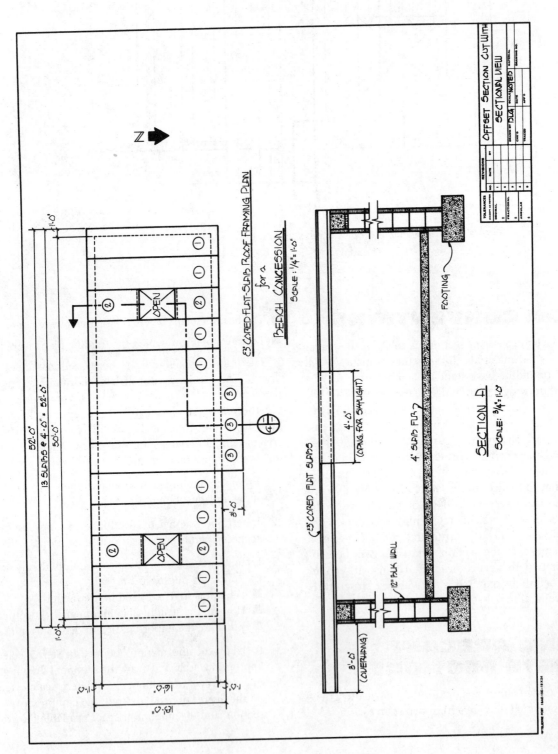

FIGURE 13.7 ■ Offset section cut with sectional view.

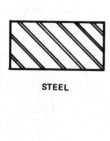

STEEL

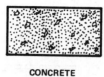

CONCRETE

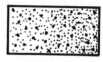

CINDER

BLOCK

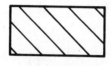

COMMON BRICK

FACE BRICK

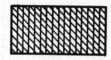

FIRE BRICK

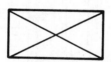

DIMENSION LUMBER

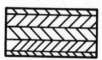

PLYWOOD

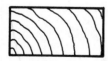

FINISH BOARD

BATT INSULATION

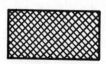

RIGID INSULATION

FIGURE 13.8 ■ Common sectioning symbols.

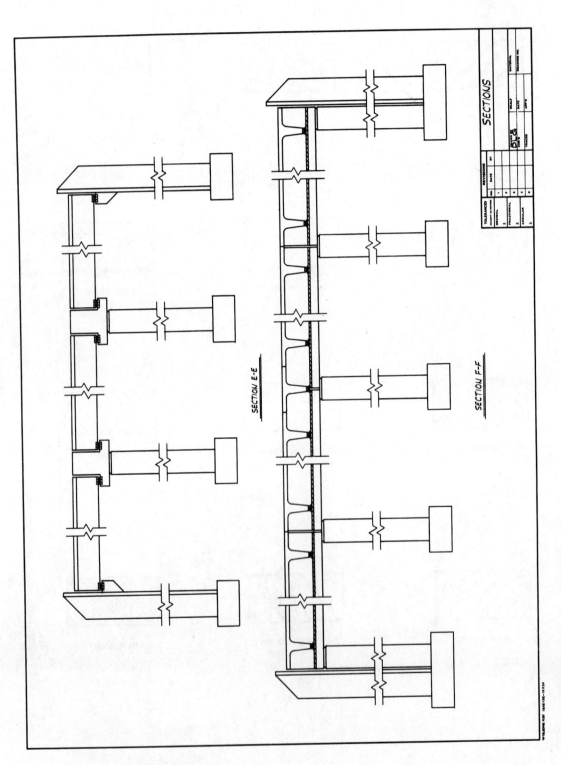

FIGURE 13.9 ■ Precast concrete full sections.

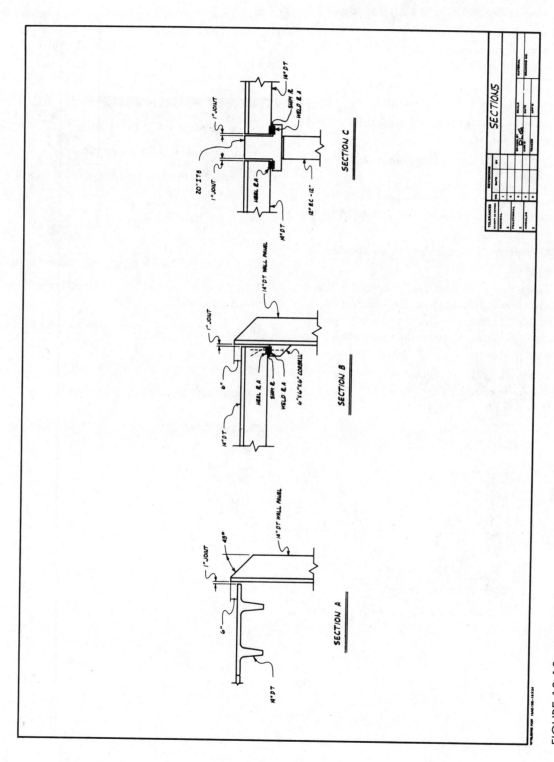

FIGURE 13.10 ■ Precast concrete partial sections.

SUMMARY

■ Sections are drawings that show the materials used in a job and how they fit together to form a completed structure.

■ Sections clarify internal relationships in a job that do not appear on a framing plan or appear hidden.

■ Sections are also important sources of height information in precast concrete working drawings.

■ The most commonly used types of sections in precast concrete drafting are full, partial, and offset sections.

■ Full sections cut through an entire building or structure along an imaginary cutting plane.

■ Full sections are of two basic types: longitudinal and cross sections.

■ Longitudinal sections cut across the entire length of a structure while cross sections cut across the entire width of a structure.

■ All sections used in precast concrete drafting are considered structural sections.

■ Structural sections differ from architectural sections in that they contain only the structural components (foundation, floors, walls, columns, beams, and roof) and exclude interior details such as carpeting, paneling, baseboards, suspended ceilings, and so on.

■ Partial sections are used to clarify isolated portions of a structure and should be used any time when a full section is not required.

■ Offset sections are used to clarify detailed interior situations in a structure that do not fall along a straight extension of the cutting plane line from its point of origin.

■ When drawing precast concrete sections, materials are symbolized by standard section conventions or by individual company symbols that are explained in a legend.

REVIEW QUESTIONS

1. In a single sentence, define *sections*.

2. What is the purpose of a section?

3. List the three most common types of sections in precast concrete drafting.

4. Define *full section*.

5. List the two types of full sections.

6. Distinguish between longitudinal and cross sections.

7. All sections in precast concrete drafting fall into what broad category?

8. How do structural sections differ from architectural sections?

9. Sketch an example of a partial section.

10. Explain how an offset section is used.

CAD ACTIVITIES

GENERAL INSTRUCTIONS

The following activities may be completed on any CAD system. Before reading the *specific instructions* for each activity (below), go through each step in the following planning checklist. The checklist applies to any CAD system and will help ensure the optimum use of your time and resources.

1. Analyze the problem carefully. Decide exactly what you are being asked to do.

2. Determine what resources and references you will need in order to complete the problem and collect them.

3. Decide if any particular standards apply to the project and have those standards available.

4. Determine what types of views will be required and how many of each.

5. Determine what the final plotted scale of the drawing will need to be, and select the appropriate paper size for plotting/printing (make sure the appropriate paper size is available).

6. Plan your drawing sequence. In what order will you develop the drawing (i.e., lines, features, dimension lines, leaders, dimensions, notes, etc.)?

7. Review the various CAD commands you will have to use in order to develop the drawing.

8. Examine your CAD system to ensure that everything is in working order, then begin the project.

SPECIFIC INSTRUCTIONS

Activity 13.1—Using the roof framing plan in Figure 13.11 as the raw data, construct a full longitudinal section.

Activity 13.2—Using the roof framing plan in Figure 13.11 as the raw data, construct a full cross section.

Activity 13.3—Figure 13.12 contains a roof framing plan of 8″ cored flat slabs for an auto repair shop and office. Construct a full longitudinal section.

Activity 13.4—Figure 13.12 contains a roof framing plan of 8″ cored flat slabs for an auto repair shop and office. Construct a full cross section through the garage area.

Activity 13.5—Figure 13.12 contains a roof framing plan of 8″ cored flat slabs for an auto repair shop and office. Construct a full cross section through the office area.

The following activities are taken from Figure 13.13:

Activity 13.6—Construct the partial section "A" showing only the structural situation at the roof.

Activity 13.7—Construct the partial section "B" showing the structural situation from the bottom of the footing to the top of the flat-slab wall panels.

Activity 13.8—Construct the partial section "C" showing the structural situation from the bottom of the footing to the top of the double-tee roof.

Activity 13.9—Construct the partial section "D" showing the structural situation from the bottom of the footing to the top of the double-tee roof.

The following activity involves major alterations to Figure 13.13:

Activity 13.10—Construct a full cross section of the structure in Figure 13.13 at a scale of 1″ = 1′-0″ with the following alterations:

a. Convert the L beam above the back wall to a rectangular beam, 12″ wide × 18″ deep. Extend the double tees to have full bearing on the rectangular beam.

b. Convert the ITB to a rectangular beam, 12″ wide × 18″ deep, and extend both bays of double tees to a half bearing on the beam less 1/2″ each to provide a 1″ joint between abutting double tees.

c. Leave the L beam at the front wall and all other structural situations intact.

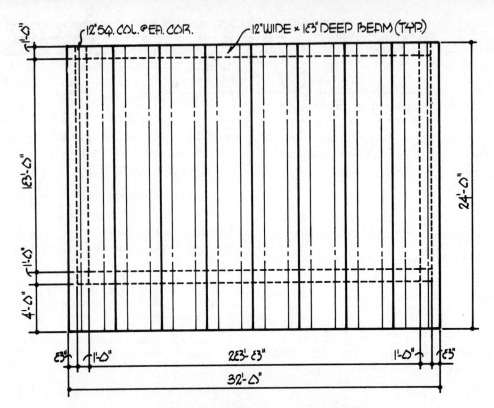

14"DT ROOF FRAMING PLAN

ADDITIONAL INFORMATION FOR SECTIONAL VIEWS
1. Columns rest on a footing of concrete 2'-0" wide x 1'-0" deep.
2. Columns are 9'-6" high.
3. Beams rest directly on top of columns.
4. 14" double tees rest directly on top of the beams.
5. The bottom of the 4" concrete slab floor is 8" above the top of footing.

FIGURE 13.11 ■ Roof framing plan for CAD activities.

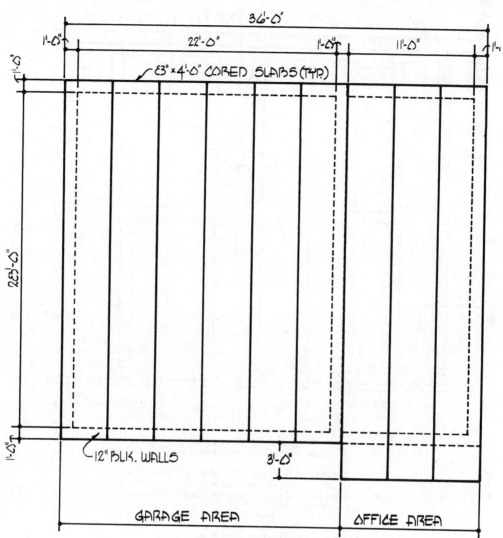

ROOF FRAMING PLAN
for an
AUTO REPAIR SHOP & OFFICE

ADDITIONAL INFORMATION FOR SECTIONAL VIEWS
1. 12″ block walls rest on 2′-0″ wide x 1′-0″ deep concrete footings.
2. Cored slabs rest directly on block lintel beams filled with concrete (See Section A — Figure 12-7).
3. The bottom of the 4″ concrete slab floor is 8″ above the top of footing.
4. The office area has an 8′-0″ high ceiling. The garage area has a 12′-0″ high ceiling.

FIGURE 13.12 ■ Roof framing plan for CAD activities.

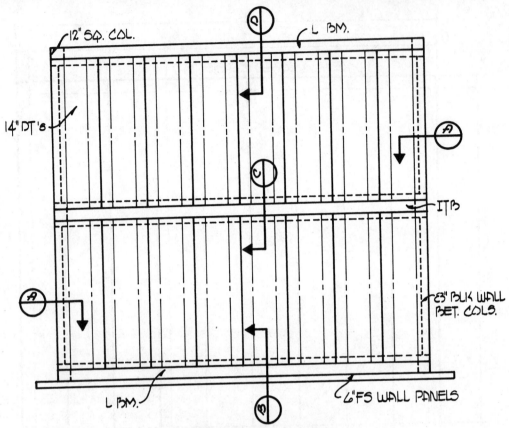

ROOF FRAMING PLAN

ADDITIONAL INFORMATION FOR SECTIONAL VIEWS

1. Left-side, right-side, and back walls are 8″ block built-up between the columns from the concrete slab floor to the bottom of the beams or double tees.
2. The 6″ flat-slab wall panels extend from the top of footing to a distance of 3′-0″ above the top of the double-tee roof.
3. Columns are 12″ square and rest on concrete footings 2′-0″ wide x 1′-0″ in depth.
4. L and ITB beam configuration dimensions may be obtained from Figure 12-2.
5. Concrete slab floor 4″ thick is located 8″ above the top of footing.
6. Bottom of double-tee elevation is 10′-0″ above the top of the concrete slab floor.

FIGURE 13.13 ■ Roof framing plan for CAD activities.

Precast Concrete Connection Details

OBJECTIVE

Upon completion of this unit, the student will be able to:

■ Construct each of the following: precast concrete baseplate connection details; precast concrete bolted beam-to-column connection details; precast concrete welded connection details; and precast concrete haunch connection details.

PRECAST CONCRETE CONNECTION DETAILS DEFINED

Precast concrete products are fabricated in a manufacturing plant and shipped to the jobsite by truck and train to be erected. Lumber that is used in building a wooden structure must be fastened together by nails, clips, or other means. Precast concrete members must also be fastened or connected. Precast concrete connections at the jobsite are achieved by bolting, welding, or a combination of both.

In order to convey the intended connection designs to readers of the plans, connection details are drawn. A *connection detail* is a plan view of a connection with a section cut through it to show an elevation view of the actual connection. Typical connections requiring details in precast concrete drafting are:

■ Column baseplate connections

■ Bolted beam-to-column connections

■ Welded connections

■ Haunch connections

COLUMN BASEPLATE CONNECTIONS

Columns are one of the most frequently used precast concrete products. So that they can be attached to a footing, columns are cast with a heavy steel baseplate in one end. This baseplate is provided with drilled holes at specific locations that fit over anchor bolts cast in the footing.

Each column is erected by placing it over its corresponding anchor bolts that have been fitted with combination washers and leveling nuts. Once the column is up, its height and degree of level can be adjusted by turning the leveling nuts. Once leveled and the top of the column is at the proper eleva-

tion, another set of washers and nuts is tightened down on each anchor bolt. This is done until the column fits firmly against the top of the baseplate permanently securing the column in place, Figure 14.1 and Figure 14.2.

Baseplates cast in columns are of two basic types. The first type is a flat plate that extends beyond the perimeter of the column. It contains holes drilled through it to fit over anchor bolts cast in the footing (Figure 14.1). The second type of baseplate fits flush with the side of the column. It contains holes for anchor bolts and has small, three-sided compartments attached to it to accommodate the anchor bolt connections (Figure 14.2).

The type of baseplate connection used depends on the design considerations of each individual job. Selecting a baseplate connection is the engineer's responsibility. Drawing baseplate connection details is the CAD technician's responsibility.

Drawing Baseplate Connection Details

When working on a job involving columns, the engineer gives the CAD technician instructions such as:

■ What type of baseplate connection is desired

■ Baseplate size and configuration specifications

■ Any other pertinent information about the baseplate

With this information known, the CAD technician may begin preparing a baseplate connection detail.

Baseplate connection details are drawn according to the following procedures:

1. Examine the baseplate and column dimensions and the amount of space available for drawing the detail. After considering these factors, select an appropriate scale for drawing the connection detail.

2. Construct an elevation view of the baseplate and a portion of the column showing the baseplate, the anchor bolts, the footing, leveling nuts with washers, and tightening nuts with washers (Figure 14.1 and Figure 14.2).

3. Cut a section horizontally through the column to show a sectional view looking down on top of the baseplate and column and draw the sectional view (Figure 14.1 and Figure 14.2).

4. Completely label all components of the connection detail.

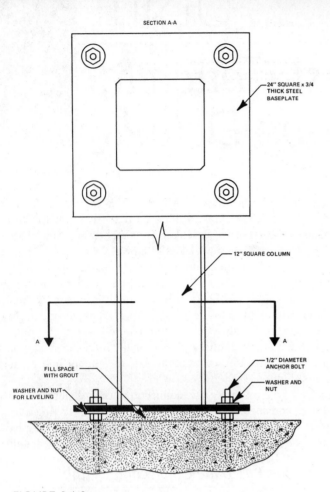

FIGURE 14.1 ■ Typical baseplate connection detail.

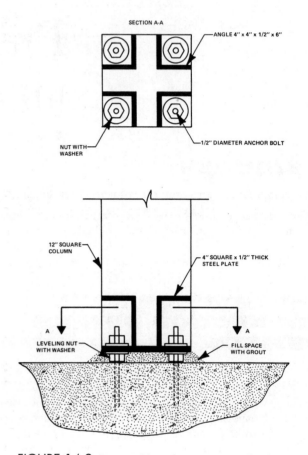

FIGURE 14.2 ■ Typical baseplate connection detail.

BOLTED BEAM-TO-COLUMN CONNECTION DETAILS

In precast concrete construction, beams are connected to columns by bolting. Each place where a beam and column are attached requires a connection detail. Connection details are first identified and then called out on the roof and floor framing plans with letter designations (Figure 14.3).

A connection detail must be drawn for each separate letter designation called out on the framing plan(s). However, in the case of connections that are exactly opposite or are very similar, the number of details required can be reduced by adding abbreviations for the words *opposite hand* or *similar.* Opposite hand is abbreviated *Opp Hd.* This means that the connection being called out is an exact mirror image of another connection that has already been called out. The callout on the framing

plan would read *Conn Det A (Opp Hd).* Similar is abbreviated *Sim.* This means that the connection being called out is substantially the same as another connection that has already been called out. The abbreviation *Sim* may be used in any situation where the beams, columns, and connection are exactly alike but other factors differ slightly (Figure 14.3).

Six basic bolted beam-to-column connections are used in precast concrete construction. In each case, the connection procedures are the same. A long threaded rod is cast into the top of the column that protrudes a distance equal to the thickness of the beam it will support. The beam is fabricated with a round, 3″ diameter blockout in the connection end(s) to fit over the threaded rod, and 4″ × 4″ × 4″ blockout(s) to provide a compartment for screwing a washer-nut combination onto the threaded rod (Figure 14.4 through Figure 14.9).

To avoid damaging the bearing surfaces of the column or the beam, a 3/4″ thick neoprene bearing pad is placed on top of the

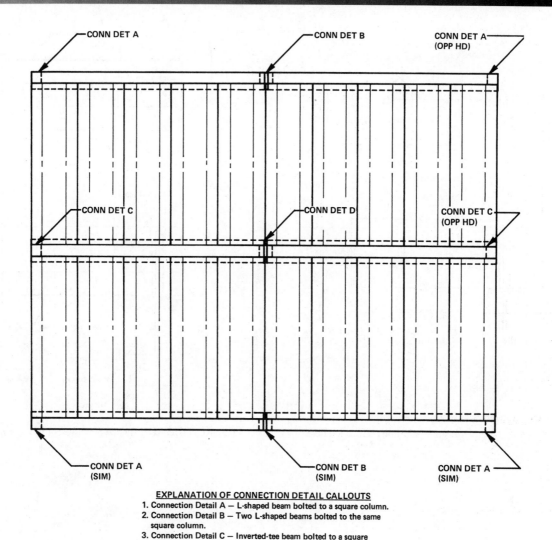

EXPLANATION OF CONNECTION DETAIL CALLOUTS

EXPLANATION OF CONNECTION DETAIL CALLOUTS
1. Connection Detail A — L-shaped beam bolted to a square column.
2. Connection Detail B — Two L-shaped beams bolted to the same square column.
3. Connection Detail C — Inverted-tee beam bolted to a square column.
4. Connection Detail D — Two inverted-tee beams bolted to the same square column.

FIGURE 14.3 ■ Connection detail callouts.

column before the beam is erected. This bearing pad cushions the bearing surface. The pad also lifts the beam just enough so that the top of the threaded rod is slightly below the top surface of the beam (Figure 14.4 through Figure 14.9).

The six basic beam-to-column connections in precast concrete construction are:

■ Single rectangular beam bearing fully on a column (Figure 14.4).

■ Two rectangular beams bearing on the same column. In this case, a 1/2″ to 1″ joint is left between the abutting beams (Figure 14.5).

■ Single L-shaped beam bearing fully on a column (Figure 14.6).

■ Two L-shaped beams bearing on the same column. In this case, a 1/2″ to 1″ joint is left between the abutting beams Figure 14.7).

■ Single inverted-tee beam bearing fully on a column (Figure 14.8).

■ Two inverted-tee beams bearing on the same column. In this case, a 1/2″ to 1″ joint is left between abutting beams (Figure 14.9).

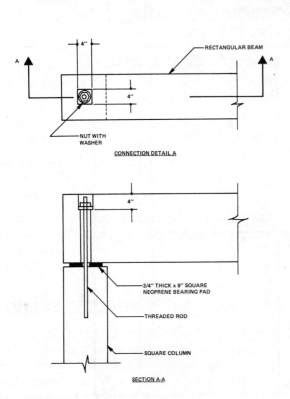

FIGURE 14.4 ■ Rectangular beam-to-column connection detail.

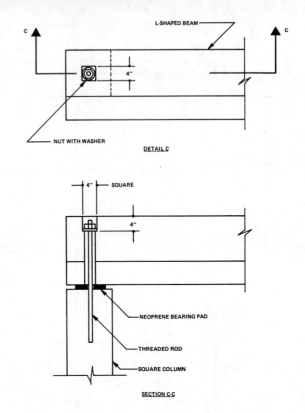

FIGURE 14.6 ■ L-shaped beam-to-column connection detail.

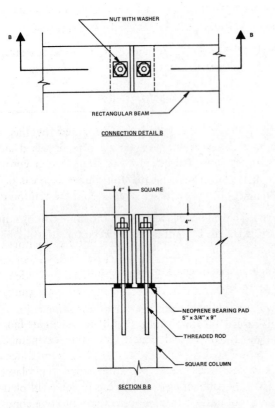

FIGURE 14.5 ■ Two rectangular beam-to-column connection details.

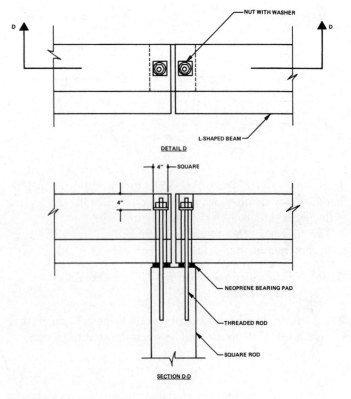

FIGURE 14.7 ■ Two L-shaped beam-to-column connection details.

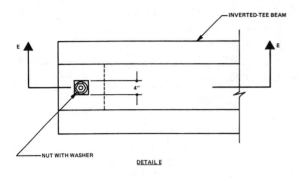

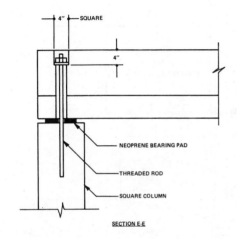

FIGURE 14.8 ■ Inverted-tee beam-to-column connection detail.

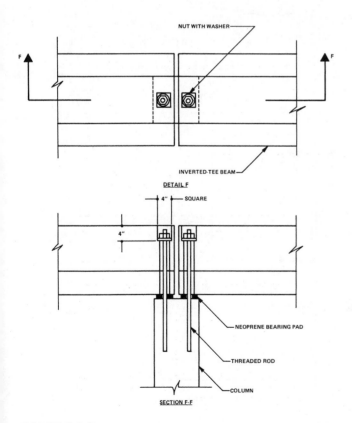

FIGURE 14.9 ■ Two inverted-tee beam-to-column connection details

Drawing Bolted Beam-to-Column Connection Details

Regardless of the types of connections used on a given precast concrete job, the procedures for drawing connection details are the same:

1. Determine from the framing plan(s) how many connection details are required and the configuration of each.

2. Draw a plan view of each connection showing only the immediate area around the connection and cut a section horizontally through it (Figure 14.4 through Figure (14.9).

3. Draw the sectional view showing an elevation of the column, beam(s), blockouts, threaded rod(s), washers/nuts, neoprene bearing pad(s), and any other information pertinent to the connection (Figure 14.4 through 14.9).

WELDED CONNECTIONS

The most frequently used connection method in precast concrete construction is welding. The primary reason for the popularity of welding connections is its versatility in the shop and in the field. By casting weld plates, weld angles, and weld strips into precast concrete members, virtually any required connection can be made. The following are the most common types of welded connections in precast concrete construction:

■ Flange connections in double-tee and single-tee members

■ Base and top connections for wall panels

■ Connections between beams and roof or floor members

■ Stem connections for double tees and single tees

Refer to Figure 14.10 through Figure 14.13 for examples of these types of welded connections.

Flange Weld Plate Connections

Roof and floor tee members must be joined together at the flanges so that several single members become fused into one solid roof or floor. This requires special welded connectors called *flange weld plates*. A flange weld plate consists of a rectangular steel plate welded to a continuous bent reinforcing bar anchor (Figure 14.10).

The steel plate provides a smooth metal surface for making a field-welded connection. The reinforcing bar anchor secures the weld plate firmly in the flange of the tee member so that, once the structure is erected and the connection welded, forces acting on the structure do not dislodge the plate causing the connection to fail.

Abutting tee members in a roof or floor system are fabricated with aligning flange weld plates. Once the tee members are erected in the field, a *single, steel strap plate* (sometimes called *connection plate*) is welded to matching flange weld plates (Figure 14.10). By connecting all matching flange weld plates, the individual roof or floor members are fused into one continuous structural unit.

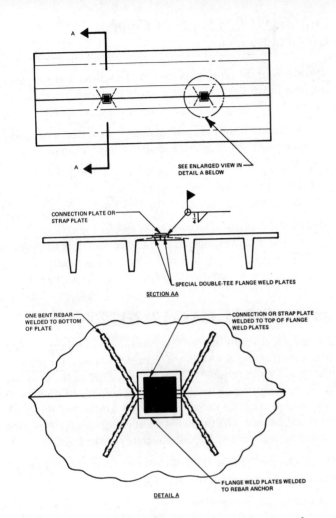

FIGURE 14.10 ■ Welded flange connections for tee members.

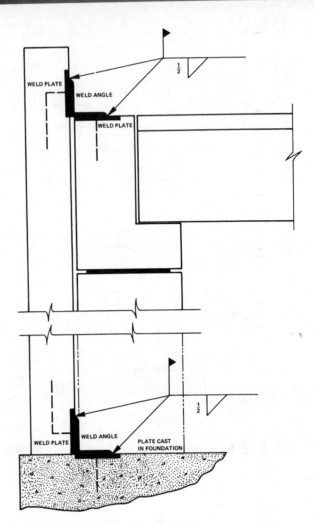

FIGURE 14.11 ■ Weld angle to weld plate connections.

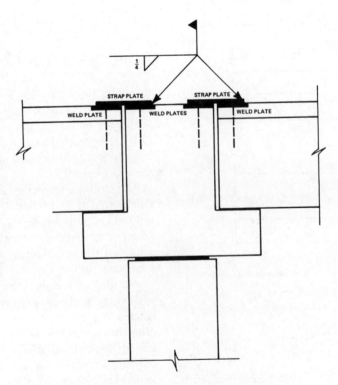

FIGURE 14.12 ■ Strap plate to weld plate connection.

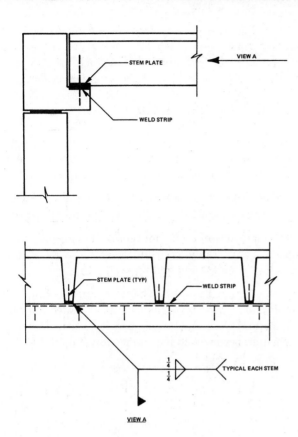

FIGURE 14.13 ■ Welded stem connections.

Weld Plate–Weld Angle Connections

An effective method of fastening precast members together or to other components of a structure involves weld plate connected by a weld angle. Weld plates are square or rectangular steel plates with reinforcing bar anchors to hold them securely in the concrete member. They are cast into precast concrete members. They are cast into precast concrete members at specific locations so as to match up, upon erection, with plates cast in other components of the structure. Matching weld plates are then connected by a weld angle (Figure 14.11).

Weld plate–weld angle connections are very versatile and can be adapted to almost any connection situation. A particular advantage of this type connection is that, if a mistake during fabrication causes weld plates to be misaligned in the field, the error can often be corrected. The error is corrected by simply shifting the weld angle in one direction or the other.

Weld Plate–Strap Plate Connections

Similar to weld plate–weld angle connections are weld plate–strap plate connections. Again, weld plates are cast into precast concrete members. They are located so that they align with plates cast in other components of the structure after erection. Matching weld plates are connected by a single strap plate welded to both (Figure 14.12). This is also a versatile connec-

tion method that can be adapted for use in numerous connection situations. Like weld plate–weld angle connections, this method allows greater tolerances in casting the weld plate in the precast members.

Stem Plate–Weld Strip Connections

A connection method specifically designed for permanently fastening double-tee and single-tee stems to beams is the stem plate–weld strip connection. A stem plate is a steel plate with two or more reinforcing bar anchors that is cut to fit into the bottom of a double-tee or single-tee stem. A weld strip is one long, continuous weld plate cut to the length of the beam in which it will be cast (Figure 14.13).

Weld strips are frequently cast in beams that support tee members instead of weld plates to avoid the difficulty of matching the many weld plates and stem plates. This results in the utilization of more steel but requires less labor time. Therefore, the cost factor is eliminated while the assurance of a connection for each tee stem is increased.

Drawing Welded Connections

Welded connections are designed by engineers or experienced CAD technicians working as designers. Before drawing welded connection details, the drafter should obtain from the engineer or designer such information as plate sizes, angle sizes, and weld information. In addition to drawing the connection details, it is the drafter's responsibility to locate and dimension all metal connectors on the appropriate framing plans.

Welded connection details differ from bolted connection details. Welded connections are drawn according to the following procedures:

1. Select an appropriate scale for drawing the details. The scale selected for drawing welded connection details depends on the number of details to be drawn, the space available in which to draw them, and individual company drafting procedures.

2. Obtain the engineer's or designer's specifications for each connection. These instructions are frequently given orally since most connections are repeated in use so often as to breed familiarity with them.

3. At the selected scale, draw only those views required to convey the connection configuration and design. A properly drawn welded connection detail will show the members being connected, all metal connectors involved, and appropriate weld symbols (Figure 14.10 through Figure 14.13).

HAUNCH CONNECTIONS

There are special situations in precast concrete construction that require special connection methods. The two most common are multistoried structures and structures in which precast concrete wall panels are designed to support roof members without involving columns or beams. The special connection method in both cases involves adding a haunch or haunches to the columns or wall panels.

A *haunch* is a bearing surface that is cast onto a column or wall panel as a secondary pour that supports beams or roof members after erection. Haunch connections on columns are usually bolted, while haunch connections on wall panels are usually welded. Figure 14.14 and Figure 14.15 show isometric illustrations of haunches on columns. Figure 14.16, Figure 14.17, and Figure 14.18 show isometric illustrations of haunches on wall panel members.

Haunched Beam-to-Column Connections

Columns in a multistory structure must support more than one system of beams. For example, a two-story structure requires a system of beams to carry the roof and another system of beams to carry the second floor. Beams that support the roof bear on top of the columns. However, beams that support the floor(s) run into the sides of columns, requiring a special connection. This special connection is accomplished by the addition of a haunch at every point where a floor beam intersects a column.

Haunch connections for beams to columns are bolted. This is accomplished by casting a threaded rod into the haunch (Fig-

ure 14.14 and Figure 14.15). Bolted connections involving haunches are drawn very much like normal beam-to-column connections, with the exception that haunch configurations and dimensions must be supplied by an engineer or designer.

Drawing Bolted Haunch Connections

Once an engineer or designer has supplied haunch configuration and dimensioning information, the procedures for drawing bolted haunch connection details are the following:

1. Determine from the framing plan(s) how many connection details are required and the configuration for each.

2. Draw a plan view of each connection showing only the immediate area around the connection and cut a section through it (Figure 14.14 and Figure 14.15).

3. Draw the sectional view showing an elevation of the column, beam(s), haunch(es), blockouts, threaded rod(s), washers/nuts, neoprene bearing pad(s), and any other information pertinent to the connection (Figure 14.14 and Figure 14.15).

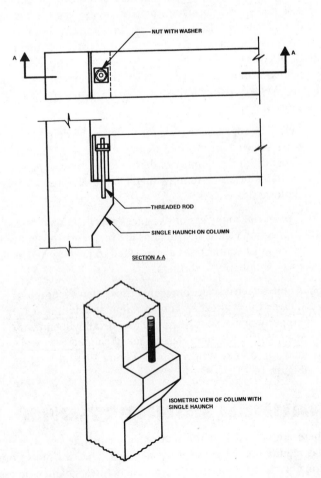

FIGURE 14.14 ■ Single-haunch bolted connection.

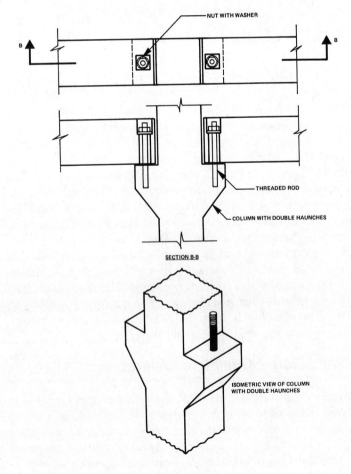

FIGURE 14.15 ■ Double-haunch bolted connection.

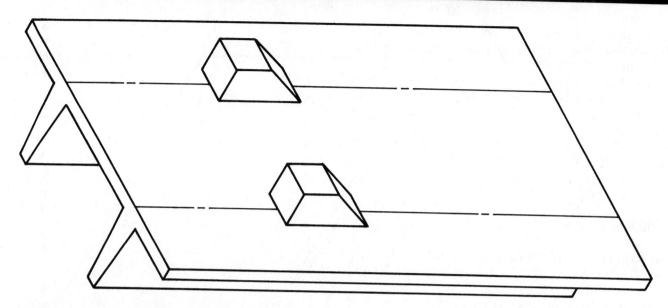

FIGURE 14.16 ■ Double-tee wall panel with two haunches.

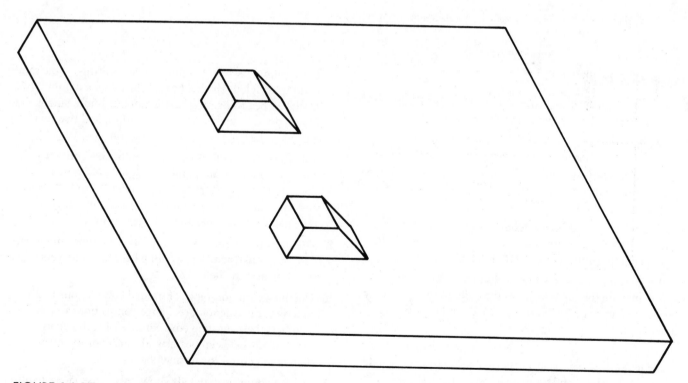

FIGURE 14.17 ■ Flat-slab wall panel with two haunches.

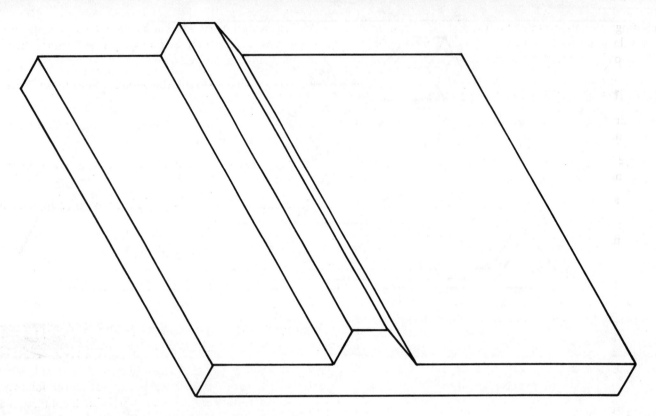

FIGURE 14.18 ■ Flat-slab wall panel with continuous haunch.

SUMMARY

■ Precast concrete members are connected at the jobsite by bolting, welding, or a combination of both.

■ The most common connection situations in precast concrete construction are column baseplate connections, bolted beam-to-column baseplate connections, bolted beam-to-column connections, welded connections, and haunch connections.

■ Columns are provided with a heavy steel baseplate cast in one end that is specially designed to fit over anchor bolts for connecting with washers and nuts.

■ Baseplates for precast concrete columns are of two types: those that are the same width as the column in which they are cast (Figure 14.2) and those that extend beyond the edges of the column (Figure 14.1).

■ Baseplate connection details contain the following information: the baseplate, anchor bolts, the footing, leveling nuts with washers, and tightening nuts with washers.

■ Connection details are called out by letter designation on the framing plans.

■ The number of connection details that must be drawn is reduced by labeling connection situations that are exactly opposite *Opp Hd*, and situations that are very similar *Sim*.

■ There are six basic beam-to-column connections in precast concrete construction: single rectangular beam bearing fully

on a column, two rectangular beams bearing on the same column, single L-shaped beam bearing on a column, two L-shaped beams bearing on the same column, single inverted-tee beam bearing on a column, and two inverted-tee beams bearing on the same column.

■ Bolted beam-to-column connection details contain a plan view of the connection and a sectional elevation showing: the column, beam(s), blockouts, threaded rod(s), washers and nuts, neoprene bearing pad(s), and any other information that is pertinent to the connection.

■ The most frequently used connection method in precast concrete construction is welding because of its versatility in the shop and the field.

■ The most common types of welded connections in precast concrete construction are: flange connections in tee members, base and top connections for wall panels, connections between beams and roof/floor members, and stem connections for tee members.

■ Welded connection details show only those views required to convey the connection design. A properly drawn welded connection contains the members being connected, all metal connectors involved, and appropriate weld symbols.

■ Additional bearing surfaces may be added to columns or wall panels through the addition of haunches.

■ Haunch connection details for bolted connections contain plan and sectional elevation views of the connection including: column, beam(s), haunch(es), blockouts, threaded rod(s), washers/nuts, neoprene bearing pad(s), and any other information pertinent to the connection.

REVIEW QUESTIONS

1. What are the two basic connection methods used in precast concrete construction?

2. List the four most common connection situations in precast concrete construction.

3. Sketch an example of the two types of column baseplates.

4. List the information that should be contained in baseplate connection details.

5. List two abbreviations that may be added to connection details called out on the framing plans to reduce the number of details required.

6. List the six basic beam-to-column connections in precast concrete construction.

7. List the information that should be contained in bolted beam-to-column connection details.

8. What are the most common welded connections in precast concrete construction?

9. List the information that should be included in haunch connection details.

CAD ACTIVITIES

GENERAL INSTRUCTIONS

The following activities may be completed on any CAD system. Before reading the *specific instructions* for each activity (below), go through each step in the following planning checklist. The checklist applies to any CAD system and will help ensure the optimum use of your time and resources.

1. Analyze the problem carefully. Decide exactly what you are being asked to do.

2. Determine what resources and references you will need in order to complete the problem and collect them.

3. Decide if any particular standards apply to the project and have those standards available.

4. Determine what types of views will be required and how many of each.

5. Determine what the final plotted scale of the drawing will need to be, and select the appropriate paper size for plotting/printing (make sure the appropriate paper size is available).

6. Plan your drawing sequence. In what order will you develop the drawing (i.e., lines, features, dimension lines, leaders, dimensions, notes, etc.)?

7. Review the various CAD commands you will have to use in order to develop the drawing.

8. Examine your CAD system to ensure that everything is in working order, then begin the project.

SPECIFIC INSTRUCTIONS

Figure 14.19 contains an example of how connection details appear when drawn as part of a set of precast concrete con-

nection details. Students should refer to this example when laying out the connection details required in the following drafting activities. Beam, column, and haunch configuration dimensions for the Unit 14 CAD Activities are contained in Figure 14.20.

Activity 14.1—Construct connection detail "A" according to the following specifications: Rectangular beam bearing fully on a square column with a 1″ diameter threaded rod and 3/4″ neoprene bearing pad.

Activity 14.2—Construct connection detail "B" according to the following specifications: Two rectangular beams bearing on the same rectangular column with a 1″ joint, 1″ diameter threaded rod, and 3/4″ neoprene bearing pad.

Activity 14.3—Construct connection detail "C" according to the following specifications: L-shaped beam bearing fully on a square column with a 1″ diameter threaded rod and 3/4″ neoprene bearing pad.

Activity 14.4—Construct connection detail "D" according to the following specifications: Two L-shaped beams bearing on the same rectangular column with a 1″ joint, 1″ diameter threaded rod, and 3/4″ neoprene bearing pad.

Activity 14.5—Construct connection detail "E" according to the following specifications: Inverted-tee beam bearing fully on a square column with a 1″ diameter threaded rod and 3/4″ neoprene bearing pad.

Activity 14.6—Construct connection detail "F" according to the following specifications: Two inverted-tee beams bearing on the same rectangular column with a 1″ joint, 1″ diameter threaded rod, and 3/4″ neoprene bearing pad.

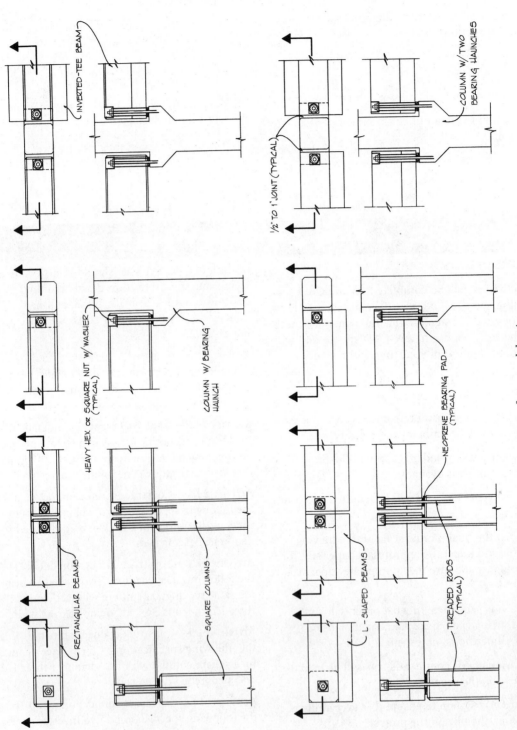

FIGURE 14.19 ■ Connection details as they appear in a set of structural drawings.

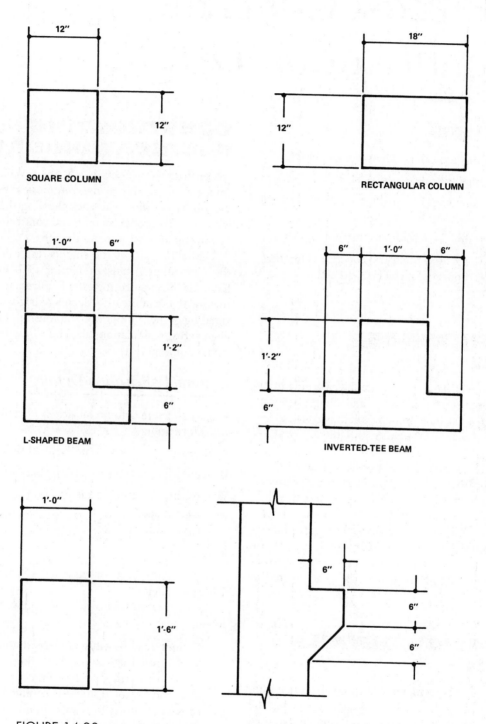

FIGURE 14.20 ■ Configuration dimensions for drafting activities.

Precast Concrete Fabrication Details

OBJECTIVES

Upon completion of this unit, the student will be able to:

- Define shop drawings and fabrication details.
- Explain how shop drawings fit into a set of precast concrete working drawings.
- Explain how fabrication details fit into the shop drawings.
- Construct fabrication details of precast concrete columns, beams, wall panels, floor/roof members, and metal connectors.

SHOP DRAWINGS DEFINED

In Unit 1, it was learned that a set of structural working drawings consists of engineering drawings and shop drawings. Engineering drawings include framing plans, sections, and connection details. These were covered in Unit 12, Unit 13, and Unit 14.

Shop drawings are comprehensive, precisely detailed drawings prepared for use by shop workers in fabricating precast concrete products and the metal connectors they contain. Shop drawings consist of fabrication details and bills of materials.

The information required in preparing shop drawings is taken from the engineering drawings. This unit deals with the preparation of fabrication details for precast concrete products and metal connectors. Bills of materials are covered in the next unit.

FABRICATION DETAILS DEFINED

Fabrication details are orthographic drawings of precast concrete beams, columns, wall panels, roof/floor members, and metal connectors. They contain all of the information required by shop workers to fabricate all of the precast products and metal connectors used in a job. Each concrete product and each metal connector requires a separate fabrication detail.

Figure 15.1 through Figure 15.4 illustrate examples of fabrication details for precast concrete products. Miscellaneous metal fabrication details are shown in Figure 15.5.

CONSTRUCTING FABRICATION DETAILS

To construct fabrication details, the CAD technician must have copies of the framing plans, sections, and connection details for the job. In addition, sketches showing the design calculations for all precast products and metal connectors must be provided by the engineer assigned to the job.

Figure 15.6 shows the engineering calculation sketches for the fabrication details in Figure 15.1 through Figure 15.4. Examine the engineering calculation sketches closely. Compare them with the fabrication details to begin developing an understanding of their important relationship. Figure 15.7 shows the engineering calculation sketches for the metal connectors detailed in Figure 15.5.

Column Fabrication Details

Precast concrete column fabrication details contain the following information:

- At least two orthographic views of the column
- Section(s) showing reinforcing bar or strand patterns
- A baseplate configuration diagram
- A special view showing the location of threaded rod(s) in relation to the top of the column
- Complete dimensions

Refer to Figure 15.1 for examples of precast concrete column fabrication details.

Column fabrication details are drawn according to the following procedures:

1. Column length, width, and depth dimensions are determined from the framing plan and products schedule, and at least two orthographic views of the column are drawn. It should be noted that it is common practice in precast concrete drafting to draw column lengths on fabrication details *not to scale*. This is because columns are often very long, making it difficult to fit them on the sheet.

2. Baseplate width, depth, and thickness are determined from the connection detail. The baseplate is drawn at the bottom end of the column, and the baseplate configuration diagram is added (Figure 15.1).

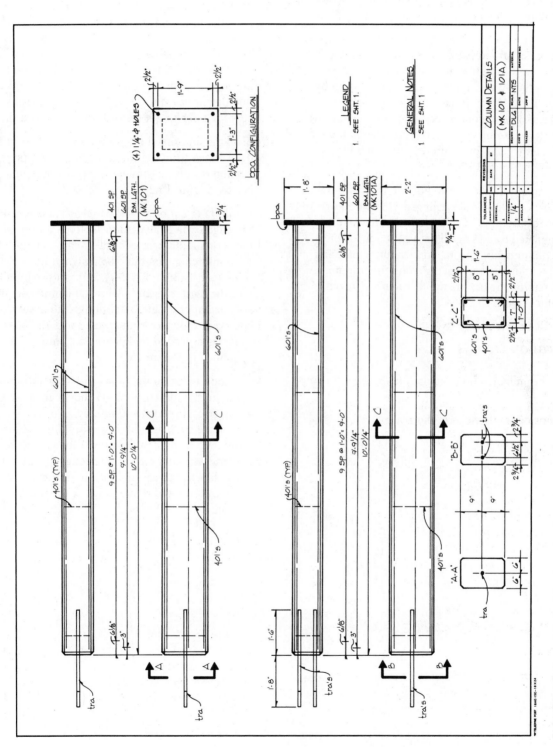

FIGURE 15.1 ■ Column fabrication details.

3. Locations of threaded rods are determined from the appropriate beam-to-column connection detail, and the rod is added to the orthographic views of the column. A special end view is then drawn to show how the threaded rod(s) fit into the top of the column. These special views are called out and drawn exactly as if they were sections (Figure 15.1).

4. Reinforcing bar or strand patterns for the column are determined from the engineering calculation sketches and represented in the orthographic views of the column by long, dashed lines. All reinforcing bars used in the column must be assigned numeric designations such as 301, 401, 501, 601, and so on and entered into the master legend. The master legend for Figure 15.1 through Figure 15.4 is shown in Figure 15.5.

5. A section is cut on the column and drawn to show how reinforcing bars or strands are arranged and spaced within the column. See Section C-C, Figure 15.1. The column detail is completed by adding all necessary dimensions including: column width, depth, and length; reinforcing bar lengths; reinforcing bar spacing along the column length; and reinforcing bar or strand spacing in a section of the column (Figure 15.1).

Beam Fabrication Details

Precast concrete beam fabrication details contain the following information:

- At least two orthographic views of the beam
- Section(s) showing reinforcing bar or strand patterns
- Blockouts
- Metal connectors
- Complete dimensions

Refer to Figure 15.2 for examples of precast concrete beam fabrication details.

Beam fabrication details are drawn according to the following procedures:

1. Beam configuration and length dimensions are determined from the framing plan, products schedule, and connection details or sections. When this information has been obtained, at least two orthographic views of the beam are drawn. It should be noted that it is common practice in precast concrete drafting to draw beam lengths on fabrication details *not to scale*. This is because beams are often very long, making it difficult to fit them on the sheet.

2. Blockout and metal connector sizes and locations are determined from the framing plan and appropriate connection details an drawn on the orthographic views of the beams (Figure 15.2).

3. Reinforcing bar or strand patterns are determined from the engineering calculation sketches and represented in the orthographic views of the beam by long, dashed lines. All reinforcing bars must be assigned numeric designations such as 301, 401, 501, 601, and so on and entered into the master legend. All metal connectors must be assigned appropriate letter designations such as wpa, spa, cpa, waa, and so on and entered into the master legend. The master legend for Figure 15.1 through Figure 15.4 is contained in Figure 15.5.

4. A section is cut on the beam and drawn to show configuration dimensions and how bars or strands are arranged and spaced within the beam. The beam detail is completed by adding all necessary dimensions including: beam length and configuration dimensions; reinforcing bar lengths; reinforcing bar spacing along the length of the beam; blockout sizes and locations; metal connectors sizes and locations; and reinforcing bar or strand spacing in the sectional view of the beam (Figure 15.2).

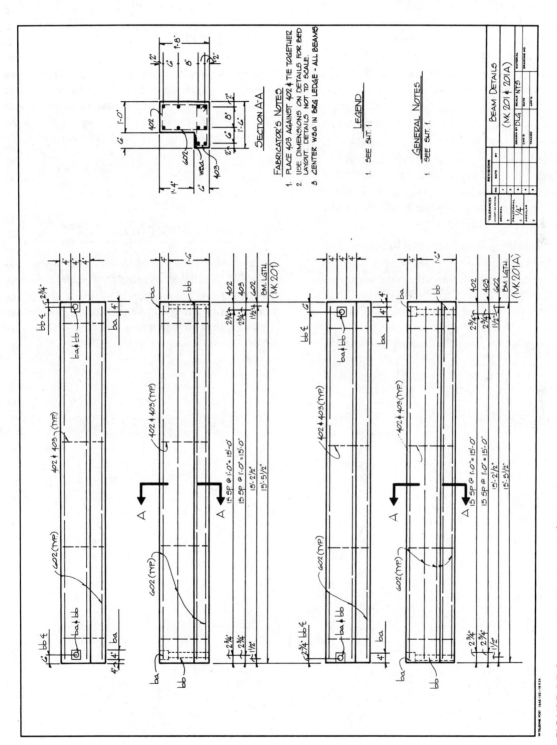

FIGURE 15.2 ■ Beam fabrication details.

Floor/Roof Member Fabrication Details

Precast concrete roof and floor members may be double tees, flat slabs, or cored flat slabs. The most common floor and roof members are double tees. However, the procedures for detailing the members are the same regardless of their type. Refer to Figure 15.3 for an example of a precast concrete floor or roof member fabrication detail.

Floor/roof member fabrication details are drawn according to the following procedures:

1. Member configuration and length dimensions are determined from the framing plan, products schedule, or illustrations of standard sizes in the case of tee members. It should be noted that it is common practice in precast concrete drafting to draw floor/roof member lengths on fabrication details *not to scale*. This is because floor/roof members are often very long, making it difficult to fit them on the sheet.

2. Blockout and metal connector sizes and locations are determined from the framing plan and appropriate connection details and drawn on the orthographic views of the floor/roof member (Figure 15.3). Blockouts and metal connectors must be assigned letter designations and entered in the master legend. The master legend for Figure 15.1 through Figure 15.4 is contained in Figure 15.5.

3. Precast concrete roof and floor members receive prestressing strands rather than reinforcing bars. Strand patterns are shown in a sectional view only. Since prestressing strands run the entire length of the precast product, they are not dimensioned in the orthographic views of the product. All stranding information required on floor/roof fabrication details is taken from the engineering calculation sketches.

4. The floor/roof member fabrication detail is completed by adding all necessary dimensions including: member length and configuration dimensions; reinforcing bar lengths and spacing (if used they must be assigned numeric designations and entered in the master legend); blockout sizes and locations; metal connector sizes and locations; strand patterns in the sectional view (Figure 15.3).

Wall Panel Fabrication Details

Like precast concrete floor and roof members, wall panels may be double tees, flat slabs, or cored flat slabs. The most common wall panels are flat slabs. Regardless of the product type, the procedures for detailing the members are the same. Refer to Figure 15.4 for an example of a precast concrete wall panel fabrication detail.

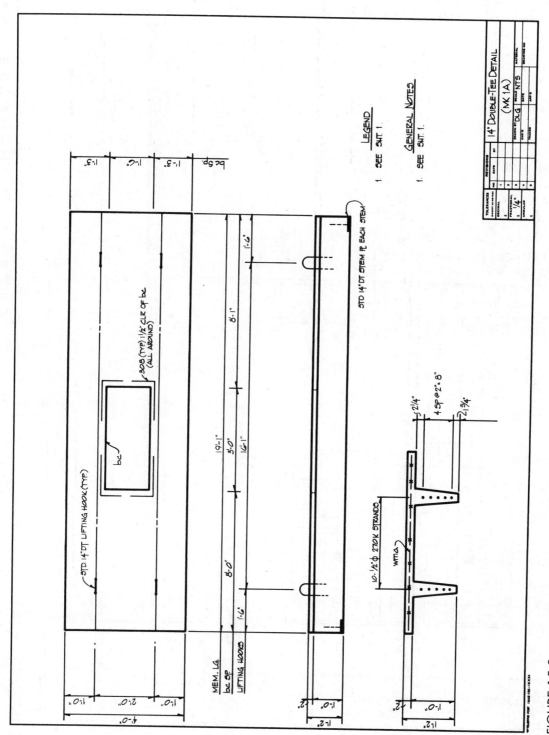

FIGURE 15.3 ■ Double-tee roof member fabrication details.

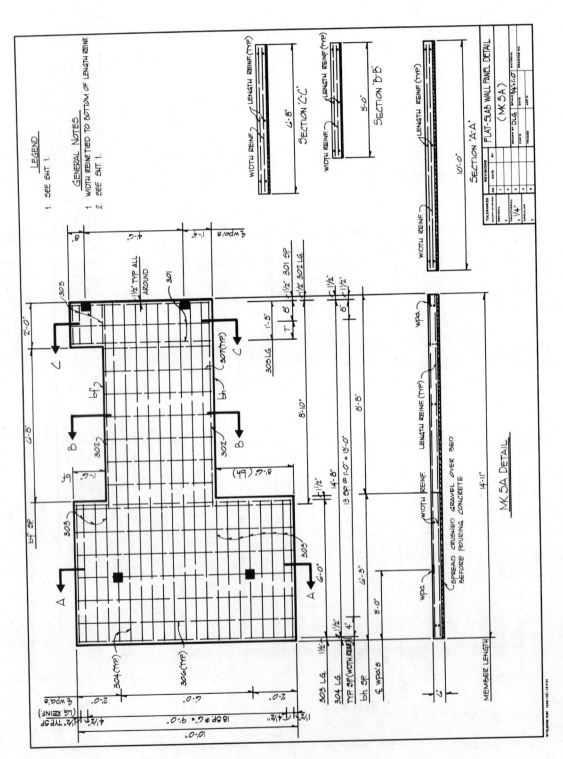

FIGURE 15.4 ■ Flat-slab wall panel fabrication detail.

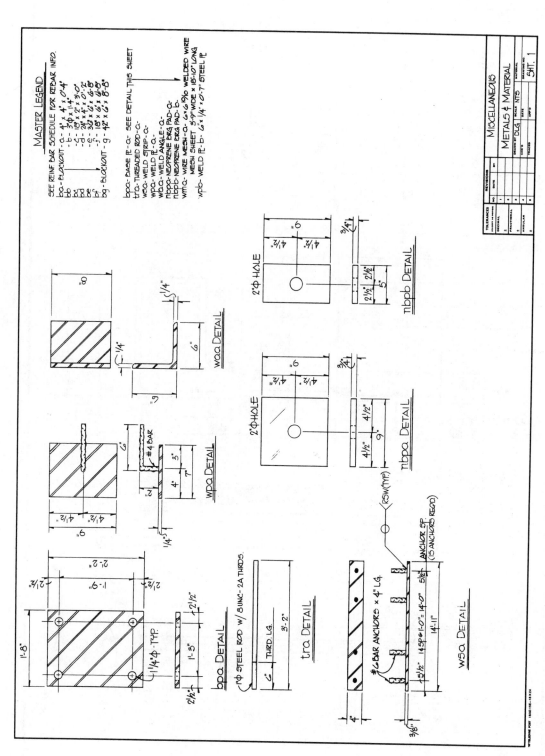

FIGURE 15.5 ■ Metal connectors fabrication details.

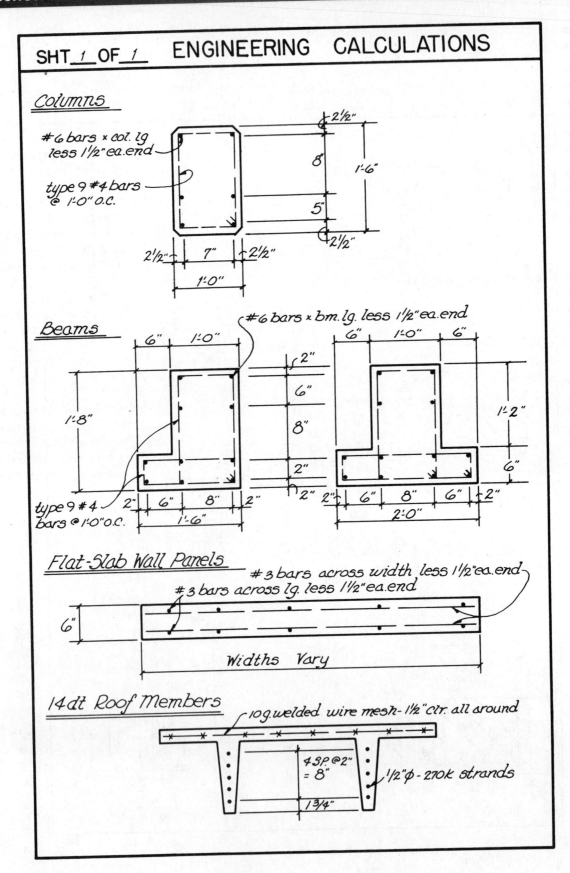

FIGURE 15.6 ■ Engineering calculation sketches.

Wall panel fabrication details are drawn according to the following procedures:

1. Wall panel length and configuration dimensions are determined from the framing plan and products schedule. It should be noted that it is common practice in precast concrete drafting to draw wall panel lengths on fabrication details *not to scale*. This is because wall panel lengths are often too long to fit comfortably on the sheet.

2. Blockout and metal connector sizes and locations are determined from the framing plan and appropriate connection details and drawn on the orthographic views of the wall panel. Later designations are assigned and entered in the master legend. The master legend for Figure 15.1 through Figure 15.4 is contained in Figure 15.5.

3. Sections are cut at each point along the wall panel where the width of the panel varies and drawn to show the arrangement and spacing of reinforcing bars within the wall panel. All reinforcing bars of differing lengths and shapes must be assigned different numeric designations and entered in the master legend.

4. The wall panel fabrication detail is completed by adding all necessary dimensions including: wall panel length and configuration dimensions; reinforcing bar lengths and spacing; blockout sizes and locations; metal connector sizes and locations; and strand patterns in the sectional views (if stranding is used) (Figure 15.4).

Metal Connector Fabrication Details

Metal connectors for a precast concrete job are designed by an engineer. The engineer may convey her design for the connectors to the drafter through sketches (Figure 15.7). The CAD technician's job is to draw orthographic fabrication details based on the engineer's design.

Metal connectors and miscellaneous materials such as neoprene bearing pads are detailed on the same sheet. This sheet then becomes the first sheet of the shop drawings. The metals sheet also contains the master legend for the job and is accompanied by a separate reinforcing bar schedule.

The *reinforcing bar schedule* is a form showing a number of bar diagrams with letters substituted for dimensions. In addition, the form has columns for listing the bar number, type, and bending dimensions (Figure 15.8). All reinforcing bars used on a job are tabulated on the reinforcing bar schedule. The reinforcing bar schedule for the members detailed in Figure 15.1 through Figure 15.4 is shown in Figure 15.9. Reinforcing bar schedule forms may be produced by the individual companies that use them or purchased from commercial producers. The format used in Figure 15.8 should be used in completing the CAD activities for this unit.

In the event that the engineering drawings are combined with the shop drawings to form a set of precast concrete working drawings, the metals sheet becomes the first sheet in the set and contains the master legend and the general notes for the job. Figure 15.5 shows an example of a metals sheet used as the first sheet in a set of precast concrete working drawings.

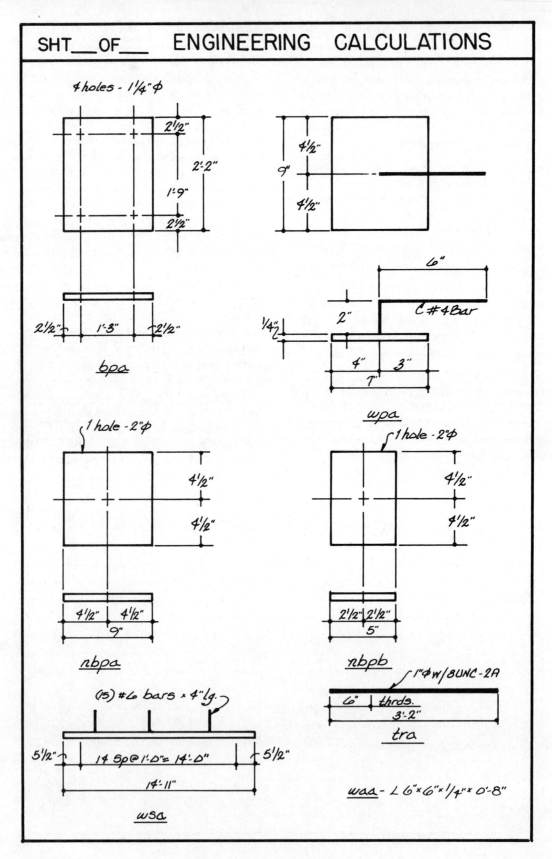

FIGURE 15.7 ■ Engineering calculations for metal connectors.

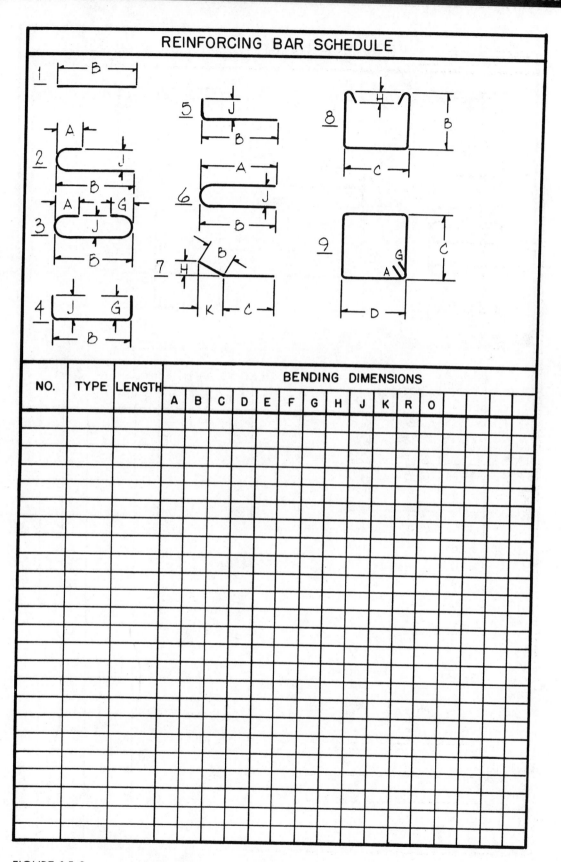

FIGURE 15.8 ■ Reinforcing bar schedule.

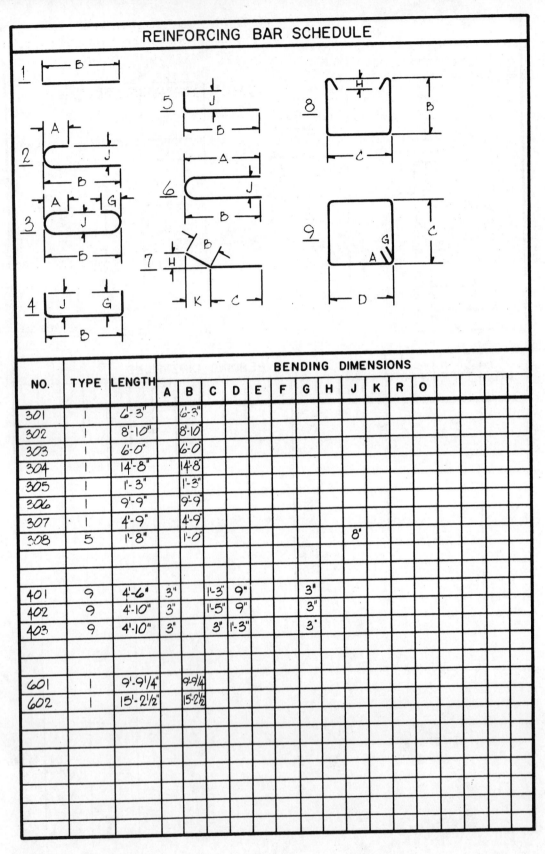

FIGURE 15.9 ■ Completed reinforcing bar schedule.

SUMMARY

■ Shop drawings are prepared for use by shop workers in fabricating precast concrete products and consist of fabrication details and bills of material.

■ Information required on shop drawings is obtained from the engineering drawings and engineering calculation sketches.

■ Fabrication details are orthographic drawings or precast concrete products or metal connectors that show all of the information needed by shop workers in fabrication.

■ In order to construct the fabrication details for a job, the drafter must have the framing plan(s), sections, connection details, and engineering calculation sketches.

■ Fabrication details are constructed according to specific procedures that include drawing at least two orthographic views of the subject; locating, sizing, and labeling all blockouts, reinforcing bars, and metal connectors; and providing complete dimensions.

■ Metal connectors for a precast concrete job are designed by an engineer. The design is conveyed to drafters by sketches.

■ The metals sheet, accompanied by a reinforcing bar schedule, contains the master legend and is the first sheet in a set of shop drawings.

■ When a set of engineering and shop drawings are combined into a set of working drawings, the metals sheet becomes the first sheet of the new set.

REVIEW QUESTIONS

1. Define *shop drawings*.

2. Where is the information that is required on shop drawings located?

3. Define *fabrication details*.

4. What must the CAD technician have in order to construct fabrication details?

5. Briefly, and in general terms, summarize the procedures for preparing precast concrete fabrication details.

6. Who designs the metal connectors in a precast concrete job?

7. Which sheet is first in a set of shop drawings? Working drawings?

CAD ACTIVITIES

GENERAL INSTRUCTIONS

The following activities may be completed on any CAD system. Before reading the *specific instructions* for each activity (below), go through each step in the following planning checklist. The checklist applies to any CAD system and will help ensure the optimum use of your time and resources.

1. Analyze the problem carefully. Decide exactly what you are being asked to do.

2. Determine what resources and references you will need in order to complete the problem and collect them.

3. Decide if any particular standards apply to the project and have those standards available.

4. Determine what types of views will be required and how many of each.

5. Determine what the final plotted scale of the drawing will need to be, and select the appropriate paper size for plotting/printing (make sure the appropriate paper size is available).

6. Plan your drawing sequence. In what order will you develop the drawing (i.e., lines, features, dimension lines, leaders, dimensions, notes, etc.)?

7. Review the various CAD commands you will have to use in order to develop the drawing.

8. Examine your CAD system to ensure that everything is in working order, then begin the project.

SPECIFIC INSTRUCTIONS

The following drafting exercises are based on the rough engineering drawings provided in Figure 15.10 through Figure 15.15 and the engineering calculation sketches in Figure 15.16. Where the use of a scale is required, a scale appropriate for the task being performed should be selected.

Column Fabrication Details

Activity 15.1—Prepare a complete fabrication detail for MK 100 in Figure 15.10. Refer to Figure 15.1 for examples.*

*The student may select the type and size of baseplate used in completing these activities.

Activity 15.2—Prepare a complete fabrication detail for MK 101 in Figure 15.10. Refer to Figure 15.1 for examples.*

Beam Fabrication Details

Activity 15.3—Prepare a complete fabrication detail for MK 200 in Figure 15.11. Refer to Figure 15.2 for examples.

Activity 15.4—Prepare a complete fabrication detail for MK 200A in Figure 15.11. Refer to Figure 15.2 for examples.

Activity 15.5—Prepare a complete fabrication detail for MK 201 in Figure 15.11. Refer to Figure 15.2 for examples.

Activity 15.6—Prepare a complete fabrication detail for MK 201A in Figure 15.11. Refer to Figure 15.2 for examples.

Wall Panel Fabrication Details

Activity 15.7—Prepare a complete fabrication detail for MK 10 in Figure 15.12. Refer to Figure 15.4 for an example.

Activity 15.8—Prepare a complete fabrication detail for MK 10A in Figure 15.12. Refer to Figure 15.4 for an example.

Activity 15.9—Prepare a complete fabrication detail for MK 10B in Figure 15.12. Refer to Figure 15.4 for an example.

Activity 15.10—Prepare a complete fabrication detail for MK 10C in Figure 15.12. Refer to Figure 15.4 for an example.

Roof Member Fabrication Details

Activity 15.11—Prepare a complete fabrication detail for MK 1 in Figure 15.13. Refer to Figure 15.3 for an example.

Activity 15.12—Prepare a complete fabrication detail for MK 1A in Figure 15.13. Refer to Figure 15.3 for an example.

Miscellaneous Metals Fabrication Details

Activity 15.13—Referring to Figure 15.10 through Figure 15.15 and Figure 15.16, prepare the legend and general notes for a metals fabrication details sheet. Refer to Figure 15.5 for an example.

Activity 15.14—Referring to Figure 15.10 through Figure 15.15 and Figure 15.16, prepare all required metal fabrication details. Refer to Figure 15.9 for an example.

Activity 15.15—Using the fabrication details completed in Activity 15.1 through Activity 15.12, prepare a complete bar reinforcing schedule. Refer to Figure 15.9 for an example.

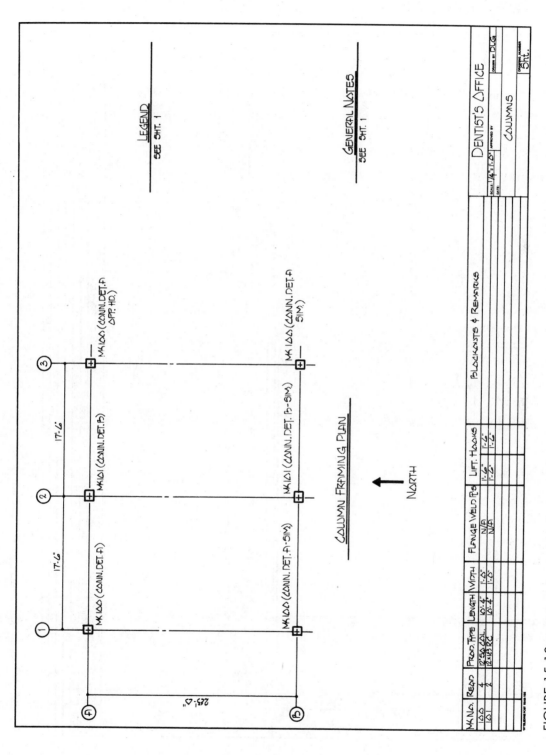

FIGURE 15.10 ■ Column framing plan for CAD activities.

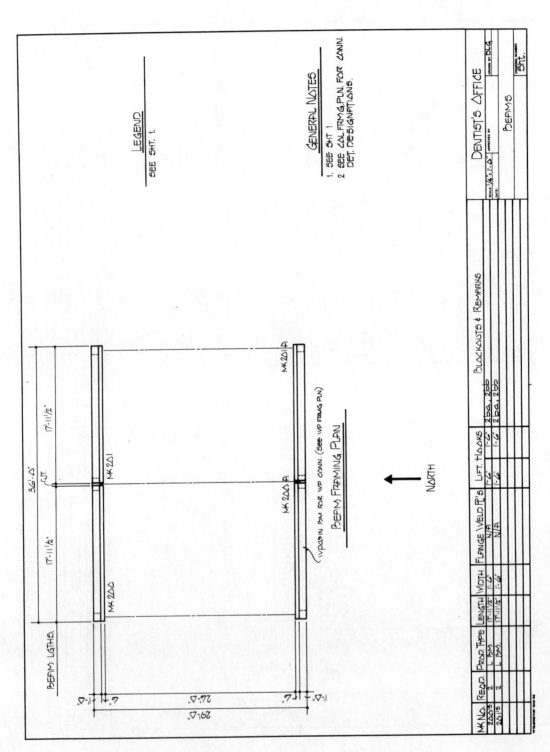

FIGURE 15.11 ■ Beam framing plan for CAD activities.

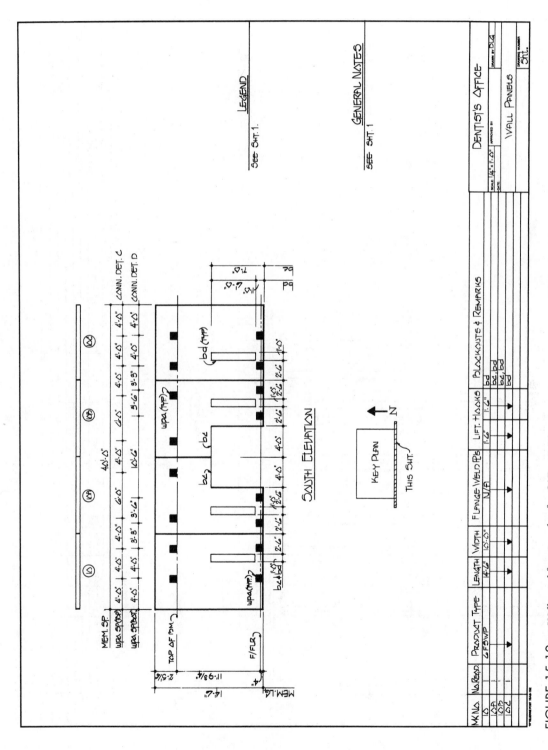

FIGURE 15.12 ■ Wall panel framing plan for CAD activities.

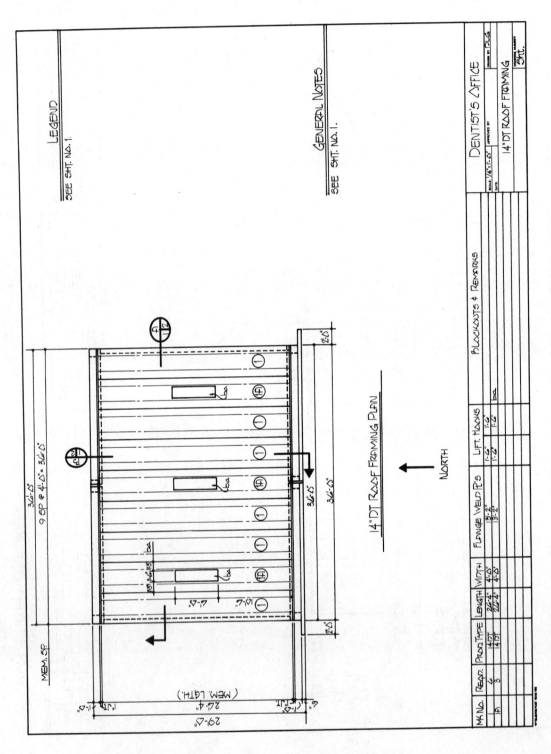

FIGURE 15.13 ■ Roof framing plan for CAD activities.

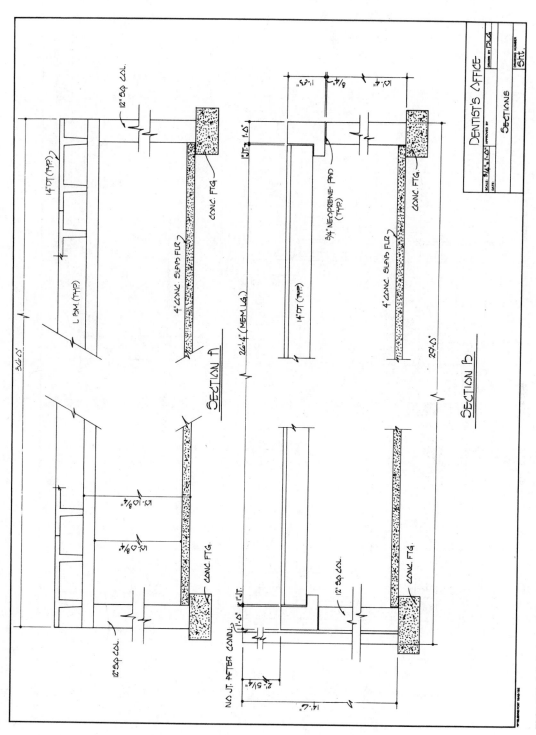

FIGURE 15.14 ■ Sections for CAD activities.

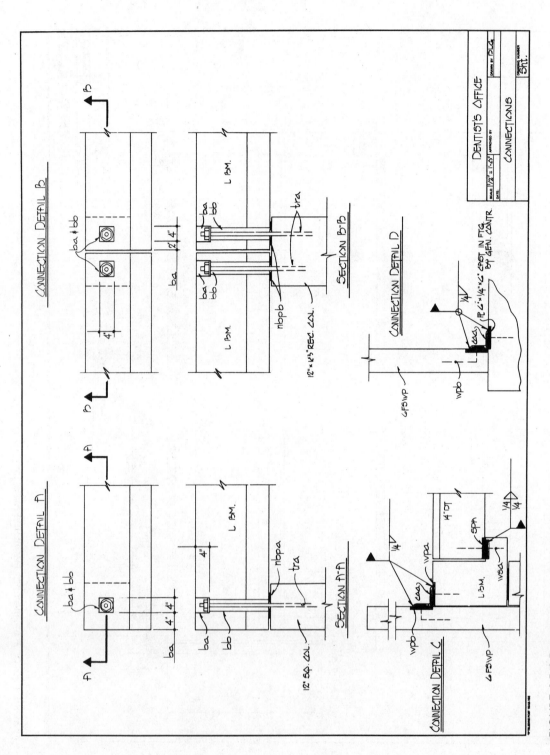

FIGURE 15.15 ■ Connection details for CAD activities.

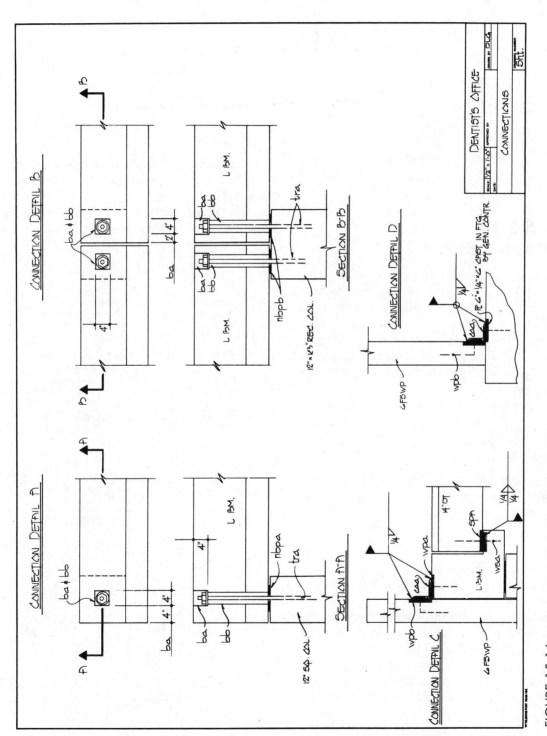

FIGURE 15.16 ■ Engineering calculation sketches for CAD activities.

Precast Concrete Bills of Materials

OBJECTIVES

Upon completion of this unit, the student will be able to:

■ Define *bill of materials* and explain its purpose.

■ Prepare bills of materials for columns, beams, floor/roof members, and wall panels.

DEFINITION AND PURPOSE

A *bill of materials* is a comprehensive listing of all materials that are used in the fabrication and erection of the precast concrete products for a given job. Bills of materials are used by shop workers. From the supply bin, the shop workers draw all of the material they need to manufacture a given member. The purchasing department also uses a bill of materials for ordering the materials that will be needed in the fabrication of the precast products for a job.

Several different procedures are used in preparing a bill of materials:

■ One complete bill of materials covering all of the precast concrete members in a job

■ Individual bills of materials for each individual precast concrete member in a job

■ One bill of materials for each separate precast concrete product type used on a job—for example, one for wall panels, and one for floor/roof members

The last method mentioned is the most popular. This is because it can serve the same purposes as both of the other two and it is more efficient. In this method, one drafter may be preparing the bill of materials for the columns while another prepares a bill for the beams. At the same time, another drafter prepares a bill for the wall panels while still another prepares a bill for the floor/roof members. When each CAD technician finishes the bill of materials in his area of responsibility, the resultant bills may be stapled together and used as one complete bill for the entire job.

Most companies either prepare or purchase standard bill of materials forms. Forms vary according to the needs of the individual company. However, all bills list the following:

■ Type of product

■ The mark numbers for all members of that product

■ All materials contained in the members

■ How many of each type of material is contained in each member

These amounts are totaled by product type and then totaled again for the entire job. Figure 16.1 shows an example of a blank bill of materials form. Bills of materials prepared for columns, beams, wall panels, and roof/floor members are shown in Figure 16.2 through Figure 16.5.

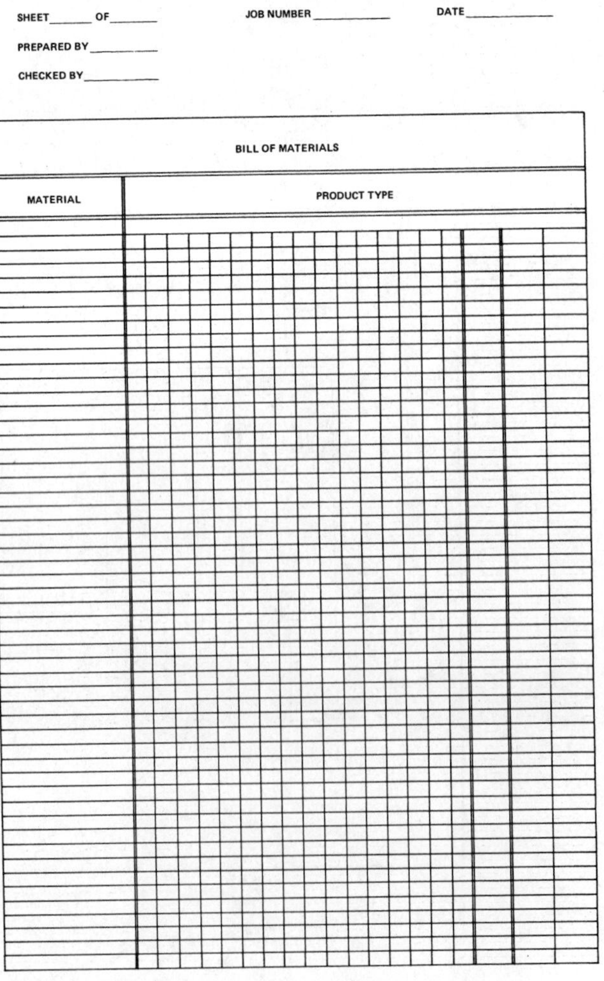

FIGURE 16.1 ■ Blank bill of materials form.

CONSTRUCTING BILLS OF MATERIALS

Bills of materials for precast concrete jobs are prepared from the fabrication details. A CAD technician assigned to prepare a bill of materials for the columns used in a job needs copies of the fabrication details of all of the columns. This is also true of the CAD technician assigned the bill of materials for any portion of a job. The procedures used in preparing the bill of materials for columns, beams, wall panels, and roof/floor members are the same:

1. Collect all of the fabrication details in the area for which the bill of materials is being prepared.

2. Examine the details and compile a list of every bar, plate, angle, bolt, washer, weld strip, rebar, bearing pad, and so on that is used in any of the members. These should be listed according to the abbreviations used on the fabrication details under the heading *material* on the bill of materials form (Figure 16.2 through Figure 16.5).

3. In the blank on the bill of materials form marked *Type of Product*, the product type should be entered: columns, beams, roof/floor members, or wall panels (Figure 16.2 through Figure 16.5).

4. A row of boxes on the bill of materials form is labeled *MK Numbers*, and each mark number for the product type being billed is entered into a box. For the convenience of the people who will use the bill of materials, the numbers should be listed in order from smallest to largest (Figure 16.2 through Figure 16.5).

5. The bill of materials is now ready to complete. The fabrication detail for each mark number listed on the bill is examined closely, the quantity of each individual material used is counted, the count for each material is recorded in the appropriate box, and the process is repeated for every mark number under the subject product type (Figure 16.2 through Figure 16.5).

SHEET 1 OF 1   JOB NUMBER 3201   DATE 2-5-80
PREPARED BY JDR
CHECKED BY DLG

BILL OF MATERIALS

MATERIAL	PRODUCT TYPE 12" SQUARE COLUMNS									Total
MK NUMBERS	100	101	102	103	104	105	106	107	108	
bpa	1	1	1	1	1	1	1	1	1	9
tra	1	2	2	4	1	4	2	1	1	18
wpa	2	-	-	3	-	1	-	-	-	6
wpb	-	2	2	2	-	1	2	-	3	12
401's	17	19	15	12	18	21	16	14	11	143
501's	-	-	-	2	-	2	-	-	-	4
502's	-	-	-	2	-	2	-	-	-	4
503's	-	-	-	2	-	2	-	-	-	4
504's	-	-	-	2	-	2	-	-	-	4
801's	4	4	4	4	4	4	4	4	4	36
802's	4	4	4	4	4	4	4	4	4	36

FIGURE 16.2 ■ Completed bill of materials for columns.

SHEET 1 OF 1   JOB NUMBER 3201   DATE 2-5-80
PREPARED BY TB
CHECKED BY DLG

BILL OF MATERIALS

MATERIAL	PRODUCT TYPE L-SHAPED BEAMS						Total
MK NUMBERS	200	201	202	203	204	205	
wsa	1	-	-	-	-	-	1
wsb	-	1	-	-	-	-	1
wsc	-	-	1	-	-	-	1
wsd	-	-	-	1	-	-	1
wse	-	-	-	-	1	-	1
wsf	-	-	-	-	-	1	1
wpc	4	-	3	2	-	5	14
wpd	-	5	-	6	-	-	11
wpe	2	2	2	2	2	2	12
301's	3	3	3	3	3	3	18
302's	3	3	3	3	3	3	18
402's	16	12	14	15	18	19	94
403's	16	12	14	15	18	19	94
601's	10	10	10	10	10	10	60

FIGURE 16.3 ■ Completed bill of materials for beams.

SHEET ___1___ OF ___1___   JOB NUMBER ___3201___   DATE _2-5-80_

PREPARED BY _DM_

CHECKED BY _DLG_

BILL OF MATERIALS

MATERIAL	PRODUCT TYPE 6" FLAT-SLAB WALL PANELS													Total
MK NUMBERS	1	1A	1B	1C	2	3	3A	3B	4	5	6	6A	6B	
wpa	2	2	2	2	2	2	2	2	2	2	2	2	2	26
wpb	2	2	2	2	2	2	2	2	2	2	2	2	2	26
404's	3	3	3	3	5	4	4	4	–	–	–	–	–	29
405's	–	–	–	–	–	–	–	3	4	5	5	5		22
406's	20	20	20	20	20	20	20	20	20	20	20	20	20	260
407's	–	–	–	–	1	1	1	–	–	1	1	1		6
408's	–	–	–	–	1	1	1	–	–	1	1	1		6
409's	–	–	–	–	1	1	1	–	–	1	1	1		6
410's	12	12	12	12	12	12	12	12	12	12	12	12		156
701's	–	–	–	2	–	–	–	2	2	–	–	–		6

FIGURE 16.4 ■ Completed bill of materials for flat-slab wall panels.

BILL OF MATERIALS

MATERIAL	PRODUCT TYPE 14" DOUBLE-TEE ROOF MEMBERS										Total
MK NUMBERS	10	10A	10B	10C	11	12	13	14	14A	15	
wpc	2	2	2	2	2	2	2	2	2		20
spa	4	4	4	4	4	4	4	4	4		40
tba	4	4	4	4	4	4	4	4	4		40
411's	–	2	2	2	–	–	–	–	–		6
412's	–	–	–	–	–	–	2	2	2		6
270K STR x L.F.	60	60	60	60	90	90	90	120	120	120	870

FIGURE 16.5 ■ Completed bill of materials for double-tee roof members.

COUNTING MATERIAL QUANTITIES

To save drafting time and for ease of reading the prints, every piece of rebar or every metal connector in a group of rebars or connectors used in a member is not shown. Rather, a few samples of the rebar or connector may be shown with one being called out. This indicates that there are more used than have been shown.

In counting material quantities, the drafter should be careful to count according to dimensional specifications and not according to the picture presented by the fabrication detail. For example, the detail for a precast member that is to be cast 20'-0" long and contain #4 bars running across its width at 6" on center would not show each individual piece of rebar. The drafter would have to read the #4 bar dimension line to determine the number of bars used (Figure 16.6).

The same rule applies in counting quantities of metal connectors or any other material used in a precast product. Quantities must be determined from the dimensions for that material with the exception of materials whose location is obvious from the picture. In these cases, the quantity of the material shown is the quantity used. An example of how metal connector quantities are counted for the bill of materials is shown in Figure 16.7.

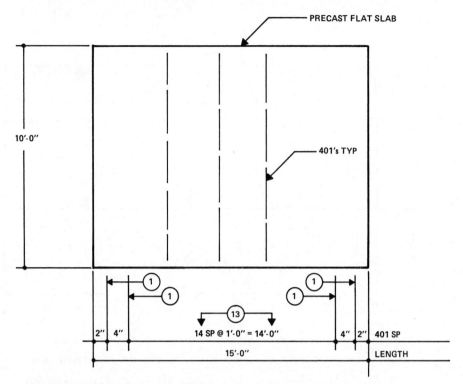

<u>PROBLEM</u>
DETERMINE THE NUMBER OF 401 BARS USED IN THE FLAT SLAB ILLUSTRATED ABOVE.

<u>SOLUTION</u>
1. LOCATE THE DIMENSION LINE THAT GIVES THE SPACING FOR 401 BARS. IN THE ILLUS-
   TRATION ABOVE, IT IS DIRECTLY ABOVE THE LENGTH DIMENSION LINE.
2. COUNT THE 401 BARS ALONG THE DIMENSION LINE. EACH LINE EXTENDED FROM THE
   FLAT SLAB TO THE DIMENSION LINE REPRESENTS ONE 401 BAR. THE DIMENSION STATE-
   MENT "14 SP @ 1'-0" = 14'-0" " CONVERTS TO 13 — 401 BARS, OR ALWAYS ONE LESS THAN
   THE NUMBER OF SPACES CALLED OUT. THIS METHOD OF DETERMINING BAR QUANTITIES
   FOR BILLS OF MATERIAL APPLIES FOR ALL PRECAST PRODUCTS.

FIGURE 16.6 ■ Counting bar quantities for the bill of materials.

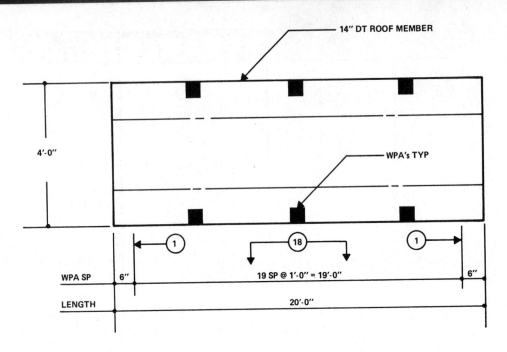

PROBLEM
DETERMINE THE NUMBER OF WPA's USED IN THE DOUBLE-TEE ROOF MEMBER ILLUSTRATED ABOVE.

SOLUTION
1. LOCATE THE DIMENSION LINE THAT GIVES THE SPACING FOR WPA's. IN THE ILLUSTRATION ABOVE, IT IS DIRECTLY ABOVE THE LENGTH DIMENSION LINE.
2. COUNT THE WPA's ALONG THE DIMENSION LINE. EACH LINE EXTENDED FROM THE FLAT SLAB TO THE DIMENSION LINE REPRESENTS ONE WPA. THE DIMENSION STATEMENT "19 SP @ 1'-0" = 19'-0" " CONVERTS TO 18 – WPA's, OR ALWAYS ONE WELD PLATE LESS THAN THE NUMBER OF SPACES CALLED OUT. THIS METHOD OF DETERMINING WELD PLATE QUANTITIES FOR BILLS OF MATERIAL APPLIES FOR ALL TYPES OF METAL CONNECTORS IN ALL TYPES OF PRECAST PRODUCTS.

FIGURE 16.7 ■ Counting plate quantities for the bill of materials.

SUMMARY

■ A bill of materials is a detailed listing of all materials used in the fabrication and erection of a job.

■ The most common method used in the preparation of a bill of materials is to prepare one bill for each product type used in a job.

■ Most companies prepare or purchase standard bill of materials forms.

■ Standard forms vary from company to company, but all contain the type of product used, mark numbers for each product being billed, a quantity count of each material used in each product, and totals for each type of material used.

■ Bills of materials are prepared from information contained in the fabrication details.

■ Material quantities are determined from dimensional specifications rather than the picture presented by the fabrication details except in cases where materials are located from the picture rather than by dimensions.

REVIEW QUESTIONS

1. Define the term *bill of materials* as it is used in precast concrete drafting.

2. Explain the most common method used in preparing precast concrete bills of materials. Why is this method the most popular?

3. List four common entries that are usually found on any bill of materials form.

4. Where do CAD technicians find the information needed in completing a bill of materials?

5. The dimensional specification *12 SP @ 1'-0"* converts to what quantity of material?

CAD **ACTIVITIES**

GENERAL INSTRUCTIONS

The following activities may be completed on any CAD system. Before reading the *specific instructions* for each activity (below), go through each step in the following planning checklist. The checklist applies to any CAD system and will help ensure the optimum use of your time and resources.

1. Analyze the problem carefully. Decide exactly what you are being asked to do.

2. Determine what resources and references you will need in order to complete the problem and collect them.

3. Decide if any particular standards apply to the project and have those standards available.

4. Determine what types of views will be required and how many of each.

5. Determine what the final plotted scale of the drawing will need to be, and select the appropriate paper size for plotting/printing (make sure the appropriate paper size is available).

6. Plan your drawing sequence. In what order will you develop the drawing (i.e., lines, features, dimension lines, leaders, dimensions, notes, etc.)?

7. Review the various CAD commands you will have to use in order to develop the drawing.

8. Examine your CAD system to ensure that everything is in working order, then begin the project.

SPECIFIC INSTRUCTIONS

In order to complete the following CAD activities, the student should prepare four blank bill of materials forms. A format similar to that shown in Figure 16.1 is recommended.

Activity 16.1—Prepare a complete bill of materials for the precast columns contained in Figure 16.8 and Figure 16.9.

Activity 16.2—Prepare a complete bill of materials for the precast beams contained in Figure 16.10 and Figure 16.11.

Activity 16.3—Prepare a complete bill of materials for the precast flat-slab wall panels contained in Figure 16.12 and Figure 16.13.

Activity 16.4—Prepare a complete bill of materials for the precast concrete double-tee roof members contained in Figure 16.14 and Figure 16.15.

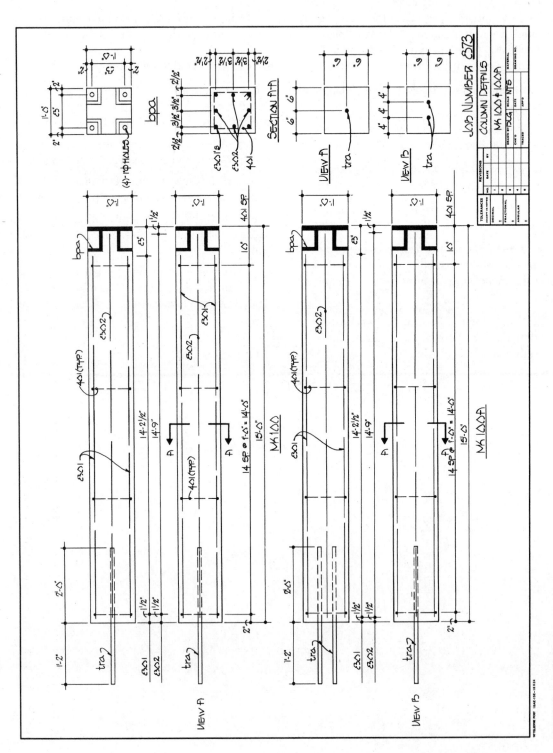

FIGURE 16.8 ■ Column details for CAD activities.

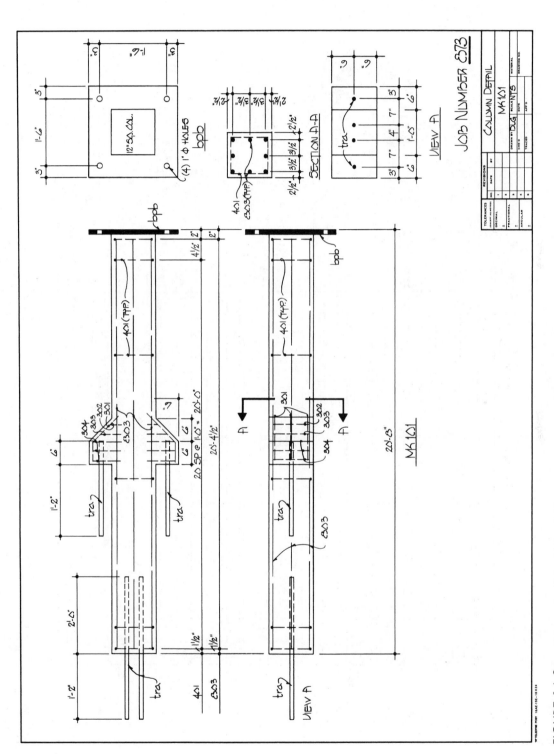

FIGURE 16.9 ■ Column detail for CAD activities.

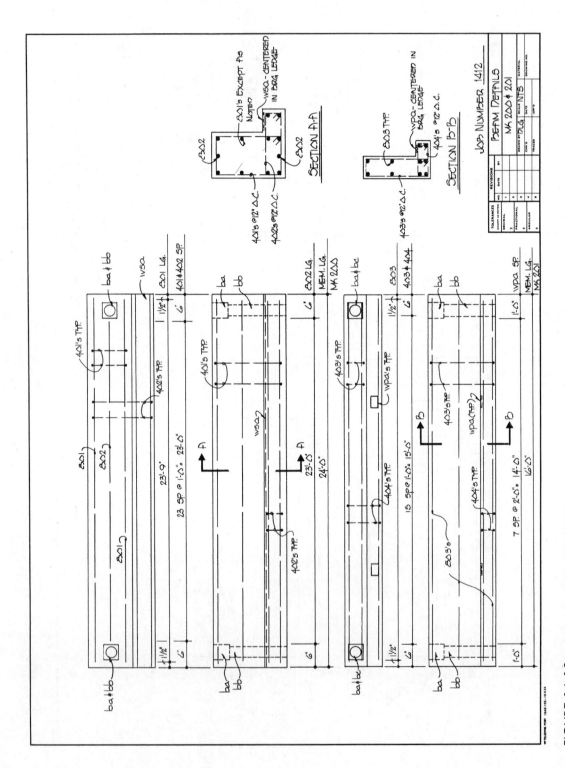

FIGURE 16.10 ■ Beam details for CAD activities.

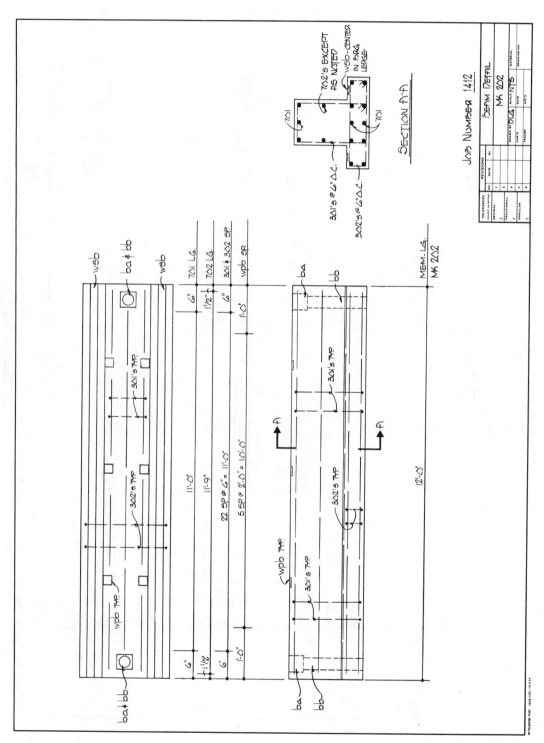

FIGURE 16.11 ■ Beam detail for CAD activities.

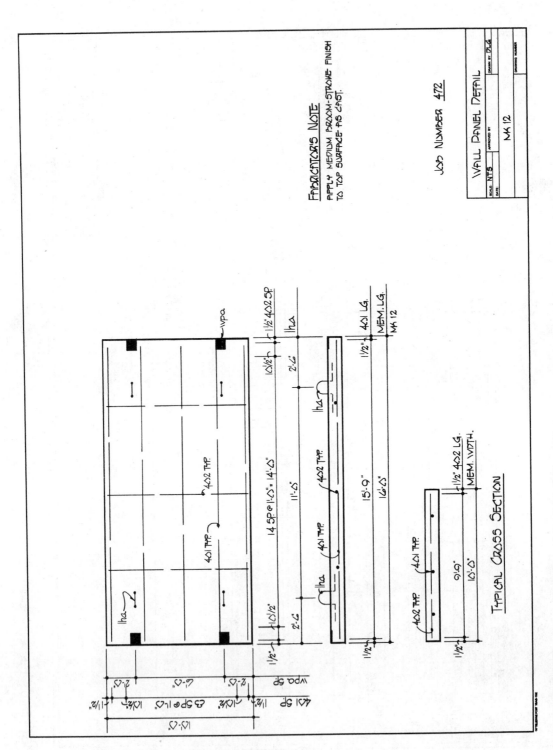

FIGURE 16.12 ■ Flat-slab wall panel detail for CAD activities.

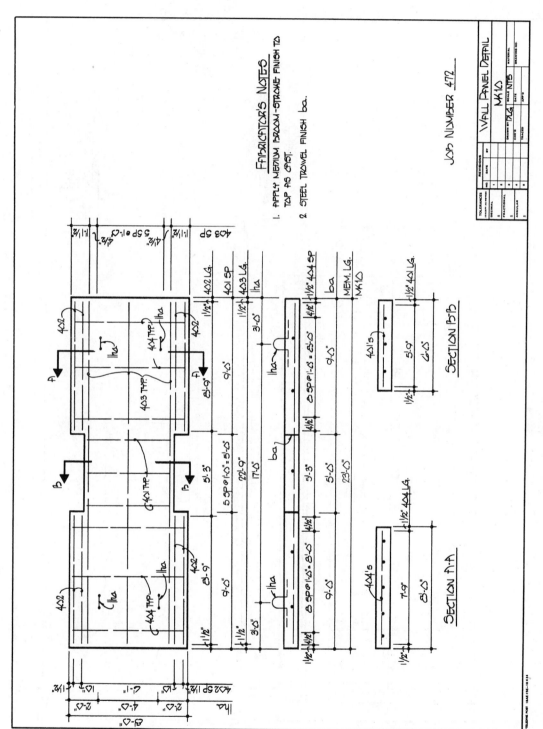

FIGURE 16.13 ■ Flat-slab wall panel detail for CAD activities.

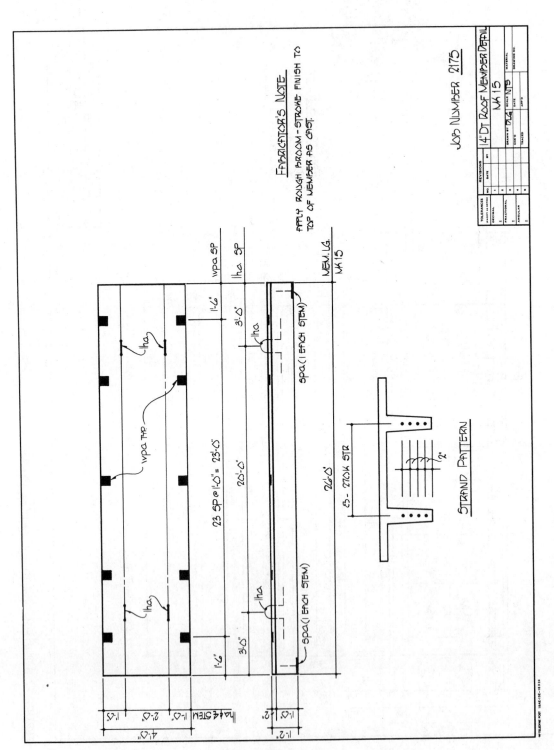

FIGURE 16.14 ■ Double-tee roof member detail for CAD activities.

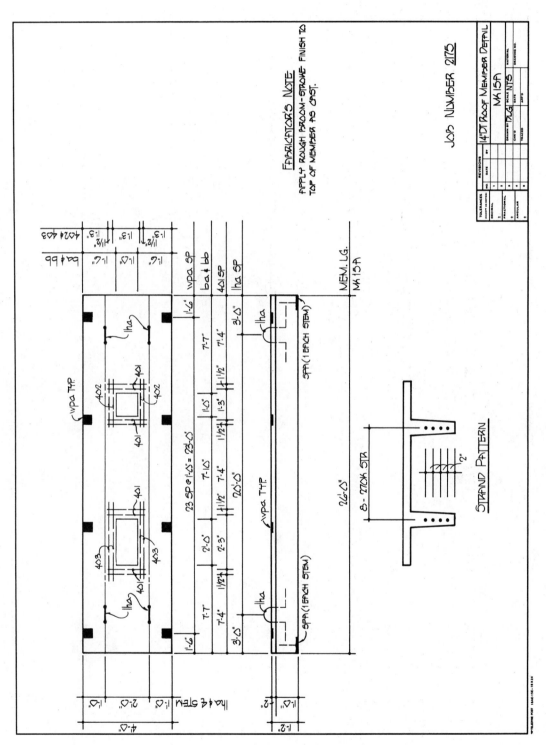

FIGURE 16.15 ■ Double-tee roof member detail for CAD activities.

SECTION IV

Structural Poured-in-Place Concrete

Poured-in-Place Concrete Foundations

OBJECTIVES

Upon completion of this unit, the student will be able to:

- Define poured-in-place concrete.
- Define engineering drawings and placing drawings as used in poured-in-place concrete drafting.
- Identify common abbreviations and symbols used in poured-in-place concrete drafting.
- Assign and interpret mark numbers for structural members.
- Prepare engineering and placing drawings for poured-in-place concrete foundations.

POURED-IN-PLACE CONCRETE CONSTRUCTION

Poured-in-place concrete construction involves the following:

- Building forms in place at the jobsite
- Placing steel reinforcing bars into the forms
- Pouring concrete into the forms around the reinforcing bars

Once the concrete has been allowed to harden sufficiently, the forms are stripped away. This leaves an in-place, fully erected structural member.

Poured-in-place indicates that the concrete is poured at the jobsite into forms that are in place. This differs from precast concrete that is poured into forms at a manufacturing plant and then shipped to the jobsite and erected in pieces. Poured-in-place concrete, with the exception of special members such as tilt-up walls, is actually erected as it is poured.

POURED-IN-PLACE CONCRETE DRAWINGS

Poured-in-place concrete drawings are of two types: engineering drawings and placing drawings. *Engineering drawings* show the general layout of the structure, size and spacing of structural components (foundations, walls, beams, columns, etc.), and various notes containing information helpful to the reader in understanding the drawings.

Placing drawings are more detailed drawings. They show the actual size, shape, and locations of all reinforcing bars in a structural member and how the bars are to be placed into the forms. Placing drawings also include comprehensive schedules containing information on every piece of reinforcing bar used in the structural members in question.

SHEET LAYOUT AND SCALES

The format chosen for laying out engineering and placing drawings tends to vary from company to company. The American Concrete Institute (ACI) recommends a standard sheet layout that is followed by many companies. However, factors such as the amount of information that must be placed on a sheet, individual company preferences, and sheet sizes used affect the layout of a sheet. Because of this, sheet layout formats are only loosely standardized. A format that represents generally what is used in industry for laying out engineering drawings is shown in Figure 17.1. A similar example for placing drawings is illustrated in Figure 17.2.

Engineering and placing drawings of poured-in-place concrete structures are drawn to scale. The scale selected for each drawing depends on several factors:

- Overall size of the structure being drawn
- Size of sheet on which the structure is being drawn
- Complexity of the drawing
- Individual company preferences

There are some industry-wide accepted guidelines for drawing the various components (plan views, elevations, sections, and details) of engineering and placing drawings:

Plan Views	Elevations	Sections and Details
1/8″ = 1′-0″	1/4″ = 1′-0″	1/4″ = 1′-0″
1/4″ = 1′-0″	3/8″ = 1′-0″	3/8″ = 1′-0″
	1/2″ = 1′-0″	1/2″ = 1′-0″
		3/4″ = 1′-0″
		1″ = 1′-0″
		1 1/2″ = 1′-0″

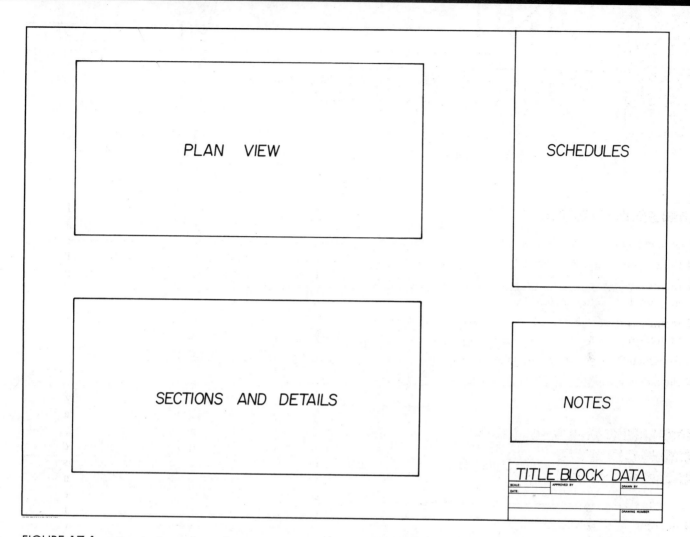

FIGURE 17.1 ■ Drawing layout format for engineering drawings. (*American Concrete Institute*)

SYMBOLS AND ABBREVIATIONS

Industry-wide standardization of symbols, abbreviations, and practices would reduce drafting time and make drawings more easily understood. However, standardization of symbols and abbreviations on poured-in-place concrete drawings is not always achieved. Some companies have their own symbols and abbreviations that have been developed over the years and are usually explained in a legend.

In spite of individual company differences, there are a number of abbreviations and symbols that have achieved industry-wide recognition, if not standardization (Figure 17.3). A similar listing of common abbreviations is shown in Figure 17.4.

MARK NUMBERING SYSTEMS

A numbering system is needed to manage the design, drawing, and construction of a structure composed of separate but interrelated parts. *Component* or *mark numbers* are used on poured-in-place concrete drawings. A well-developed mark numbering system relates not only the component number but also where the component fits into the overall job. For example, the mark number 3B4 refers to beam number 4 on the third floor.

Mark numbering systems, like other aspects of poured-in-place concrete drawings, vary from company to company. However, most systems use numbers to designate floor levels, letters to designate structural components, and numbers to identify

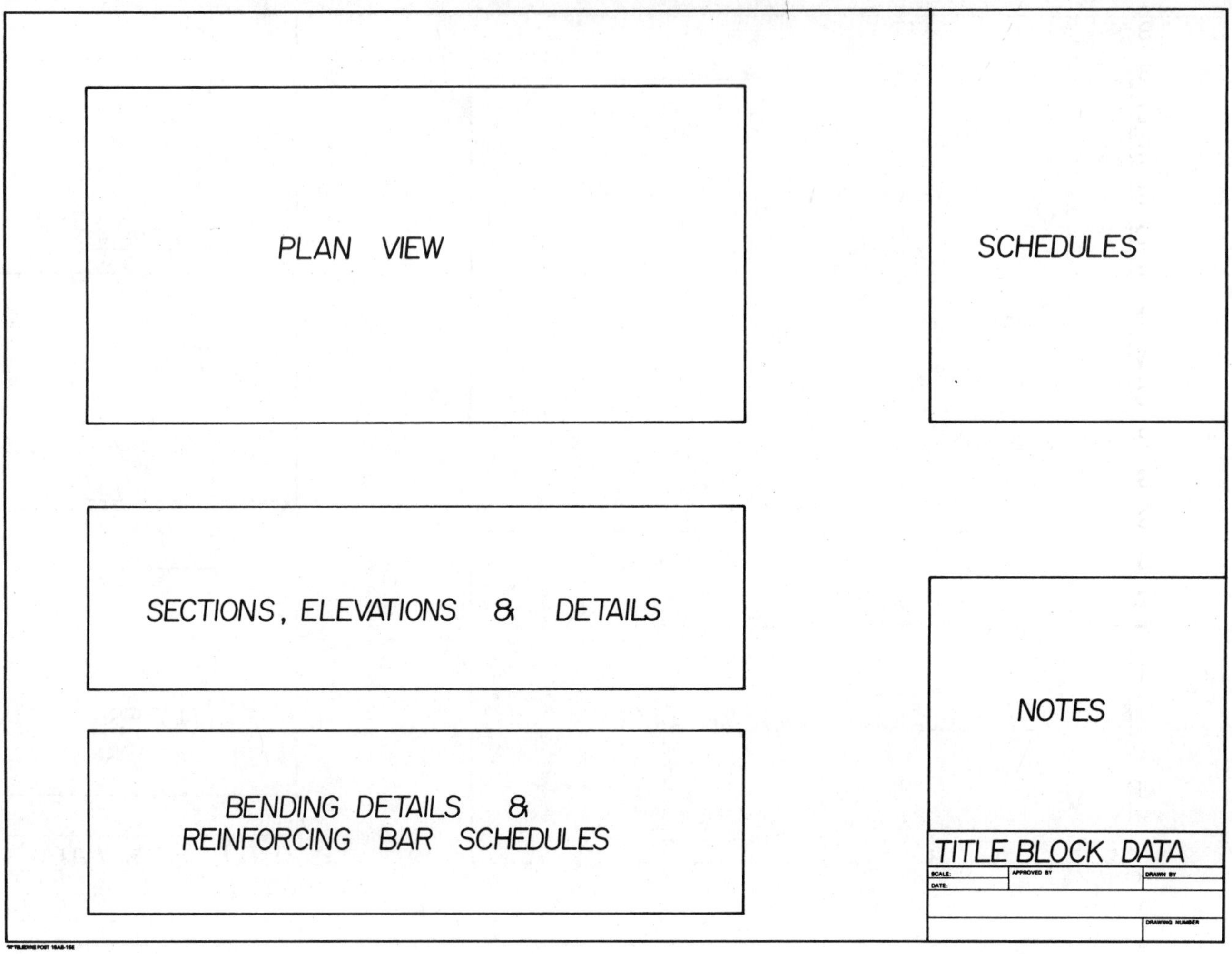

FIGURE 17.2 ■ Drawing layout format for placing drawings.

SYMBOL	DESCRIPTION
#	SYMBOL FOR NUMBER. USED TO INDICATE THE WORD NUMBER WHEN DESCRIBING BARS. EXAMPLE: #3 BAR, #4 BAR, ETC.
Ø	SYMBOL FOR DIAMETER. EXAMPLE: 1/2″ Ø BAR OR 10″ Ø COLUMN.
□	SYMBOL FOR SQUARE. EXAMPLE: 12″ □ BEAM.
@	SYMBOL FOR AT. USED IN DIMENSIONING FOR CENTER-TO-CENTER SPACING. EXAMPLE: #3 BARS @ 7″ ON CENTER.
→	INDICATES DIRECTION OF REINFORCING BARS. USED ON PLAN VIEWS.

FIGURE 17.3 ■ Common drafting symbols.

Bot	BOTTOM (AS IN BOTTOM OF MEMBER)
Bt	BENT (AS IN BENT BAR)
CT	COLUMN TIE
EF	EACH FACE
EW	EACH WAY (AS IN BARS SPACED @ 6" ON CENTER EACH WAY)
FF	FAR FACE
IF	INSIDE FACE
NF	NEAR FACE
OF	OUTSIDE FACE
PI	PLAIN BAR
St	STRAIGHT
Stir	STIRRUP
Sp	SPIRAL
T	TOP (AS IN TOP OF MEMBER)

FIGURE 17.4 ■ Commonly used abbreviations.

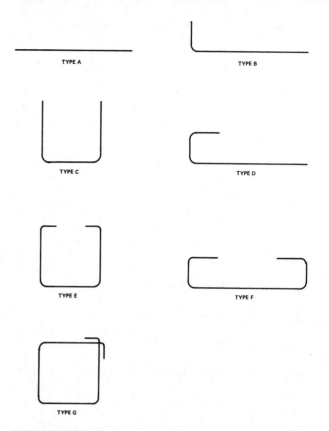

FIGURE 17.5 ■ Common bent bar types.

specific components (i.e., beam number 1, 2, 3, 4, etc.). Letter designations for beams, columns, joists, and other structural components of poured-in-place concrete jobs have achieved a high degree of industry-wide recognition. Some letter designations commonly used on poured-in-place concrete drawings are:

B—Beams

C—Columns

F—Footings or foundations

G—Girders

J—Joists

L—Lintels

S—Slabs

T—Ties

U—Stirrups (U is used to avoid conflicts with S as used for slabs)

Mark numbers are also assigned to reinforcing bars shown on poured-in-place concrete placing drawings. Various methods are used throughout the industry. Most methods use a combination of numbers and letters that designate the bar size and bar type and that specifically identify each individual length of bar. For example, the bar mark number 4A1 indicates a number 4 bar, type A, and it is the first length of bar used in the job. A bar numbered 4A2 would also be a type A, number 4 bar, but it would have a different length than a 4A1.

Designations of bar types vary from company to company, but most firms arrive at some type of standard system. Straight bars may be designated type A, J-shaped bars type B, U-shaped bars type C, and so on. Typical bent bar types found in industry with sample designations are shown in Figure 17.5.

Another common system of numbering bars uses only numbers. This method identifies the bar size and separates the bars according to lengths but does not designate the bar type. For example: Bar 601 would be a number 6 bar of some specific length. Bar 602 would also be a number 6 bar but of a different length. The bar bending schedule would be consulted to determine the bar type.

SCHEDULES

Schedules are an important part of engineering and placing drawings. Engineering drawings contain beam schedules, column schedules, and joist schedules depending on the structural members being drawn (Figure 17.6 and Figure 17.7). Placing drawings contain schedules similar to those on engineering drawings (Figure 17.8). Placing drawings also contain schedules showing the bending information for all reinforcing bars used in the job (Figure 17.9).

POURED-IN-PLACE FOUNDATION DRAWINGS

The first drawing in a set of poured-in-place concrete plans is the **foundation** plan. A complete foundation plan requires both engineering and placing drawings. The engineering drawings are used in the design and planning process and are prepared first.

Beam Schedule

Mark Numbers	Beam Width	Beam Depth	Reinforcing Bars			
			Longitudinal Reinforcing		Stirrups and Ties	
			Bar Size	Remarks	Bar Size	Bar Spacing
3B1	12″	20″	4-#6	Straight	16-#3	8 @ 12″
3B2	12″	20″	4-#6	Straight	16-#3	4 @ 12″, 4 @ 10″
3B3	10″	18″	4-#6	Straight	12-#3	6 @ 12″
3B4	10″	18″	4-#6	Straight	12-#3	3 @ 12″, 3 @ 10″
3B5	8″	14″	4-#5	Straight	8-#3	8 @ 12″
3B6	8″	14″	4-#5	Straight	9-#3	4 @ 12″, 5 @ 10″

FIGURE 17.6 ■ Sample beam schedule for engineering drawings.

Column Schedule Roof	B2, B3	C1, C4	A1, A4	Column Mark Numbers
Roof	C2, C3	A2, A3	B1, B4	
	18 × 24	18 × 18	16 × 16	Column Size
12′-0″	4-#8	4-#8	4-#8	Vertical Bars
First Floor	#3 @ 12″	#3 @ 12″	#3 @ 12″	Column Ties
	18 × 24	20 × 20	16 × 16	Column Sizes
10′-4″	6-#8	6-#8	6-#8	Vertical Bars
Basement	#4 @ 12″	#4 @ 12″	#4 @ 12″	Column Ties

FIGURE 17.7 ■ Sample column schedule.

				Longitudinal Reinforcement								
				Bottom of Beam			Top of Beam			Stirrups		
Mark No.	No. Req'd.	Width	Depth	No. Req'd.	Bar Size	Length	No. Req'd.	Bar Size	Length	No. Req'd.	Bar Size	Stirrup Spacing
2B1	3	12"	16"	4	6	21'-2"	4	5	21'-2"	22	3G1	20 @ 12" 2 @ 6"
2B2	1	12"	16"	4	6	20'-6"	4	5	20'-6"	21	3G1	19 @ 12" 2 @ 6"
2B3	1	10"	14"	3	5	16'-5"	3	4	16'-5"	17	3G2	15 @ 12" 2 @ 6"
2B4	2	10"	14"	3	5	15'-6"	3	4	15'-6"	16	3G2	14 @ 12" 2 @ 6"
2B5	1	8"	12"	3	4	10'-4"	3	4	10'-4"	11	3G3	9 @ 12" 2 @ 4"
2B6	2	8"	12"	3	4	10'-2"	3	4	10'-2"	11	3G3	9 @ 12" 2 @ 3"

(Beam Schedule)

FIGURE 17.8 ■ Sample beam schedule for placing drawings.

Bar Bending Details and Dimensions

Bar Mark Number	Bar Size	Type	Total Length	A	B	C	D	E	F	G	H
6C1	6	C	5'-2"	24"	14"						
6C2	6	C	4'-10"	22	14"						
4F3	4	F	4'-0"			4"	10"	20"			
3C4	3	C	4'-4"	20"	12"						
5F5	5	F	4'-6"			3"	12"	24"			
6F6	6	F	4'-0"			3"	10"	22"			

FIGURE 17.9 ■ Sample bar bending schedule.

Preparing Engineering Drawings for Foundations

Like so many aspects of poured-in-place concrete drafting, the procedures followed in preparing engineering drawings are not completely standardized. However, the American Concrete Institute has established guidelines for preparing engineering drawings that can simplify the process if followed.

The following are some ACI recommendations for preparing engineering drawings of poured-in-place concrete structures:

■ All information relating to the size and arrangement of concrete members and the size, positioning, and arrangement of reinforcing bars in the members must be provided. This may be accomplished through drawings, notes, or schedules.

■ Time may be saved by providing typical details and then developing a schedule containing the necessary information for members not specifically detailed.

■ All dimensions are not required, but a sufficient number should be provided to accommodate design and to indicate where bars are to be bent. More specific dimensions are left to the placing drawings.

■ Details of construction joints or splice points are helpful and should be provided to the various trades who must build the structure.

An example of an engineering drawing of a foundation for a commercial structure is shown in Figure 17.10. Examine it closely and become familiar with its layout and contents.

Preparing Placing Drawings for Foundations

Placing drawings are prepared after the engineering drawings have been completed. A common practice is to trace the outlines of the various structural components (i.e., beams, columns, walls, etc.) from the engineering drawings and then add the necessary detail. Placing drawings are prepared from information contained on the engineering drawings. They are used for fabricating the structural members for a job and ordering the materials necessary to do a job.

Placing drawings must show:

■ Size and shape of the structural members to be fabricated

■ Size, shape, and location of all bars in the structural member(s)

■ How the bars are to be placed in the forms

Figure 17.11 is an example of a placing drawing for a commercial foundation. Compare this drawing with Figure 17.10, making note of similarities and differences.

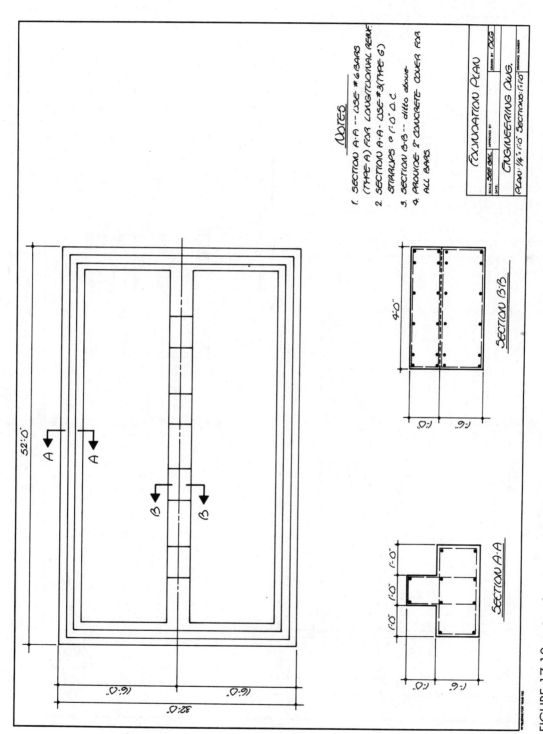

FIGURE 17.10 ■ Sample engineering drawing.

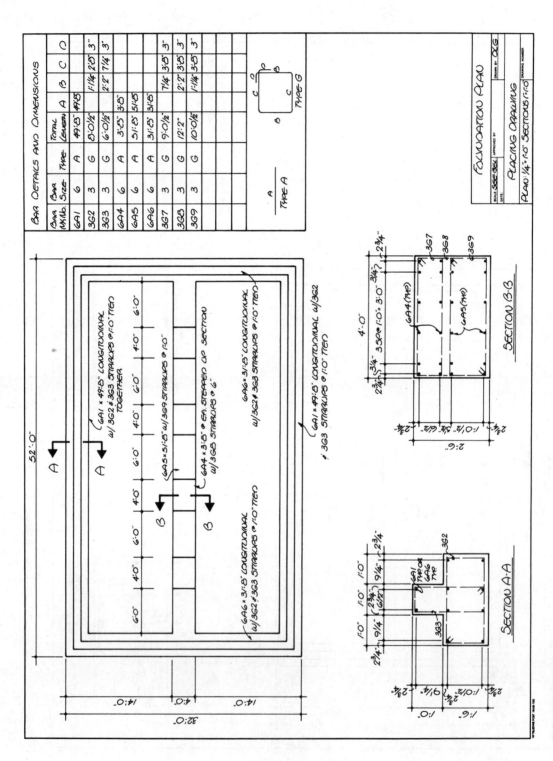

FIGURE 17.11 ■ Sample placing drawing.

SUMMARY

■ Poured-in-place concrete construction involves building forms in place at the jobsite, placing steel reinforcing bars into the forms, and pouring concrete into the forms around the reinforcing bars.

■ Poured-in-place concrete differs from precast concrete in that it is poured into temporary forms at the jobsite while precast members are poured in a manufacturing plant and shipped to the jobsite.

■ Poured-in-place concrete drawings are of two types: engineering drawings and placing drawings.

■ Engineering drawings show the general layout of the structure, size and spacing of structural components, and various notes containing information helpful in understanding the drawings.

■ Placing drawings are more detailed than engineering drawings. They show the actual size, shape, and locations of all reinforcing bars in each structural member as well as how the bars will fit into the forms.

■ Sheet layouts for poured-in-place concrete drawings are only loosely standardized, but individual companies should adhere to a general pattern of sheet layout.

■ Engineering and placing drawings are drawn to a scale selected according to the overall size of the structure being drawn, size of sheet on which the structure is being drawn, complexity of the drawing, and individual company preferences.

■ Structural CAD technicians should use poured-in-place concrete symbols that are recognized industry-wide when preparing drawings.

■ Mark numbering systems are required in order to organize and manage the design, drawing, and construction of a large structure composed of separate but interrelated parts. The mark number 3B4 refers to beam number 4 on the third floor.

■ Schedules are an important part of engineering and placing drawings. Schedules for engineering drawings are member schedules such as beam, column, and joist schedules. Schedules of replacing drawings contain bending information for reinforcing bars.

REVIEW QUESTIONS

1. Explain how poured-in-place concrete construction differs from precast concrete construction.

2. Name the two types of poured-in-place concrete drawings.

3. Make a sketch to show the general layout of a poured-in-place concrete engineering drawing.

4. Make a sketch to show the general layout of a poured-in-place concrete placing drawing.

CAD ACTIVITIES

GENERAL INSTRUCTIONS

The following activities may be completed on any CAD system. Before reading the *specific instructions* for each activity (below), go through each step in the following planning checklist. The checklist applies to any CAD system and will help ensure the optimum use of your time and resources.

1. Analyze the problem carefully. Decide exactly what you are being asked to do.

2. Determine what resources and references you will need in order to complete the problem and collect them.

3. Decide if any particular standards apply to the project and have those standards available.

4. Determine what types of views will be required and how many of each.

5. Determine what the final plotted scale of the drawing will need to be, and select the appropriate paper size for plotting/printing (make sure the appropriate paper size is available).

6. Plan your drawing sequence. In what order will you develop the drawing (i.e., lines, features, dimension lines, leaders, dimensions, notes, etc.)?

7. Review the various CAD commands you will have to use in order to develop the drawing.

8. Examine your CAD system to ensure that everything is in working order, then begin the project.

SPECIFIC INSTRUCTIONS

Activity 17.1—Prepare four, C-size sheets of paper with title blocks and borders.

Activity 17.2—On one of the sheets prepared in Activity 17.1, redraw the foundation plan in Figure 17.12, excluding the asphalt pavement area, as an engineering drawing. Refer to Figure 17.10 for an example. Revision notes and symbols should also be excluded.

Activity 17.3—On the second sheet of paper from Activity 17.1, prepare a placing drawing for the engineering drawing prepared in Activity 17.2. Refer to Figure 17.11 for an example.

Activity 17.4—Redraw an engineering drawing of Figure 17.12, making the following changes:
   a. Change the cross-sectional dimension of Section A-A to 3′-0″ and adjust the longitudinal reinforcement spacing accordingly.
   b. Add an identical row of reinforcing bars 2″ down from the top of Section A-A and surround both rows with Type G stirrups at 1′-0″ on center.

Activity 17.5—Prepare a complete placing drawing for the engineering drawing developed in Activity 17.4.

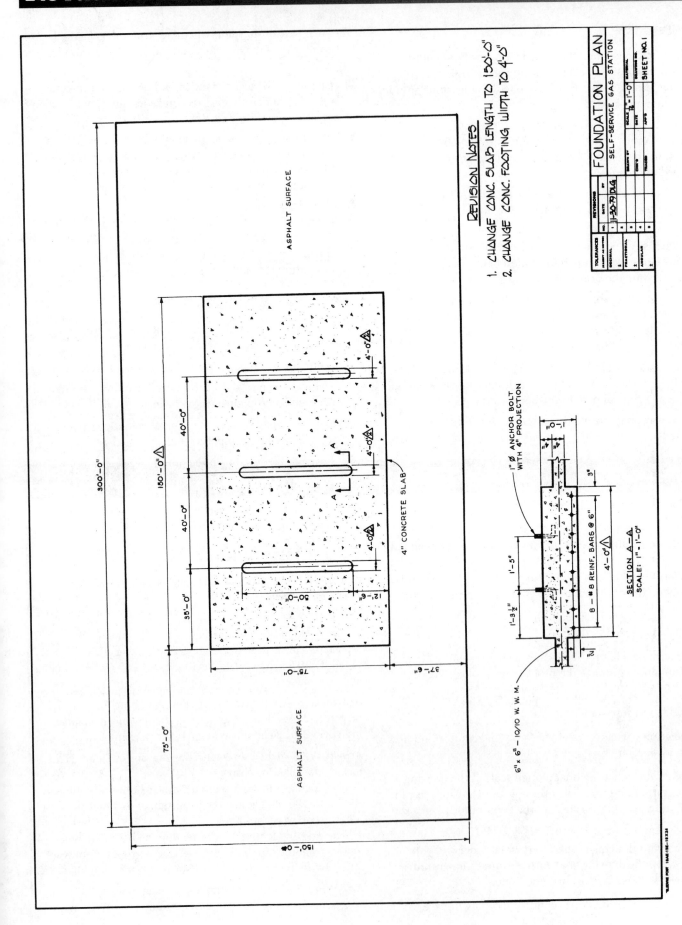

FIGURE 17.12 ■ Foundation plan for CAD activities.

Poured-in-Place Concrete Walls and Columns

OBJECTIVES

Upon completion of this unit, the student will be able to:

■ List the four basic categories of poured-in-place concrete walls and give a definition of each.

■ Define and distinguish between the two basic types of poured-in-place concrete columns.

■ Prepare complete engineering drawings of poured-in-place concrete wall systems and drawings of poured-in-place concrete columns.

■ Prepare complete placing drawings of poured-in-place concrete wall components and drawings of poured-in-place concrete columns.

POURED-IN-PLACE CONCRETE WALLS AND COLUMNS

Two major components of most poured-in-place concrete structures are walls and columns. Walls fall into four basic categories: security walls, shear walls, tilt-up walls, and retaining walls. Columns fall into two categories: tied columns and spiral columns. The terms *tied* and *spiral* refer to the type of reinforcing used in the columns.

Security Walls

Security walls may be used for a number of different applications. As the name implies, their primary function is usually to provide some type of security. Common applications are to provide security from fire, break-ins, or noise. Bank vaults are very often composed of poured-in-place concrete security walls. Firewalls in apartments and other commercial structures are also a common application.

Shear Walls

Shear walls can be made of poured-in-place concrete. They are used to brace tall structures against lateral loading from wind and swaying motions. Elevator shafts and stairwells may be designed as shear walls (Figure 18.1). Poured-in-place concrete shear walls, such as those illustrated in Figure 18.1, provide support for the floors of the building as well as overall lateral support for the building.

Tilt-Up Walls

Exterior walls of a structure are commonly constructed of **tilt-up walls.** A tilt-up wall may be precast or poured in place, as may all structural concrete walls listed in this unit. A poured-in-place tilt-up wall breaks the rule of poured-in-place concrete construction. This rule says that members are poured into forms that have been built in place at the jobsite, thereby speeding up pouring and erection concurrently.

Tilt-up wall components are poured into forms that have been built at the jobsite, but they are not in place. The forms for tilt-up walls are built on the ground or on the floor of the building. Concrete is poured over the reinforcing bars in the forms and allowed to harden. The wall panels are then lifted into place by a hoist or crane (Figure 18.2).

Retaining Walls

A **retaining wall** is a freestanding, self-supporting wall designed to hold back earth, water, or other material. Basement walls hold back earth or water and must be reinforced to do so. However, since a structure rests on basement walls, they are not freestanding and should not be classified as retaining walls. True retaining walls fall into two categories: gravity walls and cantilever walls (Figure 18.3).

Gravity walls are very heavy, poured-in-place concrete walls. Their use is usually restricted to applications requiring less than 6′ of height. Applications requiring more height have a tendency to cause the wall to tip over. For these applications, the cantilever wall is more appropriate.

The *cantilever wall* is tapered from top to bottom with the thickest portion at the bottom resting on a wide footing. The footing is poured in the ground and heavily reinforced. Its purpose is to prevent the forces of pressure from the material being held back from tipping the wall over. The vertical portion of the cantilever wall, known as the *stem*, is also heavily reinforced. A cantilever wall in section showing the steel reinforcing bars is shown in Figure 18.4.

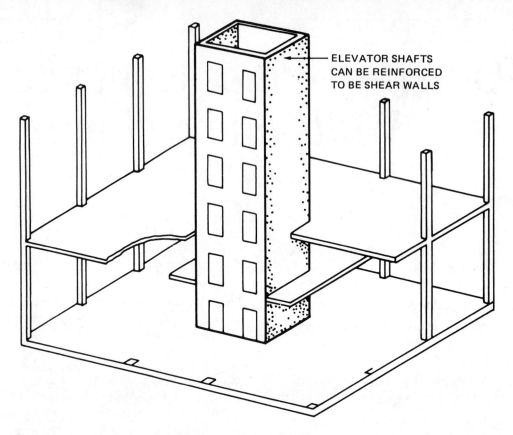

FIGURE 18.1 ■ Elevator shaft as shear walls. (*Charles D. Willis*)

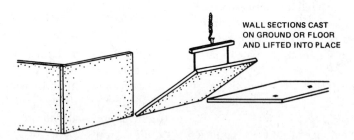

FIGURE 18.2 ■ Poured-in-place concrete tilt-up walls. (*Charles D. Willis*)

Tied and Spiral Columns

Columns in poured-in-place concrete construction are classified as either tied or spiral according to the method of reinforcing used in them. *Tied* columns are square or rectangular in cross section, while *spiral* columns are round (Figure 18.5).

Tied columns are reinforced with long, straight reinforcing bars tied off with wire ties or smaller, bent reinforcing bars (Figure 18.6). Spiral columns are reinforced with long, straight reinforcing bars surrounded by spiral wire or reinforcing bars (Figure 18.6).

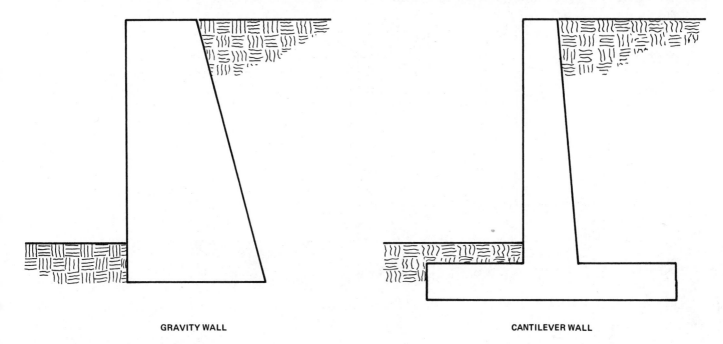

GRAVITY WALL

CANTILEVER WALL

FIGURE 18.3 ■ Retaining walls.

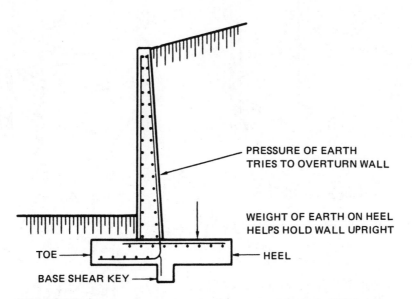

PRESSURE OF EARTH
TRIES TO OVERTURN WALL

WEIGHT OF EARTH ON HEEL
HELPS HOLD WALL UPRIGHT

TOE —

— HEEL

BASE SHEAR KEY —

FIGURE 18.4 ■ Cantilever retaining wall in section. (*Charles D. Willis*)

Poured-in-place columns are often used in multistory buildings. When this is the case, the columns must be spliced. Column splicing usually occurs at floor levels. Splicing a poured-in-place column involves extending the main reinforcing steel out of the bottom column so that it can be spliced to the main reinforcing of the column that will rest on top of it (Figure 18.7).

Most poured-in-place columns are square, round, or rectangular. However, on occasion, design and appearance may dictate the need for a different shape of column. When this is the case, several special column shapes are available (Figure 18.8).

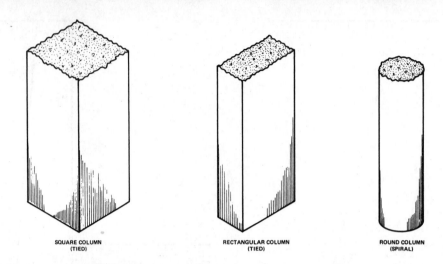

SQUARE COLUMN
(TIED)

RECTANGULAR COLUMN
(TIED)

ROUND COLUMN
(SPIRAL)

FIGURE 18.5 ■ Tied and spiral column configurations.

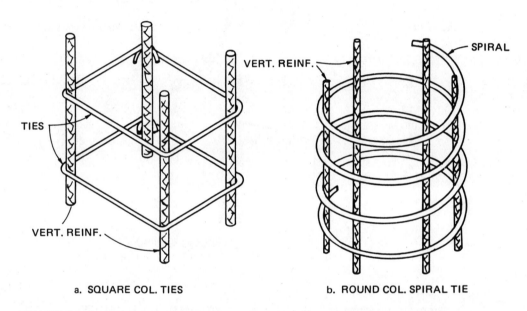

TIES

VERT. REINF.

VERT. REINF.

SPIRAL

a. SQUARE COL. TIES

b. ROUND COL. SPIRAL TIE

FIGURE 18.6 ■ Typical column reinforcing. (*Charles D. Willis*)

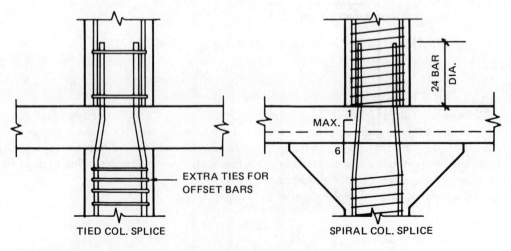

EXTRA TIES FOR
OFFSET BARS

24 BAR DIA.

MAX.

1

6

TIED COL. SPLICE

SPIRAL COL. SPLICE

FIGURE 18.7 ■ Column splicing details. (*Charles D. Willis*)

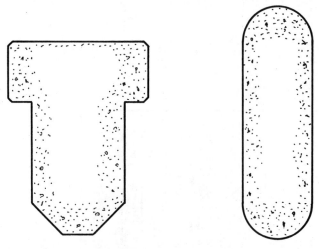
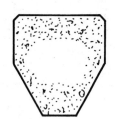

FIGURE 18.8 ■ Special column configurations.

WALL AND COLUMN ENGINEERING DRAWINGS

CAD technicians prepare engineering drawings of column layouts and wall systems from sketches and specifications supplied by structural engineers or architects. Engineering drawings contain the following:

■ A plan view of the walls or columns

■ General dimensions

■ An indication of the types and sizes of reinforcing bars to be used

■ Sections and/or details

■ Necessary notes

■ Wall component or column schedules

Figure 18.9 through Figure 18.11 illustrate wall and column engineering drawings.

WALL AND COLUMN PLACING DRAWINGS

CAD technicians prepare placing drawings from engineering drawings and sketches and specifications provided by engineers. Placing drawings contain:

■ Plan views

■ Sections and/or details

■ Schedules

■ Bending details

■ Notes

The placing drawings for a job are usually drawn to the same scale as the engineering drawings so that the outlines of structural components can be traced to save drafting time. Poured-in-place concrete wall and column placing drawings are shown in Figure 18.12 and Figure 18.13.

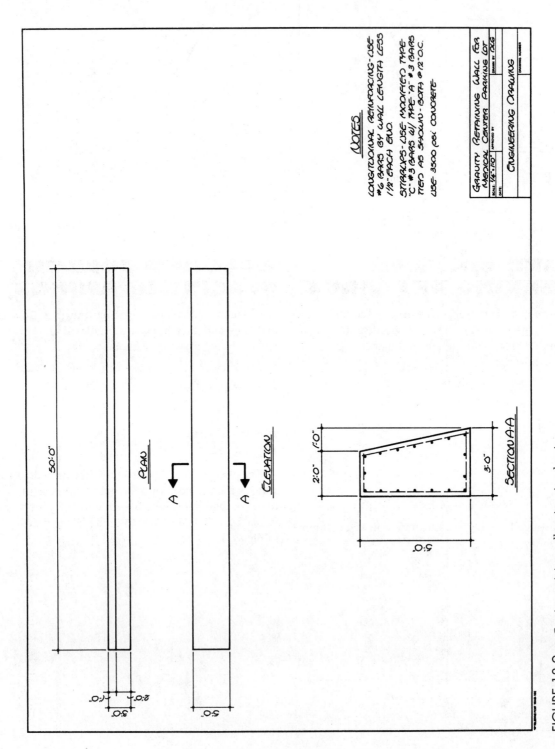

FIGURE 18.9 ■ Gravity retaining wall engineering drawing.

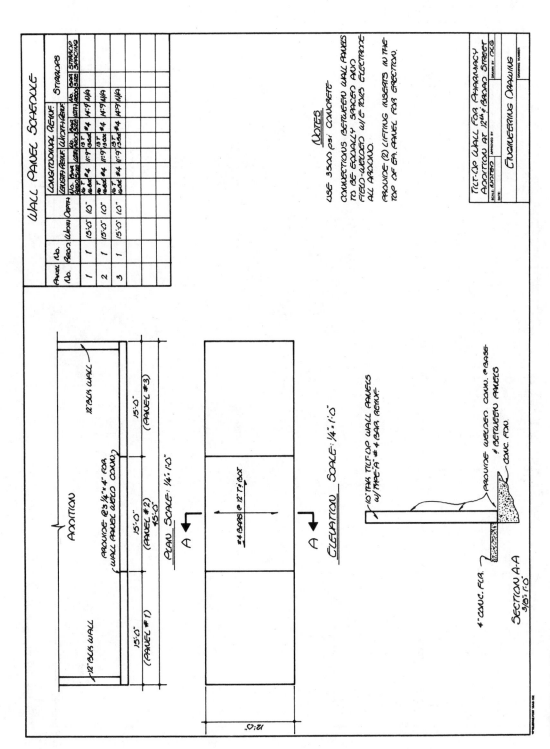

FIGURE 18.10 ■ Tilt-up wall engineering drawing.

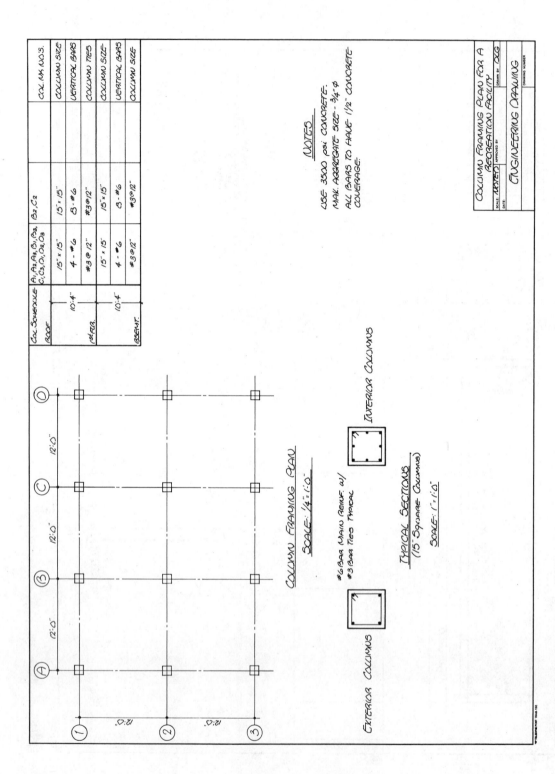

FIGURE 18.11 ■ Column engineering drawing.

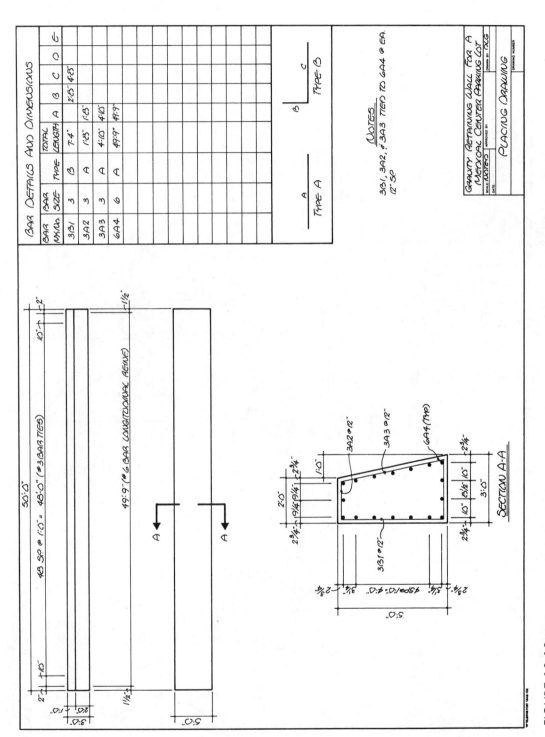

FIGURE 18.12 ■ Retaining wall placing drawing.

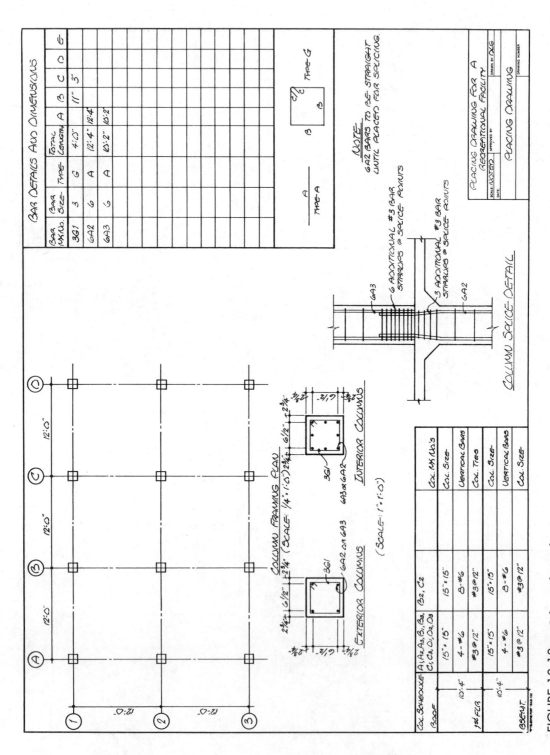

FIGURE 18.13 ■ Column placing drawing.

SUMMARY

■ Walls are a major component of most poured-in-place concrete structures. Walls fall into four categories: security walls, shear walls, retaining walls, and tilt-up walls.

■ Columns are another major component of most poured-in-place concrete structures and fall into two categories: tied columns and spiral columns.

■ Security walls are used to provide security from fire, break-ins, or noise.

■ Shear walls are used to brace tall structures against lateral loading from wind or other forces.

■ Exterior walls of a structure are commonly tilt-up walls that are poured in temporary forms on the ground or the floor of the building and erected with a hoist or crane.

■ Retaining walls are freestanding, self-supporting walls designed to hold back earth, water, or other material. There are two types: gravity walls and cantilever walls.

■ Gravity walls are very heavy, poured-in-place concrete walls usually restricted to applications requiring less than 6′ of height.

■ Cantilever walls are poured on a wide footing that prevents them from tipping over. Consequently, they can be used for applications requiring more than 6′ of height.

■ Tied columns are square or rectangular in cross section and are reinforced with stirrup ties wrapped around long, straight longitudinal reinforcing bars.

■ Spiral columns are usually round and are reinforced with spiral wire wrapped around long, straight, longitudinal reinforcing bars.

REVIEW QUESTIONS

1. List the four basic categories of walls.

2. Sketch a simple illustration of each type of wall listed in Question 1.

3. Make a sketch that illustrates the different reinforcing methods for tied and spiral columns.

4. List three uses of security walls.

5. What is the primary function of shear walls?

6. What is a tilt-up wall?

7. What are the two types of retaining walls?

8. Sketch an example of a cross section of each of the two types of retaining walls listed in Question 7.

CAD ACTIVITIES

GENERAL INSTRUCTIONS

The following activities may be completed on any CAD system. Before reading the *specific instructions* for each activity (below), go through each step in the following planning checklist. The checklist applies to any CAD system and will help ensure the optimum use of your time and resources.

1. Analyze the problem carefully. Decide exactly what you are being asked to do.

2. Determine what resources and references you will need in order to complete the problem and collect them.

3. Decide if any particular standards apply to the project and have those standards available.

4. Determine what types of views will be required and how many of each.

5. Determine what the final plotted scale of the drawing will need to be and select the appropriate paper size for plotting/printing (make sure the appropriate paper size is available).

6. Plan your drawing sequence. In what order will you develop the drawing (i.e., lines, features, dimension lines, leaders, dimensions, notes, etc.)?

7. Review the various CAD commands you will have to use in order to develop the drawing.

8. Examine your CAD system to ensure that everything is in working order, then begin the project.

SPECIFIC INSTRUCTIONS

Activity 18.1.—Figure 18.14 shows a foundation plan for a self-service gas station. The concrete slab area is 75'-0" × 150'-0" in the center of a larger area. Draw an engineering drawing for a gravity retaining wall for the left side of the concrete slab area. The wall should be 4'-0" tall, 3'-0" wide at the top, and 4'-0" wide at the bottom. The plan should be similar to the example in Figure 18.9.

Activity 18.2.—Prepare a placing drawing for the retaining wall drawn in Activity 18.4. The plan should be similar to the example in Figure 18.12.

Activity 18.3.—Prepare an engineering drawing of a cantilever retaining wall for the right side of the concrete slab in Figure 18.14. Refer to Figure 18.4 for a sample cross-sectional illustration of a similar wall.

Activity 18.4.—Prepare a placing drawing for the retaining wall drawn in Activity 18.3. Refer to Figure 18.12 for a similar plan that can be used as an example.

Activity 18.5.—Figure 18.14 contains three, 50'-0" long concrete foundations equally spaced within the concrete slab area. Using the dimensions available, place three, 12" square columns (equally spaced) on each of the foundations. Using this as a starting point, prepare an engineering drawing for the columns. Each column is to be 15'-0" high, reinforced with four #5 bars longitudinally, and #3 bar stirrups @ 12" on center. Refer to Figure 18.11 for an example.

Activity 18.6.—Prepare a placing drawing for the column plan drawn in Activity 18.5. Refer to Figure 18.13 for an example.

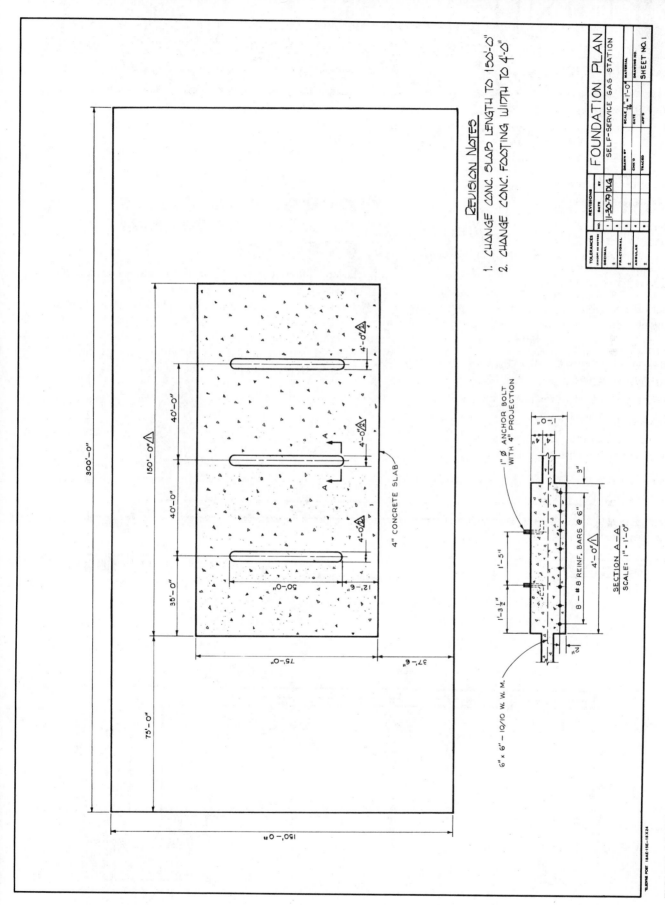

FIGURE 18.14 ■ Poured-in-place concrete drawing for CAD activity.

Poured-in-Place Concrete Floor Systems

OBJECTIVES

Upon completion of this unit, the student will be able to:

- Distinguish between ground-supported and suspended floor systems.

- Define and recognize one-way solid slab-and-beam floor systems; one-way ribbed or joist-slab floor systems; two-way solid slab-and-beam floor systems; two-way flat-plate floor systems; and waffle-slab floor systems.

- Prepare complete engineering drawings of poured-in-place concrete floor systems.

- Prepare complete placing drawings of poured-in-place concrete floor systems.

POURED-IN-PLACE CONCRETE FLOOR SYSTEMS

Several different types of poured-in-place concrete floor systems are used in the heavy construction industry. One-way solid slab and beam, one-way ribbed or joist slab, two-way solid slab and beam, two-way flat plate, and waffle-slab floor systems are the most common. Floor systems in poured-in-place concrete construction are either ground-supported systems or suspended systems (Figure 19.1).

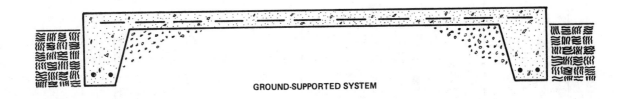

GROUND-SUPPORTED SYSTEM

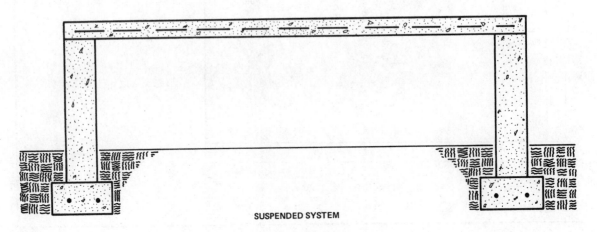

SUSPENDED SYSTEM

FIGURE 19.1 ■ Two classifications of poured-in-place concrete floor systems.

One-Way Solid Slab and Beam

One of the most commonly used floor systems in poured-in-place concrete is the one-way solid slab-and-beam system. This type of system consists of a 4″ to 6″ thick **slab** of concrete supported by parallel beams or walls. It is used primarily in short-span situations of less than 12′.

As in poured-in-place columns, the one-way slab-and-beam system derives its classification from the method of reinforcement used. The *one-way* in the name refers to the fact that the main slab reinforcing runs in only one direction, from beam to beam. A poured-in-place concrete one-way solid slab-and-beam floor system is shown in Figure 19.2.

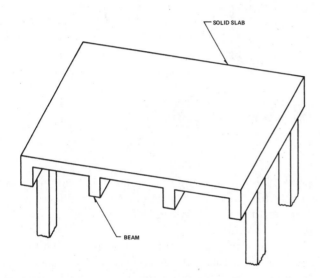

FIGURE 19.2 ■ One-way solid slab-and-beam floor system.

One-Way Ribbed or Joist Slab

One disadvantage of the one-way solid slab-and-beam floor system is that it can be very heavy, particularly when used in situations involving spans of over 12′. The weight problem can be solved by using a modified version of the one-way solid slab-and-beam system called the *one-way ribbed* or *joist slab*. This system combines a thinner slab with concurrently poured joists or ribs to create a floor system very similar to a precast concrete double-tee system.

The joist or ribs are usually spaced 24″ to 35″ on center and are reinforced with two or more reinforcing bars. The slab between and above the joists is reinforced in the same manner as the one-way solid slab-and-beam system. Figure 19.3 shows an example of a one-way ribbed or joist system. Lateral stability may be increased by adding bridging at intervals between the joists or ribs (Figure 19.4).

Two-Way Solid Slab and Beam

In construction situations that call for square slabs or rectangular slabs that are almost square, the two-way solid slab-and-beam floor system is used. As the name implies, the slab has beams running in two directions and is reinforced in two directions. This configuration allows for greater floor area with fewer beams and no girders. This type of floor system is slightly more difficult to construct. However, it is relatively economical in that it usually requires less building materials.

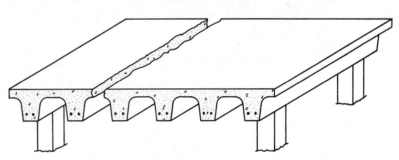

FIGURE 19.3 ■ One-way ribbed or joist-slab system.

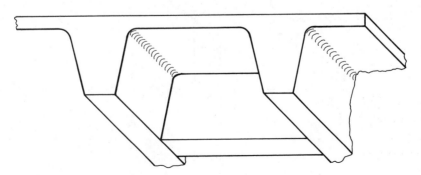

FIGURE 19.4 ■ Bridging between joists.

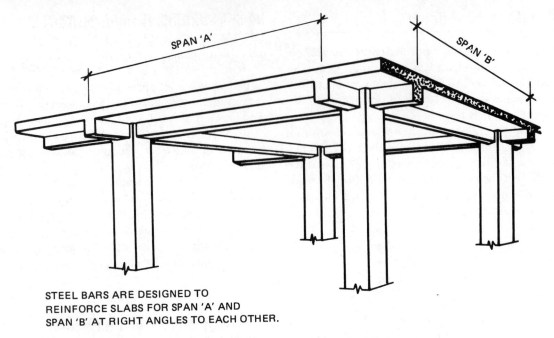

STEEL BARS ARE DESIGNED TO
REINFORCE SLABS FOR SPAN 'A' AND
SPAN 'B' AT RIGHT ANGLES TO EACH OTHER.

FIGURE 19.5 ■ Two-way solid slab-and-beam floor system. (*Charles D. Willis*)

A two-way solid slab-and-beam floor system is shown in Figure 19.5.

Two-Way Flat-Plate Floor Systems

The three floor systems just discussed are actually variations of the two-way solid-slab system. The flat-plate system is the most simple. It consists of a floor slab of uniform thickness supported by columns only. There are no beams or girders (Figure 19.6). The absence of beams and girders and the flat floor make this type of system ideal in applications where headroom is a problem.

The *two-way* designation in the title indicates that the reinforcement is placed in both directions. Common applications of this type of floor system include hotels, motels, apartment buildings, and condominiums.

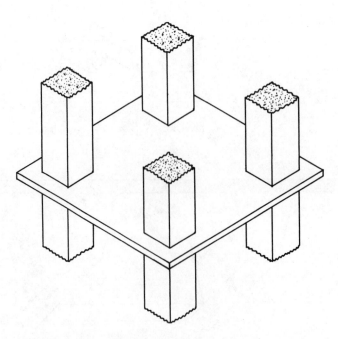

FIGURE 19.6 ■ Two-way flat-plate system.

Waffle Slab

In construction applications that are to be heavily loaded, extra thick slabs are required. However, the thicker the slab, the heavier it is, the more expensive it is, and, often, the less attractive it is. When a thick slab is required, these disadvantages may be overcome through the use of a waffle slab.

The *waffle-slab floor system* (Figure 19.7) is a thick slab with a series of geometric recesses formed into it. These recesses are formed with removable fillers placed in the forms before pouring. They cut down on the weight and cost of the slab without substantially decreasing its structural capabilities. These recesses also add a pleasing appearance to the slab.

POURED-IN-PLACE CONCRETE FLOOR SYSTEM DRAWINGS

Poured-in-place concrete floor systems require both engineering and placing drawings. The engineering drawings consist of a plan view of the floor, sections, slab schedules, and notes (Figure 19.8). Placing drawings contain a plan view of the floor, sections showing precise placement of the reinforcement bars, a slab schedule, and a bar details schedule (Figure 19.9).

Slabs shown in the plan view must be crossed to indicate the breadth of their length and width (Figure 19.8 and Figure 19.9). They are also given mark number designations. For example, the slab designation 1S2 would be interpreted slab number 2 on the first floor. Slab designation 3S4 would be interpreted as slab number 4 on the third floor.

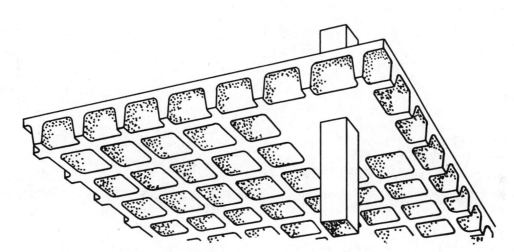

FIGURE 19.7 ■ Waffle-slab floor system.

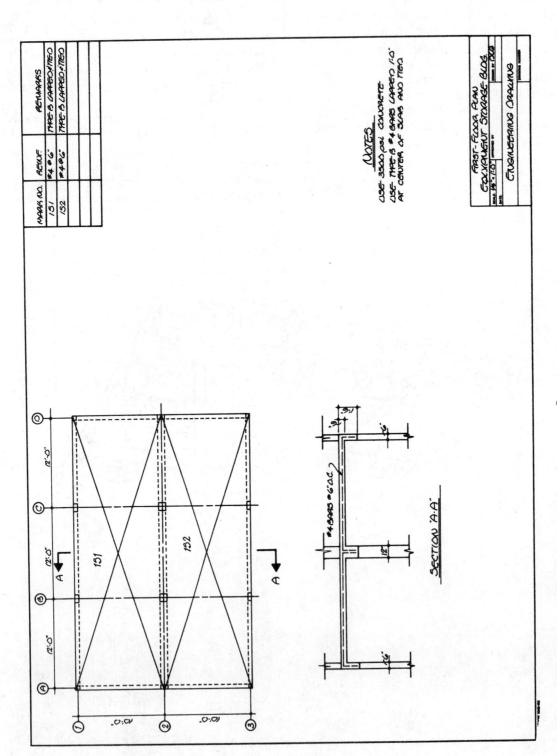

FIGURE 19.8 ■ Engineering drawing for poured-in-place concrete floor system.

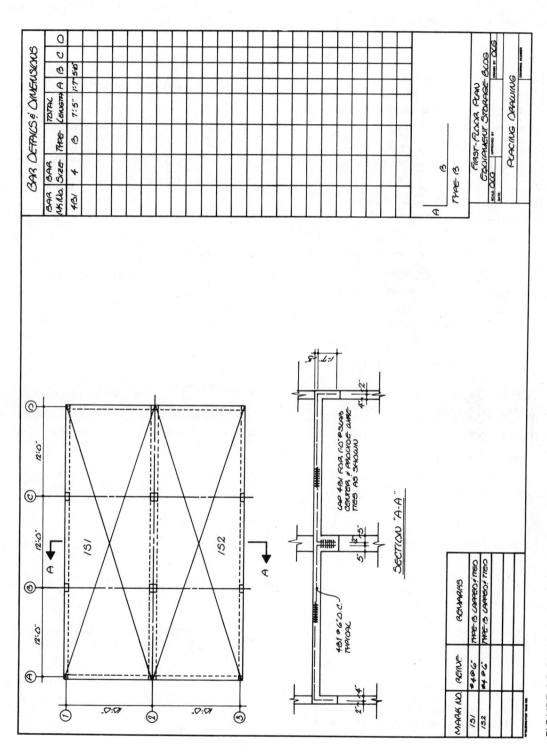

FIGURE 19.9 ■ Placing drawing for a poured-in-place concrete floor system.

SUMMARY

■ The most common types of poured-in-place concrete floors are: one-way solid slab and beam, one-way ribbed or joist slab, two-way solid slab and beam.

■ Floor systems in poured-in-place concrete construction are either ground-supported or suspended systems.

■ The one-way solid-slab system consists of a slab of uniform thickness supported by parallel beams or walls.

■ The one-way ribbed or joist-slab system combines a thin slab with concurrently poured joists to create a floor system that is similar to the precast concrete double-tee system.

■ Lateral stability in one-way ribbed or joist floor systems may be increased by adding bridging at intervals between the joists.

■ The two-way solid slab-and-beam floor system is used in applications requiring square or rectangular slabs.

■ The two-way solid-slab system is more difficult to construct than the one-way systems, but it is economical because it cuts down on the number of beams and requires no girders.

■ The two-way flat-plate floor system is one of the simplest poured-in-place concrete floor systems to construct and is ideal in applications where headroom is a problem.

■ Common applications of the two-way flat-plate system are hotels, motels, apartment buildings, and condominiums.

■ The waffle slab may be used in applications requiring a thick slab.

■ The waffle slab is reduced in cost and weight by a series of recesses formed into it with removable fillers. These necessary reductions do not substantially decrease its structural capabilities.

REVIEW QUESTIONS

1. List the most common types of poured-in-place concrete slabs.

2. Make a sketch that illustrates suspended floor systems and ground-supported systems.

3. Make a sketch to illustrate the one-way solid-slab system.

4. Make a sketch to illustrate the one-way ribbed or joist system.

5. How can the lateral stability of ribs or joists be increased?

6. Make a sketch to illustrate the two-way solid slab-and-beam system.

7. Make a sketch to illustrate the two flat-plate system(s).

8. List three common applications of the two-way flat-plate system.

9. Make a sketch to illustrate the waffle-slab system.

10. How are the recesses in a waffle slab formed?

CAD ACTIVITIES

GENERAL INSTRUCTIONS

The following activities may be completed on any CAD system. Before reading the *specific instructions* for each activity (below), go through each step in the following planning checklist. The checklist applies to any CAD system and will help ensure the optimum use of your time and resources.

1. Analyze the problem carefully. Decide exactly what you are being asked to do.

2. Determine what resources and references you will need in order to complete the problem and collect them.

3. Decide if any particular standards apply to the project and have those standards available.

4. Determine what types of views will be required and how many of each.

5. Determine what the final plotted scale of the drawing will need to be, and select the appropriate paper size for plotting/printing (make sure the appropriate paper size is available).

6. Plan your drawing sequence. In what order will you develop the drawing (i.e., lines, features, dimension lines, leaders, dimensions, notes, etc.)?

7. Review the various CAD commands you will have to use in order to develop the drawing.

8. Examine your CAD system to ensure that everything is in working order, then begin the project.

SPECIFIC INSTRUCTIONS

Activity 19.1—Redraw the engineering drawing in Figure 19.8 with the following changes:
   a. Increase the 12'-0" dimensions to 15'-0".
   b. Increase the 10'-0" dimensions to 12'-0".
   c. Make all columns 12" square and use number 5 bars.

Activity 19.2—Prepare a complete placing drawing for the engineering drawing completed in Activity 19.1.

Activity 19.3—Convert the engineering drawing of the one-way solid slab-and-beam system to a two-way solid slab-and-beam system by adding 12" wide beams between columns B1 and B2, B2 and B3, C1 and C2, and C2 and C3. Provide type B #4 bars at 6" on center in both directions. Use the same dimensions used in Figure 19.8.

Activity 19.4—Prepare a complete placing drawing for the engineering drawing completed in Activity 19.3.

Poured-in-Place Stairs and Ramps

OBJECTIVES

Upon completion of this unit, the student will be able to:

■ Sketch examples of the various types of stairs.

■ Perform stair design computations.

■ Develop engineering and placing drawings of stairs and ramps.

■ Explain the purpose of a ramp.

TYPES OF STAIRS

Stairs are a set of steps of various designs that provide easy access between the floors of a building. Four types of stairs are very common in heavy construction: straight stairs, L-shaped stairs, double-L stairs, and U-shaped stairs. Straight stairs are the most common and the simplest to construct. As the name implies, they run straight from one elevation to the next with no turns (Figure 20.1).

L-shaped stairs have one 90-degree turn between elevations. The two sides of the L are connected by a landing (Figure 20.2). Double-L stairs are the same as L-shaped stairs except they make two turns and have two landings (Figure 20.3). U-shaped stairs traverse part of the distance between floors, usually half, in one direction and traverse the remaining distance in the opposite direction (Figure 20.4).

STAIR DESIGN

The structural CAD technician may be called upon to perform stair design computations from time to time. In order to do so, one must be familiar with the critical dimensions required in designing stairs.

The first is the total rise. The *total rise* is the distance from one floor to the next or, in other words, the height that the stairs must span. The next important term is the *total run* or horizontal distance the stairs cover. These two terms are often given or known before the computations are performed. The

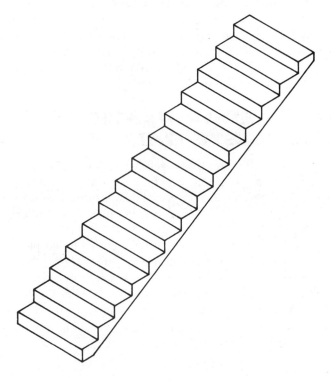

FIGURE 20.1 ■ Straight stairs.

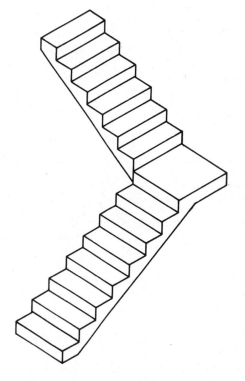

FIGURE 20.2 ■ L-shaped stairs.

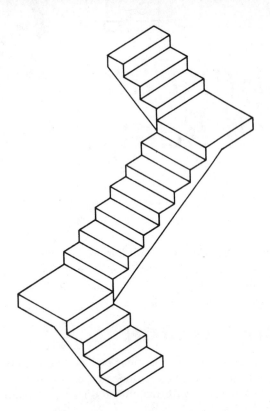

FIGURE 20.3 ■ Double-L stairs.

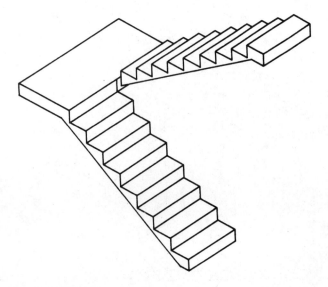

FIGURE 20.4 ■ U-shaped stairs.

two critical dimensions that must be computed are the *riser height* and the *tread width* (Figure 20.5).

When performing stair design computations, the structural CAD technician must work within certain parameters of proper design. The following are rules of thumb that ensure proper design of stairs:

■ There is always one less tread than risers.

■ All risers are the same height and all treads are the same width.

■ The sum of one riser height and one tread width should equal between 17″ and 18″.

Stair Design Computations

Stair design computations are not difficult. However, they should proceed in a certain predetermined order to ensure correct results:

1. Determine the total height that the stairs will have to span and convert the distance to inches.

2. Divide the total height by 7″ (the ideal riser height). This number should be rounded according to standard mathematical practices. For example, 14.57″ would be rounded to 15″; 14.47″ would be rounded to 14″.

3. The final number from Step 2 is the number of risers that this set of stairs will require. This rounded number is then divided into the total height dimension (stated in inches). The answer determined by this division is the exact height of the risers.

4. Since there is always one less tread than risers, the tread width can be easily computed. The number of risers less one is divided into the total run dimension stated in inches. The result is the exact width of each tread. Figure 20.6 shows how the result of stair computations are indicated on a structural drawing.

POURED-IN-PLACE CONCRETE RAMPS

Ramps are slightly sloped passages that may be used by persons with handicaps or persons who for any number of reasons are unable to use stairs. Ramps serve the same purpose as stairs in that they provide access from one elevation in a building to another.

Public awareness of the problems incurred by people with handicaps in transporting themselves in and out of public buildings has rapidly increased the number of ramps being added to areas that have historically had only stairs. Figure 20.7 shows how a ramp is shown in section on a structural drawing.

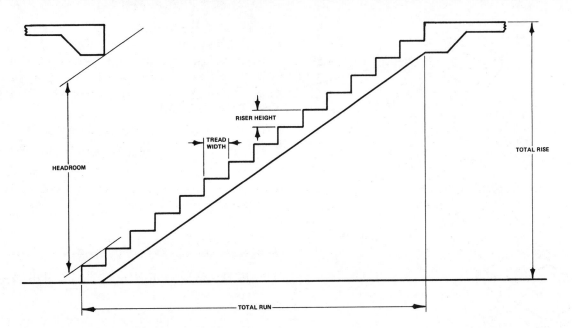

FIGURE 20.5 ■ Critical stair dimensions.

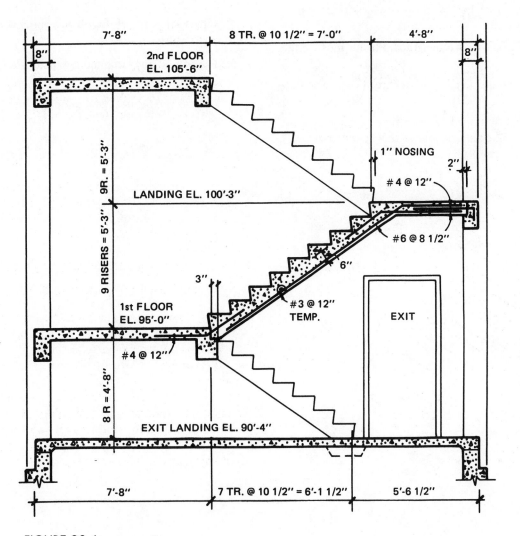

FIGURE 20.6 ■ Stairwell in section.

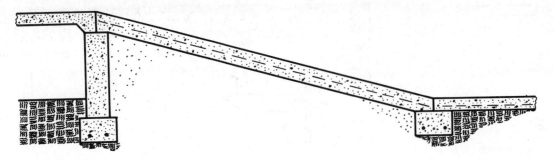

FIGURE 20.7 ■ Poured-in-place concrete ramp

SUMMARY

■ Four types of stairs are commonly used in heavy construction: straight stairs, L-shaped stairs, double-L stairs, and U-shaped stairs.

■ The structural drafter must be aware of certain critical dimensions in stair design: total rise, total run, tread width, and riser height.

■ In good stair design, there is always one less tread than risers.

■ All risers must be the same height and treads must be the same width.

■ The sum of one riser height and one tread width should equal between 17″ and 18″.

■ Ramps are slightly sloped passages that serve the same purpose for people with handicaps that stairs serve for people without handicaps.

REVIEW QUESTIONS

1. Sketch and label an example of each of the four common types of stairs.

2. Sketch an example of a set of straight stairs and label the dimensions that are critical.

3. When designing stairs, what is the proper ratio of risers to treads?

4. What is the rule of thumb governing riser height and tread width?

5. The sum of any riser and any tread should equal

   _____.

6. What is the purpose of a ramp?

CAD ACTIVITIES

GENERAL INSTRUCTIONS

The following activities may be completed on any CAD system. Before reading the *specific instructions* for each activity (below), go through each step in the following planning checklist. The checklist applies to any CAD system and will help ensure the optimum use of your time and resources.

1. Analyze the problem carefully. Decide exactly what you are being asked to do.

2. Determine what resources and references you will need in order to complete the problem and collect them.

3. Decide if any particular standards apply to the project and have those standards available.

4. Determine what types of views will be required and how many of each.

5. Determine what the final plotted scale of the drawing will need to be, and select the appropriate paper size for plotting/printing (make sure the appropriate paper size is available).

6. Plan your drawing sequence. In what order will you develop the drawing (i.e., lines, features, dimension lines, leaders, dimensions, notes, etc.)?

7. Review the various CAD commands you will have to use in order to develop the drawing.

8. Examine your CAD system to ensure that everything is in working order, then begin the project.

SPECIFIC INSTRUCTIONS

Activity 20.1—Using the following specifications, design a set of stairs for a building: total rise = 9'-0" and total run = 12'-4".

Using Figure 20.5 as an example, draw a side elevation of your stairs at a scale of 1" = 1'-0". Completely label the drawing and indicate the number of treads and risers.

Activity 20.2—Repeat the instructions in Activity 20.1 for the following specifications: total rise = 9'-2" and total run = 12'-6".

Activity 20.3—Repeat the instructions in Activity 20.1 for the following specifications: total rise = 9'-11 1/4" and total run = 13'-3".

Activity 20.4—Repeat the instructions in Activity 20.1 for the following specifications: total rise = 10'-4 1/2" and total run = 13'-8 1/2".

Activity 20.5—Redraw Figure 20.6 at a scale of 3/4" = 1'-0" after adding 6" to each rise and run shown and recomputing all numbers.

SECTION V

Structural Wood Drafting

Structural Wood Floor Systems

OBJECTIVES

Upon completion of this unit, the student will be able to:

■ Sketch an example of the following types of floor systems: joist and girder systems, plywood systems, and trussed floor systems.

■ Select floor joists from span data tables according to design requirements.

■ Prepare framing plans, sections, and details for structural wood floor systems.

JOIST AND GIRDER FLOOR SYSTEMS

Floor systems of commercial buildings are of two types: on grade and off grade. *On-grade floor systems* are made of poured-in-place concrete. *Off-grade floor systems* are usually made of wooden joists, girders, and planks. Off-grade floors are often used in commercial construction. This is because of the maintenance advantages they offer, such as easy access to plumbing and ductwork for heating and cooling.

Framing plans for off-grade floors may be drawn in either one of two ways:

■ Joists may be shown on the floor plan with an arrow symbol indicating the direction of the span and a note indicating the joist size and spacing (Figure 21.1).

■ A floor joist framing plan showing the joists and girders in the plan may be used (Figure 21.2).

Girders in wooden floor systems are of two types: built-up wooden girders and steel girders. In either case, joists frame into or over girders and are supported by them. Unless span and loading conditions are too great, built-up wooden girders are used because they are easier to work with and more economical than steel girders.

Several construction methods are used for connecting joists to wooden girders. Four of the most common are illustrated in Figure 21.3.

In certain heavy-loading or long-span situations, it may become necessary to use steel girders. Wooden joists may be framed on top of steel girders by first attaching a wooden nailer to the girder and then the joists to it (Figure 21.4). However, since framing over the top of a girder decreases the

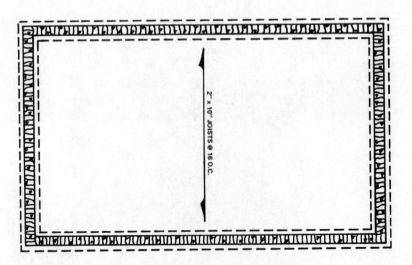

FIGURE 21.1 ■ Joists shown on foundation plan.

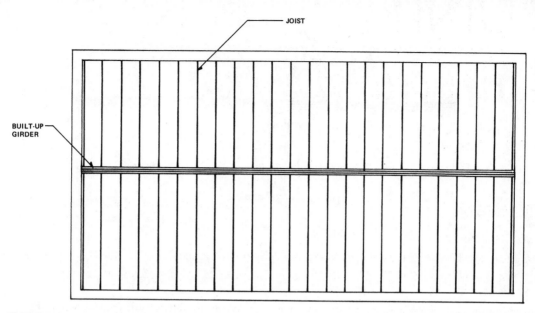

FIGURE 21.2 ■ Floor joist framing plan.

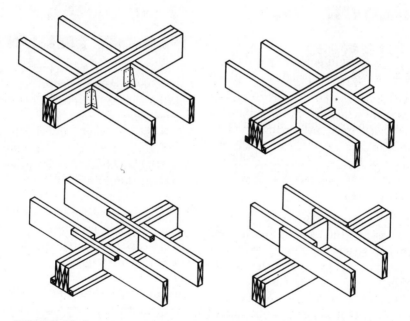

FIGURE 21.3 ■ Joist-to-girder connections. (*Goodheart-Wilcox Company, Inc.*)

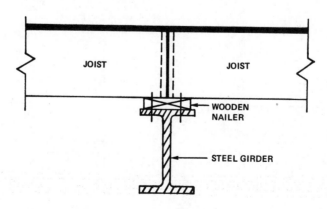

FIGURE 21.4 ■ Wood joist bearing on steel girder.

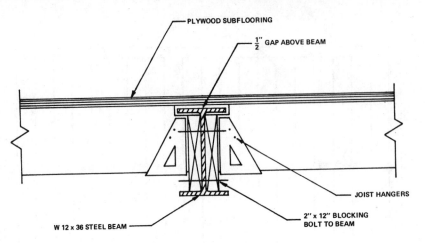

FIGURE 21.5 ■ Wood joist bearing into side of steel girders.

amount of crawl space under the floor, a more popular method is to frame into the side of the girder (Figure 21.5).

PLYWOOD FLOOR SYSTEMS

A wooden floor system that is gaining in popularity is the plywood system. Tongue-and-groove plywood panels 1 1/8″ thick have been developed for use simultaneously with wooden girders spaced at 4′-0″ on center to form a sound, economical floor (Figure 21.6). Since joists are not required, material costs are decreased. Because the large 4′ × 8′ plywood panels are quickly and easily installed, labor costs are also decreased (Figure 21.7).

BRIDGING

Bridging is a term used to describe the construction method whereby joists in a wooden floor system are stiffened to prevent misalignment, twisting, or warping. Bridging also helps distribute the structural load on the floor over a larger floor area.

The two basic types of bridging are: solid bridging and cross bridging. Solid bridging consists of solid cuts of lumber placed between the joists and secured by nailing (Figure 21.8). Cross bridging may be made of wood; but, in recent years, it is more commonly accomplished with two diagonal metal braces attached at intervals between the joists (Figure 21.9).

FLOOR TRUSS SYSTEMS

A floor system that is sometimes used instead of the joist and girder system is the *floor truss system*. Floor trusses are made of wooden *chords* (top and bottom members) with wooden or metal diagonal cross braces in between. The braces that are used vary in length and depth, depending on the span and load requirements of the individual job (Figure 21.10 and Figure 21.11). Another type of floor truss is actually a wooden reproduction of a steel beam. It is made of plywood flanges separated by a deep plywood web (Figure 21.12).

Floor trusses have several advantages over joists and girders. Since they span from one wall to the next wall, they require no intermediate supports. In addition, their ease of installation makes them more economical in some situations. The spaces between braces form convenient avenues for passing plumbing and ductwork through the floor. Trusses are usually spaced at 24″ on center and are shown on a floor truss framing plan.

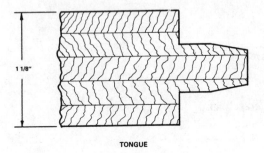

TONGUE

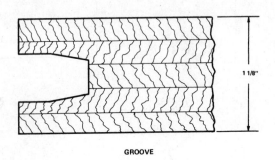

GROOVE

FIGURE 21.6 ■ Tongue-and-groove plywood.

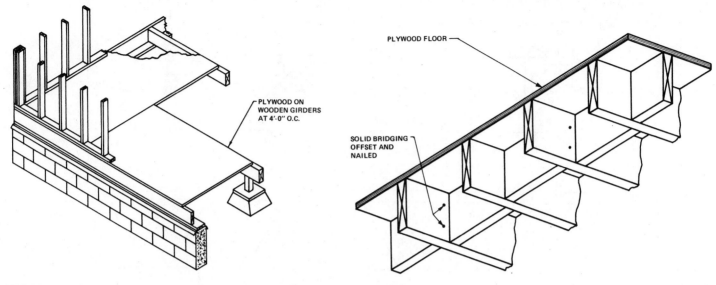

FIGURE 21.7 ■ Plywood floor system. (*American Technical Society*)

FIGURE 21.8 ■ Solid bridging between joists.

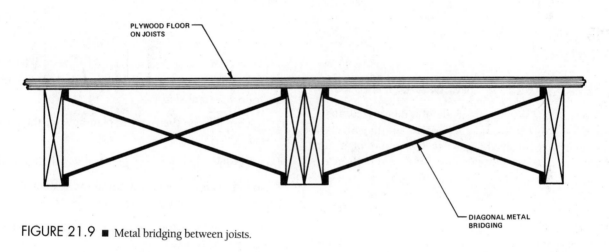

FIGURE 21.9 ■ Metal bridging between joists.

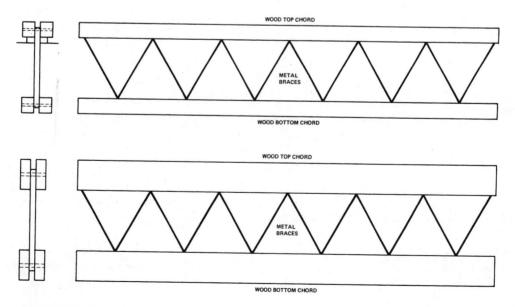

FIGURE 21.10 ■ Floor trusses—wood chords and metal braces.

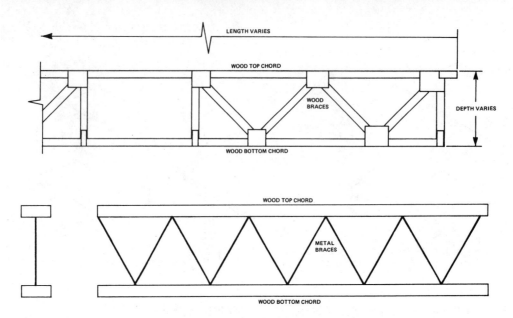

FIGURE 21.11 ■ Floor trusses—all wood and wood chords with metal braces.

FLOOR JOIST SELECTION

Joists in wooden floor systems are made of lumber sized nominally as: 2 × 6, 2 × 8, 2 × 10, and 2 × 12 (refer to Appendix A). The structural drafter involved in the development of plans for a wooden floor system must be able to select the properly sized joist to meet the design requirements of the job. To assist the drafter in this task, the National Forest Products Association provides span data tables for floor joists (refer to Appendix B in the back of this text).

These tables provide span data for joists made of all of the commonly used types and grades of wood. Lumber grades, joist sizes, joist spacings, and span capabilities for each size and spacing are also listed.

In order to select a particular joist from the table, the drafter must follow several steps:

1. Determine from the framing plan how far the joist has to span.

2. Determine what type and grade of wood are specified.

3. Go to the appropriate column in the span table and locate a joist and spacing that meet the span requirements.

Appendix B contains applicable span tables as well as directions explaining how to use them.

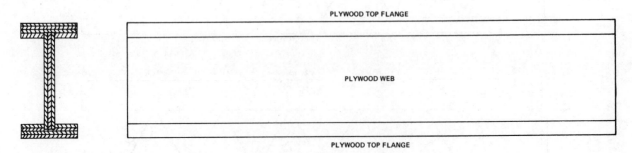

FIGURE 21.12 ■ Plywood composite floor truss.

SUMMARY

■ The most common structural wood floor systems are: joist and girder systems, plywood systems, and trussed floor systems.

■ Joists are shown on the plans in one of two ways: by an arrow symbol on the foundation plan or by drawing of a complete joist framing plan.

■ Girders in wooden floor systems are of two types: built-up wooden girders and steel girders.

■ Plywood floor systems combine tongue-and-groove plywood panels 1 1/8″ thick and wooden girders spaced at 4′-0″ on center to make a sound, economical floor system.

■ Bridging is used between joists to prevent misalignment, twisting, or warping.

■ Bridging may be wood or metal and falls into two basic categories: solid bridging and cross bridging.

■ Floor truss systems are sometimes used instead of joists and girders.

■ Floor truss systems have several advantages over joist and girder systems: longer spanning capabilities, easier installation, and convenience for installation of plumbing and ductwork.

■ Structural drafters should be able to select joist sizes and spacings from span data tables.

■ Wooden joists are nominally sized as: 2 × 6, 2 × 8, 2 × 10, and 2 × 12.

REVIEW QUESTIONS

1. List the three most common types of structural wood floor systems.

2. Sketch an example of how joists may be indicated on a foundation plan.

3. List the two types of girders used in wooden floor systems.

4. Sketch an example of each of the four most common construction methods for joining joists and girders.

5. Sketch an example of how joists are attached to the side of a steel girder.

6. What is bridging? What are the two types? Sketch an example of each.

7. Sketch an example of four different types of floor trusses.

8. List the four different sizes of wooden floor joists.

CAD ACTIVITIES

GENERAL INSTRUCTIONS

The following activities may be completed on any CAD system. Before reading the *specific instructions* for each activity (below), go through each step in the following planning checklist. The checklist applies to any CAD system and will help ensure the optimum use of your time and resources.

1. Analyze the problem carefully. Decide exactly what you are being asked to do.

2. Determine what resources and references you will need in order to complete the problem and collect them.

3. Decide if any particular standards apply to the project and have those standards available.

4. Determine what types of views will be required and how many of each.

5. Determine what the final plotted scale of the drawing will need to be, and select the appropriate paper size for plotting/printing (make sure the appropriate paper size is available).

6. Plan your drawing sequence. In what order will you develop the drawing (i.e., lines, features, dimension lines, leaders, dimensions, notes, etc.)?

7. Review the various CAD commands you will have to use in order to develop the drawing.

8. Examine your CAD system to ensure that everything is in working order, then begin the project.

SPECIFIC INSTRUCTIONS

The title block as shown in Figure 1.21 and 1/2″ borders are required for all activities. The following activities should be completed on C-size paper.

Activity 21.1—Figure 21.13A contains the outline of a floor system that is to make use of joists to span from wall to wall.

Convert this figure to a foundation plan such as the one shown in Figure 21.1, using an 8″ foundation wall centered on a 16″ wide footing. Select the proper joist size and spacing from the span tables in Appendix A and place the proper indications on the foundation plan.

Activity 21.2—Figure 21.13B contains the outline of a floor system that is to make use of joists to span from wall to wall. Following the same instructions set forth in Activity 21.1, draw the foundation plan and place the proper indications on it.

Activity 21.3—Figure 21.13C contains the outline of a floor system that is to make use of floor joists to span from wall to wall. Select the proper joist size and spacing from the span tables in Appendix A. Draw a joist framing plan similar to the one shown in Figure 21.2.

Activity 21.4—Figure 21.13D contains the outline of a floor system that is to be split into two bays by 6 × 10 built-up wooden girders. Appropriate joist size and spacing are to be chosen. The joists are to span into the side of the girders and be connected with metal connectors. Select the joists from the span data tables in Appendix A and draw a floor joist framing plan.

Activity 21.5—Figure 21.13E contains the outline of a floor system that is to be split into three equal bays by 6 × 10 built-up girders. Appropriate joist size and spacing must be chosen. The joists are to span on top of the girders and be braced in between with solid bridging. Select the joists from the span tables in Appendix A and draw a floor joist framing plan.

Activity 21.6—Cut a section through the girder in Activity 21.4 to show how the joists are connected to the girder. Place this connection detail on the same sheet as the framing plan.

Activity 21.7—Cut a section through one of the girders in Activity 21.5 to show how the joists are connected to the girder. Place this connection detail on the same sheet as the framing plan.

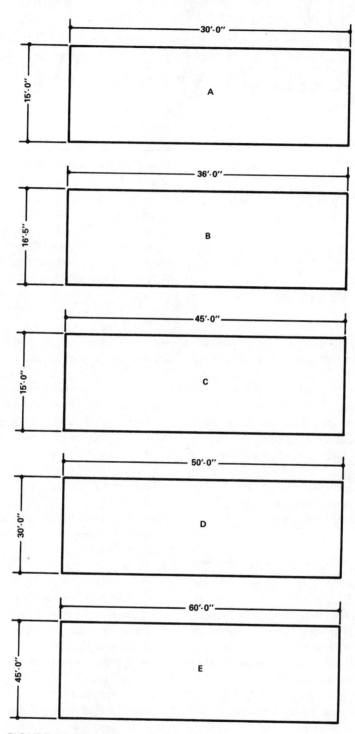

FIGURE 21.13 ■ Figures for CAD activities.

22 UNIT

Structural Wood Walls

OBJECTIVES

Upon completion of this unit, the student will be able to:

■ Define and illustrate platform framing, balloon framing, and MOD 24 framing.

■ Define and illustrate bracing.

■ Prepare structural drawings of wall details and sections.

STRUCTURAL WOOD WALLS

One of the major structural components of a wooden structure is the wall. Walls are either load bearing or non–load bearing. *Load-bearing walls* are those that actually carry the load imposed by the floor(s) and roof. Exterior walls of a structure are almost always load bearing.

Non-load-bearing walls carry no load. They are simply built up from the floor to the ceiling and serve the purpose of partitioning space into rooms, closets, and so on. Only load-bearing walls are considered structural.

Historically, structural wood walls were of two types: platform framed walls and balloon framed walls. A recent addition to these two framing methods is the MOD 24 wall.

Platform Framing

Platform framing is the most common framing method used for building structural wood walls. Also called *western framing,* it begins with a 2 × 6 sill that is bolted to the foundation wall. Floor joists of the proper size and spacing (Unit 21) are nailed to the sill. Headers are attached to the ends of the joists and the subfloor to the tops of the joists. A 2 × 4 soleplate is then attached to the tops of the joists through the subfloor. Wall **studs** are nailed to the soleplate and capped with a double top plate composed of two 2 × 4s. A typical wall section that illustrates this method of wall framing is shown in Figure 22.1. Figure 22.2 shows an elevation of a platform wall showing the various components.

Balloon Framing

Balloon framing is an older method than platform framing. However, it is seldom used any longer as the primary framing method in structural wood walls. Like platform framing, balloon framing begins with a 2 × 6 sill anchored to the top of the foundation wall. However, from this point on, the two types of framing differ.

In balloon framing, the floor joists rest on the sill with headers or braces placed between the joists. The wall studs also rest on the sill and are placed against the floor joists. Second floors are accomplished by bearing the floor joist on ribbon boards that are nailed into slots cut in the wall studs. Again, braces are placed between the floor joists. These braces serve both as a source of lateral stability and as a fire stop.

There are two important disadvantages to balloon framing that have decreased its popularity. The first is that the 18′ to 20′ long studs required are not readily available. Those that can be found are very expensive due to the additional length. A more practical problem with balloon framing is that the carpenters have no subfloor to stand on when building the stud wall. This method of wall framing is illustrated in Figure 22.1. An elevation of a balloon wall showing the various components is shown in Figure 22.3.

Mod 24 Framing

The ever-increasing price of labor and materials prompted the advent of a new framing method, **MOD 24.** The name comes from the 24″ modular spacing of wall studs. Floor and roof joists are also spaced on 24″ centers. The most structurally sound MOD 24 framing arrangement has floor joists, wall studs, and roof joists aligned.

In addition to using less wood and requiring less labor time, MOD 24 framing produces less waste. This is because most building materials such as lumber, sheathing, plywood, and so on are produced in 8′ lengths or widths. Since 8′ can be equally divided into 24″ modules, less waste results when a piece of lumber must be cut to fit.

Structural designers and architects with an interest in reducing cost design a structure so that it is completely modular in both length and width. For example, rather than designing a building that would be 43′-5″ long × 25′-7″ wide, the cost-minded designer would make the building 44′-0″ × 26′-0″. The latter dimensions are modules of 24″. Therefore, the client would get a larger building for the same or a lesser price because labor time and material waste are reduced. An example of MOD 24 framing is shown in Figure 22.4.

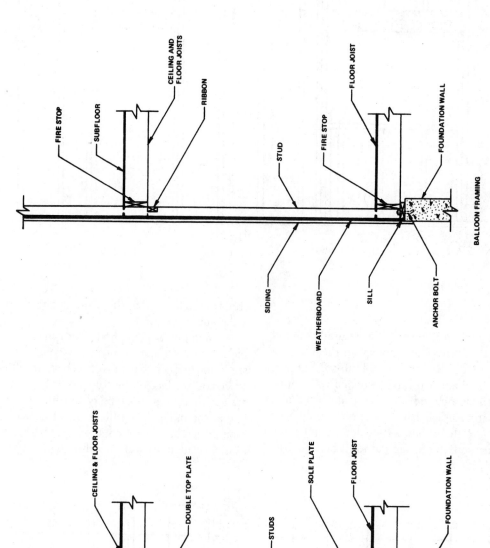

FIGURE 22. 1 ■ Typical wall sections for platform and balloon framing.

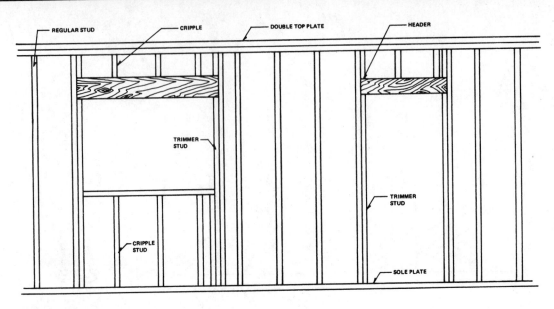

FIGURE 22.2 ■ Elevation of platform wall.

BRACING

Wooden walls by themselves are not structurally sound enough to withstand the many forces acting on them after a structure has been erected. To provide additional support against bending, buckling, and wracking, wooden walls must be braced. Several methods are used. One of the most common methods has been to form a brace that extends diagonally from the top of the soleplate to the bottom of the double top plate, Figure 22.5. In this method, 2 × 4s are cut to fit between the studs at angles so that the result is one continuous brace from soleplate to top plate.

A similar bracing method is known as the *let-in brace*. In this method, a 1 × 6 brace is placed diagonally across the face of the studs in grooves that have been cut to allow the face of the brace and the face of the studs to be flush (Figure 22.6). There are advantages to this type of bracing. The primary advantage is that the brace, being flush with the face of the studs rather than placed between them, does not interfere with the installation of insulation or electrical wiring.

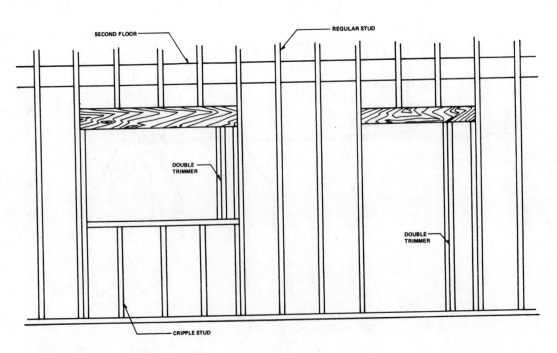

FIGURE 22.3 ■ Elevation of balloon wall.

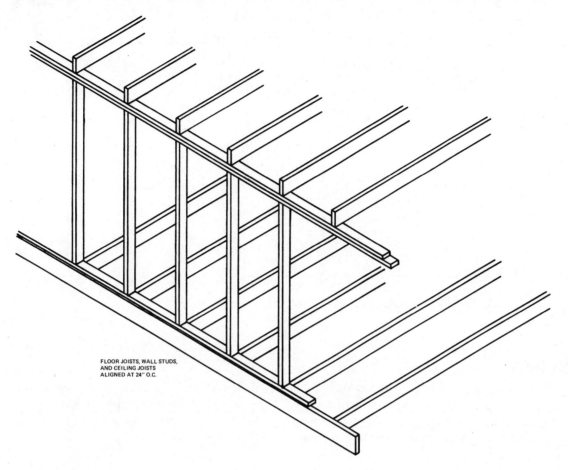

FIGURE 22.4 ■ MOD 24 wall framing.

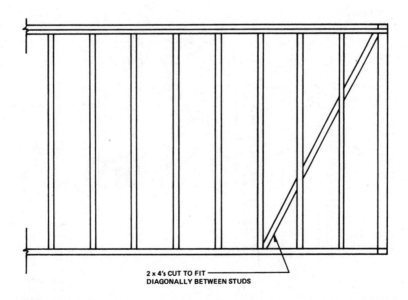

FIGURE 22.5 ■ Diagonally cut wall braces.

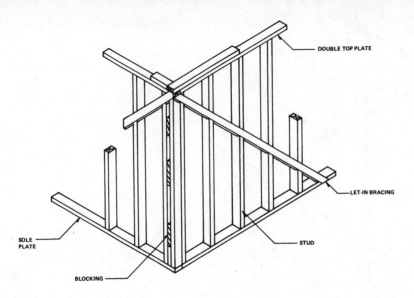

FIGURE 22.6 ■ Let-in wall bracing.

Another effective method of bracing wooden walls is to attach a sheet of plywood to the exterior of the wall at each corner (Figure 22.7). This is a very effective, structurally sound bracing method. It has the same advantages as the let-in brac-ing method and to an even greater degree. However, since the cost of plywood is prohibitive, only the corners receive a ply-wood brace. The remainder of the wall is braced with sheath-ing or weatherboard.

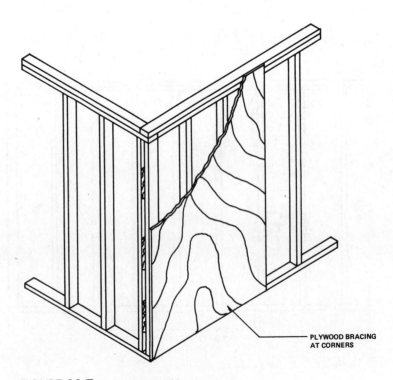

FIGURE 22.7 ■ Plywood wall bracing.

SUMMARY

■ Walls in a structure are either load bearing or non–load bearing.

■ Load-bearing walls carry the weight of the floors in multistory buildings and the weight of the roof in multistory and single-story buildings.

■ Non-load-bearing walls carry no weight (except their own) and are built up from the floor to the roof.

■ Exterior walls are usually load bearing. Interior walls may be load bearing or non–load bearing, depending on the design circumstances of the individual job.

■ Platform or western framing is the most common method of framing wooden walls.

■ Balloon framing, though it is an older method than platform framing, is seldom used any longer due to the increased cost of the 18′ to 20′ studs required and practical building difficulties inherent in this method.

■ MOD 24 framing came about in an attempt to cut material and labor costs.

■ The most structurally sound MOD 24 framing arrangement has the joists, studs, and roof joists aligned.

■ Bracing is used in wooden walls to prevent bending, buckling, or wracking.

■ The three bracing methods commonly used in wooden wall construction are diagonal bracing between studs with specially cut 2 × 4s, let-in bracing with 1 × 6s, and plywood corner bracing.

REVIEW QUESTIONS

1. Which two of the following are the two types of structural wood walls?
   a. partition walls
   b. load-bearing walls
   c. interior walls
   d. non-load-bearing walls

2. Which of the following types of walls carry the weight of the floors and the roof in a building?
   a. partition walls
   b. balloon walls
   c. load-bearing walls
   d. MOD 24 walls

3. Which type of wall is more likely to be load bearing, an interior wall or an exterior wall?

4. Make a sketch that illustrates the difference between platform and balloon framing.

5. Make a sketch that illustrates MOD 24 framing.

6. Make a sketch that illustrates the diagonal-cut 2 × 4 method of wall bracing.

7. Make a sketch that illustrates the let-in method of wall bracing.

8. Make a sketch that illustrates the plywood method of wall bracing.

CAD ACTIVITIES

GENERAL INSTRUCTIONS

The following activities may be completed on any CAD system. Before reading the *specific instructions* for each activity (below), go through each step in the following planning checklist. The checklist applies to any CAD system and will help ensure the optimum use of your time and resources.

1. Analyze the problem carefully. Decide exactly what you are being asked to do.

2. Determine what resources and references you will need in order to complete the problem and collect them.

3. Decide if any particular standards apply to the project and have those standards available.

4. Determine what types of views will be required and how many of each.

5. Determine what the final plotted scale of the drawing will need to be, and select the appropriate paper size for plotting/printing (make sure the appropriate paper size is available).

6. Plan your drawing sequence. In what order will you develop the drawing (i.e., lines, features, dimension lines, leaders, dimensions, notes, etc.)?

7. Review the various CAD commands you will have to use in order to develop the drawing.

8. Examine your CAD system to ensure that everything is in working order, then begin the project.

SPECIFIC INSTRUCTIONS

The activities for this unit are based on the wall section shown in Figure 22.8.

Activity 22.1—Begin to understand how to draw typical wall sections for structural wood walls by redrawing Figure 22.8 to a scale of 3/4″ = 1′-0″.

Activity 22.2—Redraw Figure 22.8 to a scale of 3/4″ = 1′-0″ with the following changes:
    a. first-floor joists 2 × 10 @ 16″ on center
    b. second-floor joists 2 × 10 @ 16″ on center
    c. ceiling joists 2 × 10 @ 16″ on center

Activity 22.3—Redraw Figure 22.8 to a scale of 3/4″ = 1′-0″ with the following changes:
    a. all joists 2 × 8 @ 12″ on center
    b. all studs 2 × 4 @ 16″ on center

Activity 22.4—Using the same size and spacing for all structural components of the wall, convert Figure 22.8 to balloon framing.

Activity 22.5—Convert Figure 22.8 to balloon spacing and redraw it at a scale of 3/4″ = 1′-0″ with the following changes:
    a. all joists 2 × 10 @ 16″ on center
    b. All studs 2 × 4

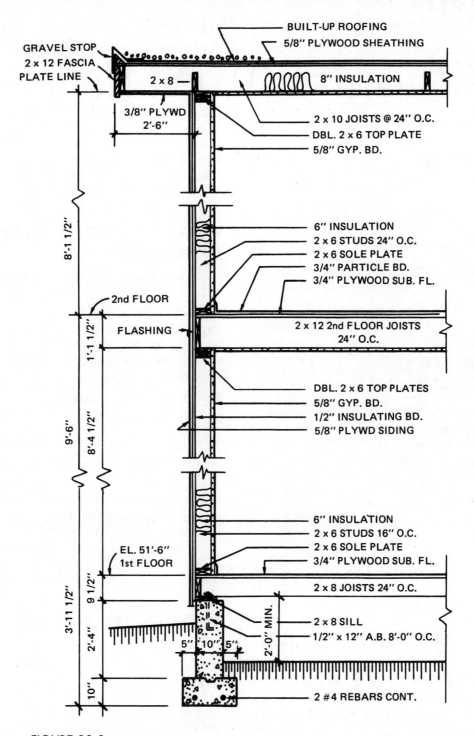

GRAVEL STOP
2 x 12 FASCIA
PLATE LINE

BUILT-UP ROOFING
5/8'' PLYWOOD SHEATHING

2 x 8

8'' INSULATION

3/8'' PLYWD
2'-6''

8'-1 1/2''

2nd FLOOR

FLASHING

1'-1 1/2''

9'-6''

8'-4 1/2''

EL. 51'-6''
1st FLOOR

9 1/2''

3'-11 1/2''

2'-4''

10''

5'' 10'' 5''

2'-0'' MIN.

2 x 10 JOISTS @ 24'' O.C.
DBL. 2 x 6 TOP PLATE
5/8'' GYP. BD.

6'' INSULATION
2 x 6 STUDS 24'' O.C.
2 x 6 SOLE PLATE
3/4'' PARTICLE BD.
3/4'' PLYWOOD SUB. FL.

2 x 12 2nd FLOOR JOISTS
24'' O.C.

DBL. 2 x 6 TOP PLATES
5/8'' GYP. BD.
1/2'' INSULATING BD.
5/8'' PLYWD SIDING

6'' INSULATION
2 x 6 STUDS 16'' O.C.
2 x 6 SOLE PLATE
3/4'' PLYWOOD SUB. FL.

2 x 8 JOISTS 24'' O.C.

2 x 8 SILL
1/2'' x 12'' A.B. 8'-0'' O.C.

2 #4 REBARS CONT.

FIGURE 22.8 ■ Typical wall section for CAD activities.

23 UNIT

Structural Wood Roofs

OBJECTIVES

Upon completion of this unit, the student will be able to:

■ Identify the most common classification of roof configurations and draw roof configuration diagrams.

■ Calculate the slope and pitch of a roof.

■ Draw eave and ridge details.

■ Select ceiling joist sizes and spaces from span tables.

■ Identify the various types of roof trusses and select roof trusses from span tables according to design requirements.

COMMON ROOF CLASSIFICATIONS

Structural wood roofs are divided into two broad classifications: flat roofs and sloped roofs. *Flat roofs* encompass those that are actually flat as well as those that have a very gradual slope. *Slope roofs* range from medium slopes to very steep slopes. Another important distinction between these two classifications of roofs is that flat roofs are usually built up, while sloped roofs are shingled.

Sloped roofs are the most popular in structural wood construction. This is because they provide for water runoff, snow runoff, additional attic or storage space, and an appearance that is pleasing to the eye. Some of the most popular sloped wooden roofs are the gable, the hip, the gambrel, the dutch hip, the mansard, and the A frame (Figure 23.1).

ROOF SLOPE AND PITCH

The two most important terms when studying wooden roofs are slope and pitch. The **slope** of a roof is the amount of rise measured in inches that occurs for every 12″ of run. The term *run* means a distance equal to half of the total span (Figure 23.2). A roof slope that rises at a rate of 3″ per foot of run has a slope of 3:12. Slopes ranging from 3:12 to 5:12 are considered medium slopes. Slopes that are 6:12 or greater are considered steep slopes.

The **pitch** of a roof is a figure arrived at by dividing the total span into the total rise (Figure 23.2). A chart showing several of the most commonly used roof pitches and their correspond-

ing slopes is shown in Figure 23.3. Roof slopes are indicated on structural drawings with a slope symbol that lists the units of rise for every 12″ of run (Figure 23.2).

EAVE AND RIDGE DETAILS

Most wooden roofs are designed to overhang the exterior walls a certain specified distance. This overhang area is called the *eave* or *cornice*. Structural drafters draw sectional views cut through the eaves of a building to guide the contractor in constructing the roof. The following are the four most common eave or cornice configurations used in wood construction (refer to Figure 23.4):

■ Wide-box eave configuration

■ Open eave configuration

■ Wide-box eave configuration without lookouts

■ Narrow-box eave configuration

Eave configuration details are drawn to show how the ceiling joists or roof trusses connect to the bearing wall, how far the rafters or trusses overhang the wall, and the construction of the eave or cornice (Figure 23.4).

A *ridge detail,* which is a section cut through the highest point on the roof, must also be drawn. Ridge details show the ridge board in section and how the rafters connect to it (Figure 23.5).

WOODEN ROOF FRAMING

Wooden roofs are framed by conventional methods or with roof trusses. *Conventional* or *stick framing* is accomplished by spanning from wall to wall with ceiling joists and from double top plate to ridge board with rafters. When stick framing is to be used in constructing a roof, the structural CAD technician must select ceiling joists that will meet the design and span requirements. These joists are selected from a ceiling joist span table (see Appendix B) in the same manner as floor joists are selected.

Occasionally, the span in a building is so long that stick building the roof is not feasible. In these cases, roof trusses may be used. A *roof truss* is a prefabricated structural member

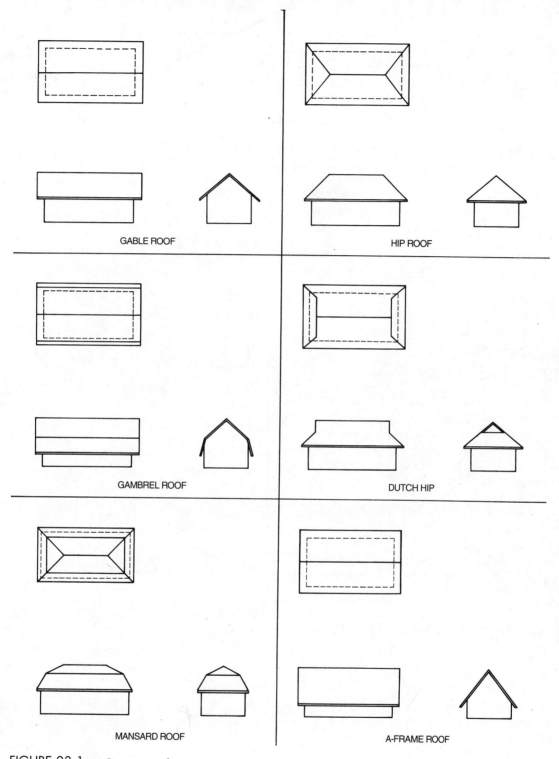

FIGURE 23.1 ■ Common roof types.

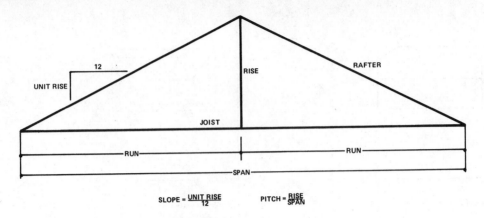

$$SLOPE = \frac{UNIT\ RISE}{12} \qquad PITCH = \frac{RISE}{SPAN}$$

FIGURE 23.2 ■ Roof slope and pitch.

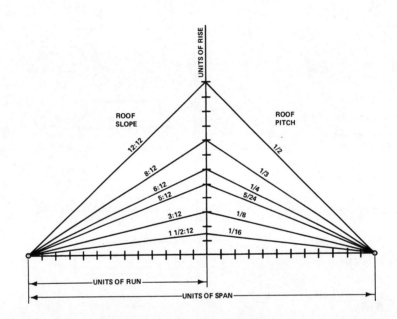

FIGURE 23.3 ■ Common roof slopes and pitches. *(Goodheart-Wilcox Company, Inc.)*

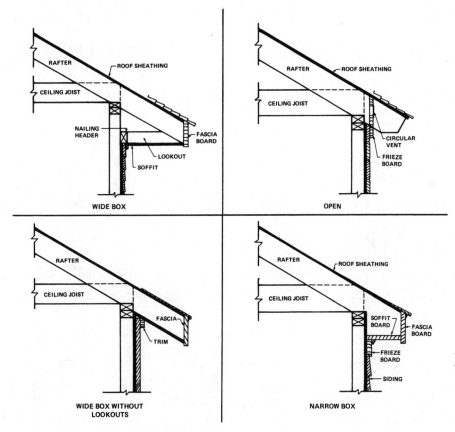

FIGURE 23.4 ■ Eave configuration details.

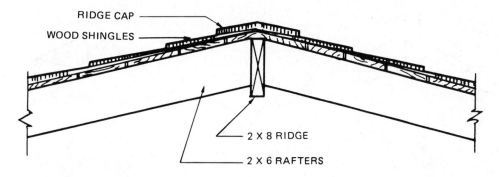

FIGURE 23.5 ■ Ridge detail. (*Charles D. Willis*)

composed of a joist member, rafter members, and various types of support members between the joist and rafters. Because of the additional support members, a roof truss spans farther than a conventional joist-rafter combination.

There are numerous different types of trusses with several different configurations for each type. The four most common types of trusses are the common, the scissors, the mono pitch, and the flat (Figure 23.6 through Figure 23.9). Trusses span great distances, but they too have limitations. Trusses must be carefully selected to meet design requirements. Structural CAD technicians select roof trusses from span tables in much the same manner as they select joists. The span tables for common, scissors, mono pitch, and flat roof trusses are shown in Figure 23.10.

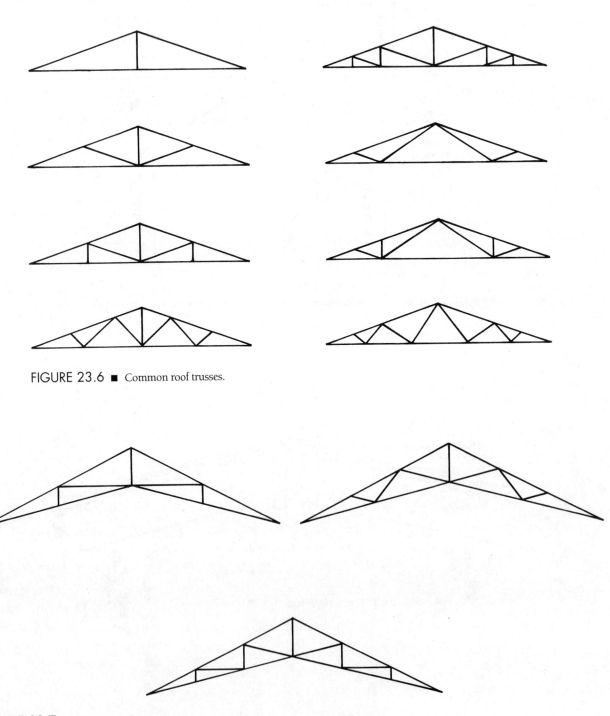

FIGURE 23.6 ■ ■ Common roof trusses.

FIGURE 23.7 ■ ■ Scissors roof trusses.

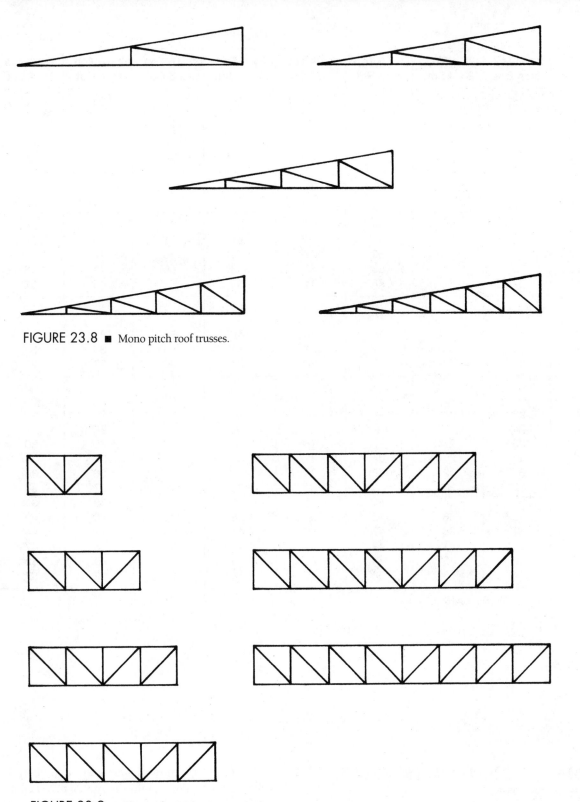

FIGURE 23.8 ■ Mono pitch roof trusses.

FIGURE 23.9 ■ Flat roof trusses.

	Pitch	Chord Size			Chord Size			Chord Size		
		2 × 4 Top 2 × 4 Bot.	2 × 6 Top 2 × 4 Bot.	2 × 6 Top 2 × 6 Bot.	2 × 4 Top 2 × 4 Bot.	2 × 6 Top 2 × 4 Bot.	2 × 6 Top 2 × 6 Bot.	2 × 4 Top 2 × 4 Bot.	2 × 6 Top 2 × 4 Bot.	2 × 6 Top 2 × 6 Bot.
Common	2/12	22'	23'	34'	25'	25'	39'	28'	28'	44'
	3/12	29'	31'	44'	33'	34'	50'	37'	38'	56'
	4/12	33'	39'	49'	37'	42'	55'	41'	46'	62'
	5/12	35'	45'	53'	39'	48'	59'	44'	52'	66'
	6/12	37'	51'	55'	41'	53'	62'	44'	57'	68'
Mono Pitch	Pitch									
	2/12	22'	23'	34'	25'	25'	38'	28'	28'	44'
	3/12	30'	31'	45'	33'	35'	51'	38'	39'	57'
	4/12	33'	39'	50'	37'	42'	56'	42'	46'	63'
	5/12	35'	45'	53'	40'	48'	60'	44'	53'	67'
	6/12	37'	51'	56'	41'	53'	62'	45'	57'	68'
Scissors	6/2*	32'	38'	48'	36'	42'	54'	40'	46'	61'
	6/3	28'	30'	42'	31'	34'	48'	35'	38'	54'
	6/4	21'	22'	32'	24'	24'	36'	27'	27'	42'
Flat	16"	23'	—	—	24'	—	—	26'	—	—
	18"	24'	—	—	26'	—	—	28'	—	—
	20"	26'	27'	—	28'	28'	—	30'	30'	—
	24"	29'	30'	34'	31'	31'	37'	33'	33'	39'
	28"	31'	32'	38'	33'	34'	40'	36'	36'	43'
	30"	32'	33'	39'	34'	35'	42'	37'	37'	45'
	32"	33'	34'	41'	35'	36'	43'	38'	38'	47'
	36"	35'	36'	43'	37'	38'	46'	40'	40'	50'
	42"	37'	38'	47'	39'	41'	50'	43'	44'	54'
	48"	39'	41'	49'	41'	43'	53'	45'	47'	57'
	60"	42'	46'	54'	45'	49'	59'	49'	52'	64'
	72"	44'	50'	58'	48'	53'	63'	52'	57'	68'

*6/12 = Top chord pitch, 2/12 = bottom chord pitch

FIGURE 23.10 ■ Roof truss span table. (*Alpine Engineered Products, Inc.*)

ROOF CONFIGURATION DIAGRAMS

Structural CAD technicians involved in a job using a wooden roof are required to draw roof configuration diagrams. These are plan views of the completed roof showing the configurations of the roof. Roof diagrams may be drawn to scale or simply drawn out in the proper proportions but not to scale.

Roof configuration diagrams are extremely helpful to the carpenter who must build the roof. The diagrams give a completed picture of what the carpenter is trying to construct. Figure 23.11 is an example of a roof configuration diagram for a hip roof showing the proper proportions for each roof component. Examples of roof configuration diagrams for several other common types of sloped roofs are shown in Figure 23.12.

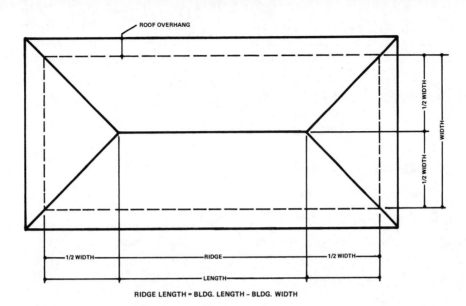

FIGURE 23.11 ■ Hip roof configuration diagram.

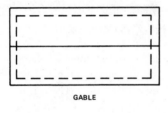

GABLE

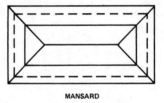

MANSARD

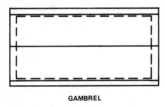

GAMBREL

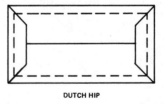

DUTCH HIP

FIGURE 23.12 ■ Roof configuration diagrams.

SUMMARY

■ Structural wood roofs are divided into two broad classifications: flat roofs and sloped roofs.

■ Flat roofs are those that range from actually flat to a very gradual slope.

■ Sloped roofs range from medium slopes to very steep slopes.

■ Flat roofs are built up, while sloped roofs are usually shingled.

■ The most popular sloped roofs are the gable, hip, gambrel, dutch hip, mansard, and A frame.

■ The slope of a roof is the amount of rise measured in inches that occurs for every 12″ of run.

■ The pitch of a roof is arrived at by dividing the rise by the span.

■ The four most common eave configurations are the wide box, the open eave, the wide box without lookouts, and the narrow box.

■ Conventional or stick framing involves joists bearing on the double top plate and rafters spanning from double top plate to the ridge board.

■ The four most common types of roof trusses are the common, the scissors, the mono pitch, and the flat.

■ Roof configuration diagrams provide a completed picture to guide the carpenter in constructing a wooden roof.

REVIEW QUESTIONS

1 What are the two broad classifications of wooden roofs?

2. List the six most popular types of sloped roofs.

3. Sketch a plan view of each of the six roof types listed in Question 2.

4. Sketch an example that shows how a 3:12 roof slope would be indicated on a drawing.

5. Calculate the roof pitches, given the following data:
   Roof A—3′ rise with 24′ span
   Roof B—6′ rise with 24′ span

6. Sketch an example of the four most common eave configurations.

7. List the four most common types of roof trusses.

8. Sketch an example of each of the four types of roof trusses listed in Question 7.

9. Sketch an example of a roof configuration diagram for a gable roof.

10. Sketch an example of a roof configuration diagram for a dutch hip roof.

CAD ACTIVITIES

GENERAL INSTRUCTIONS

The following activities may be completed on any CAD system. Before reading the *specific instructions* for each activity (below), go through each step in the following planning checklist. The checklist applies to any CAD system and will help ensure the optimum use of your time and resources.

1. Analyze the problem carefully. Decide exactly what you are being asked to do.

2. Determine what resources and references you will need in order to complete the problem and collect them.

3. Decide if any particular standards apply to the project and have those standards available.

4. Determine what types of views will be required and how many of each.

5. Determine what the final plotted scale of the drawing will need to be and select the appropriate paper size for plotting/printing (make sure the appropriate paper size is available).

6. Plan your drawing sequence. In what order will you develop the drawing (i.e., lines, features, dimension lines, leaders, dimensions, notes, etc.)?

7. Review the various CAD commands you will have to use in order to develop the drawing.

8. Examine your CAD system to ensure that everything is in working order, then begin the project.

SPECIFIC INSTRUCTIONS

Activity 23.1—In the upper left-hand corner of your sheet draw a single-line representation of a plan view of a building that is 24′-0″ long × 12′-0″ wide. Use a 1/4″ = 1′-0″ scale.
   a. Select the proper ceiling joist size and spacing from the span tables in Figure 23.10, and enter them on the plan view.
   b. Assume a gable roof with 2 × 4 rafters, a 2′ overhang, and a wide-box eave configuration. At a scale of 1″ = 1′-0″, draw an eave configuration detail.
   c. Use a 2 × 6 ridge board, and, at a scale of 1″ = 1′-0″, draw a ridge detail.
   d. Place a roof configuration diagram drawn not to scale immediately over the title block.

Activity 23.2—Repeat the instructions in Activity 23.1 for a building that is 36′-0″ long × 15′-0″ wide, has a 1′ overhang, a narrow-box eave configuration, and a hip roof.

Activity 23.3—Repeat the instructions in Activity 23.1 for a building that is 36′-0″ long × 18′-0″ wide, has a 1′ overhang, an open eave configuration, and a dutch hip roof.

Activity 23.4—Repeat the instructions in Activity 23.1 for a building that is 50′-0″ long × 25′-0″ wide, has a 2′ overhang, a wide-box eave configuration without lookouts, a gable roof, and common roof trusses.

Activity 23.5—Repeat the instructions in Activity 23.1 for a building that is 48′-0″ long × 24′-0″ wide, has a 1′ overhang, an open eave configuration, a gable roof, and scissors roof trusses.

Structural Wood Posts, Beams, Girders, and Arches

OBJECTIVES

Upon completion of this unit, the student will be able to:

■ Sketch examples of post-and-beam construction details, structural wood laminated arches, and various types of laminated beams and girders.

■ Prepare post, beam, girder, and arch drawings including: framing plans, sections, and connection details.

POST-AND-BEAM CONSTRUCTION

Post-and-beam wood construction is a popular method used for constructing commercial and large residential buildings. It involves three structural members: posts, beams, and planks (Figure 24.1).

There are several advantages to post-and-beam construction. In addition to its aesthetic qualities, post-and-beam construction allows for large glass areas in walls and wide overhangs. This is due to the longer spanning capabilities of the beams as compared to conventional wooden joists.

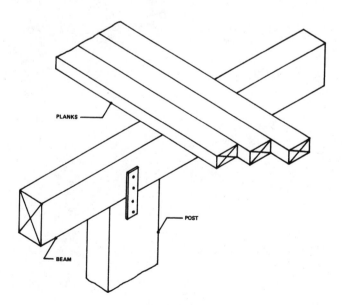

FIGURE 24.1 ■ Three components of post-and-beam construction. (*Goodheart-Wilcox Company, Inc.*)

Several types of beams are used in post-and-beam construction. The most popular have been developed because of the lack of availability of solid beams cut from large trees. These include laminated beams, steel reinforced beams, and box beams (Figure 24.2).

The most problematic area in post-and-beam construction involves connecting the posts to the footings and the beams to the posts. Several special metal connectors have been developed to solve the problem. Some of the most commonly used connectors and connection situations are illustrated in Figure 24.3.

LAMINATED ARCHES

Another popular construction method for building large commercial buildings of wood involves laminated arches. The term *laminated* means that the arch is composed of several layers of wooden planks glued together under intense pressure to form one larger, continuous structural member.

Laminated arches are a popular construction material for buildings requiring large, open areas such as churches, gymnasiums, and hangars. They are especially popular where the structural members used to span and support the building are to be exposed and must have a pleasing appearance.

Several types of laminated arches are used. Three of the most popular are the three-hinged arch, the tudor arch, and the A-frame arch (Figure 24.4). An example of one, a half section of an arch with dimensions relating it to the completed building, is shown in Figure 24.5. Notice that the roof slope can be varied to suit the needs of the individual job.

Laminated arches are attached to the footings in a manner similar to that used for connecting posts in post-and-beam construction. Figure 24.6 illustrates two commonly used connection methods.

LAMINATED BEAMS AND GIRDERS

Beams and girders may also be laminated in much the same way as arches. Figure 24.2 illustrates the two types of laminated beams and girders, horizontal laminated and vertical laminated. Connecting laminated beams poses the same problems encountered with posts, beams, and arches.

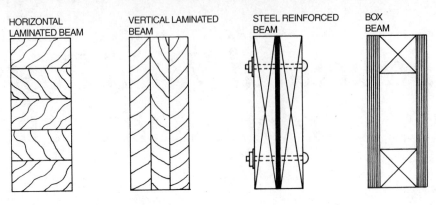

FIGURE 24.2 ■ Wooden beams. (*Goodheart-Wilcox Company, Inc.*)

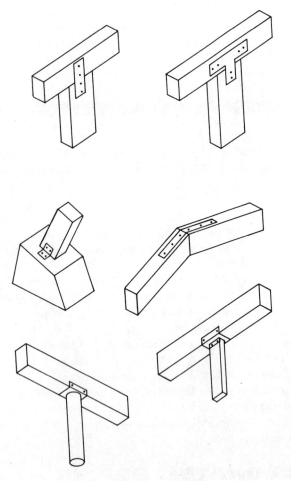

FIGURE 24.3 ■ Metal fasteners for post-and-beam connections (*Goodheart-Wilcox Company, Inc.*)

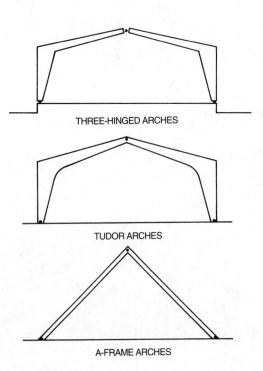

FIGURE 24.4 ■ Common wooden arches.

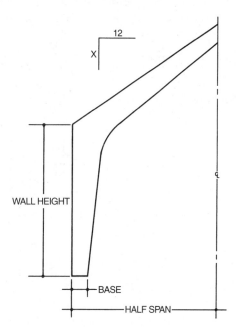

FIGURE 24.5 ▪ Half-arch detail.

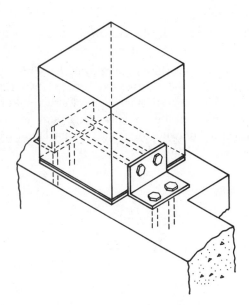

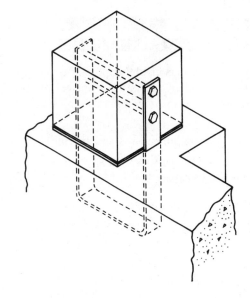

FIGURE 24.6 ▪ Base connections—posts and arches.

Special metal connectors, similar to those shown in Figure 24.3, have been developed for connecting laminated beams and girders to other structural components and for connecting other structural members to them. Two connection methods commonly used for connecting laminated beams and girders are shown in Figure 24.7.

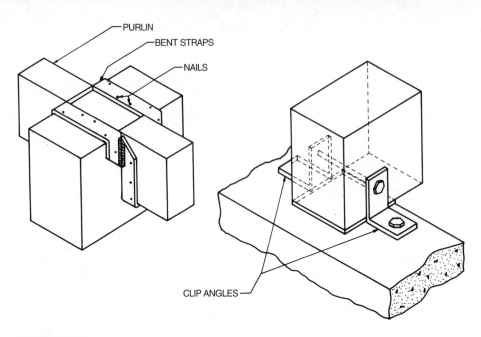

FIGURE 24.7 ■ Laminated beam connections.

POST, BEAM, GIRDER, AND ARCH DRAWINGS

Structural drawings for buildings involving posts, beams, girders, and arches made of wood are very similar to steel and pre-cast concrete drawings. The primary components of the drawings are:

■ A plan view showing overall dimension and spacings

■ Sections showing height information and how the structural members fit together

■ Connection details showing how the structural members are to be connected during erection

Figure 24.8 shows the drawing for a post, beam, and plank walkway to be constructed between two existing buildings. Note that as in all other types of structural drawings the connection details are drawn to a larger scale to emphasize detail. A structural drawing for a church using laminated tudor arches is shown in Figure 24.9.

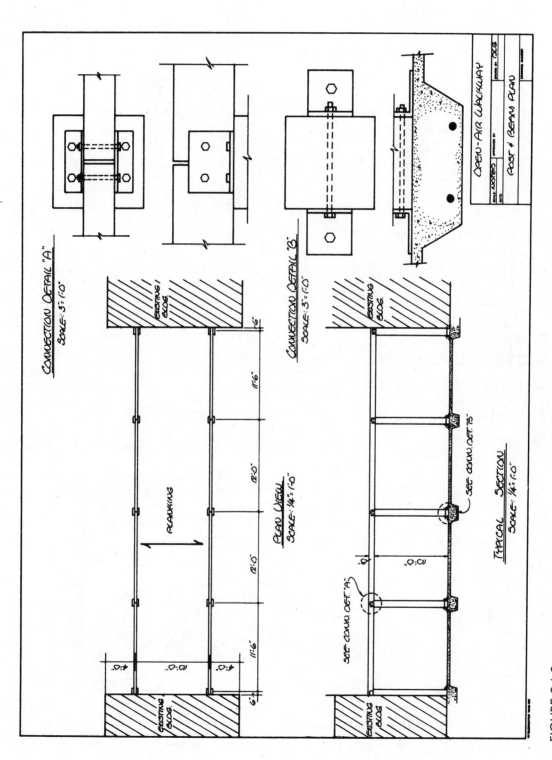

FIGURE 24.8 ■ Post-and-beam drawing.

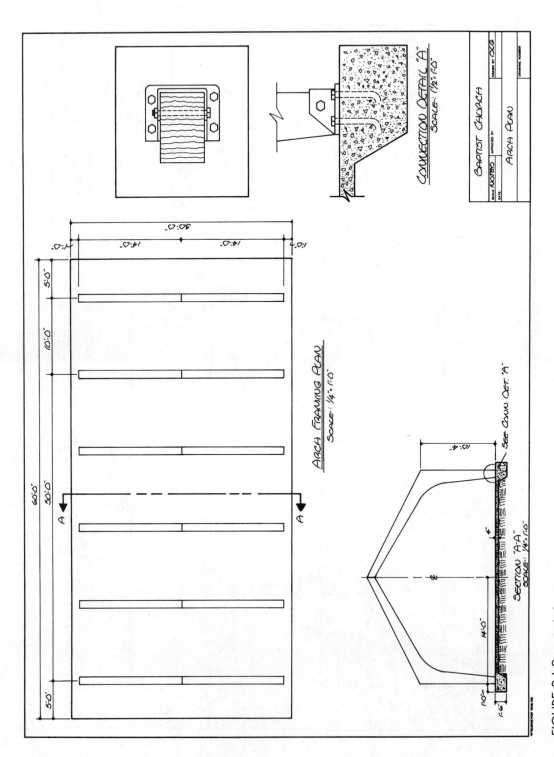

FIGURE 24.9 ■ Arch drawings.

SUMMARY

■ Post-and-beam construction involves three structural components: posts, beams, and planks.

■ Post-and-beam construction is advantageous in that it allows for large glass-wall areas and greater overhangs.

■ In order to circumvent the lack of solid wood structural beams, several types of built-up wooden beams are used: laminated beams, steel reinforced beams, and box beams.

■ The most problematic area in post, beam, girder, and arch construction is connections. These problems have been overcome through the development of special metal connectors for virtually every connection situation that has been identified.

■ Laminated arches are a popular structural building component in the construction of churches, gymnasiums, hangars, etc.

■ Three of the most popular types of laminated arches are: the three-hinged arch, the tudor arch, and the A-frame arch.

■ Laminated beams and girders are either horizontal or vertical laminated.

■ Structural drawings for post, beam, girder, and arch construction contain plan view(s), section(s), and connection details.

REVIEW QUESTIONS

1. Make a sketch and label it to illustrate the three basic components of post-and-beam construction.

2. List two advantages of post-and-beam construction over conventional wall and joist construction.

3. Make a sketch to illustrate four beam configurations that are used instead of solid wood construction.

4. How have problems in connecting wooden beams, girders, posts, and arches been solved?

5. List the three most common types of laminated arches.

6. Sketch an example of each of the three types of arches listed in Question 5.

7. Sketch an example that illustrates the difference between a vertical and a horizontal laminated beam.

8. List the primary components of structural wood drawings.

CAD ACTIVITIES

GENERAL INSTRUCTIONS

The following activities may be completed on any CAD system. Before reading the *specific instructions* for each activity (below), go through each step in the following planning checklist. The checklist applies to any CAD system and will help ensure the optimum use of your time and resources.

1. Analyze the problem carefully. Decide exactly what you are being asked to do.

2. Determine what resources and references you will need in order to complete the problem and collect them.

3. Decide if any particular standards apply to the project and have those standards available.

4. Determine what types of views will be required and how many of each.

5. Determine what the final plotted scale of the drawing will need to be, and select the appropriate paper size for plotting/printing (make sure the appropriate paper size is available).

6. Plan your drawing sequence. In what order will you develop the drawing (i.e., lines, features, dimension lines, leaders, dimensions, notes, etc.)?

7. Review the various CAD commands you will have to use in order to develop the drawing.

8. Examine your CAD system to ensure that everything is in working order, then begin the project.

SPECIFIC INSTRUCTIONS

Activity 24.1—Redraw the isometric connection detail shown on the left in Figure 24.6 so that it is an actual connection detail showing a plan view with a section cut through it to provide a complete elevation view. The bottom of the arch resting on the footing should be 12″ square, the angles 4″ × 4″ × 1/4″ × 6″, and the anchor bolts 3/4" in diameter.

Activity 24.2—Repeat the instructions in Activity 24.1 for the isometric connection detail on the right in Figure 24.6. The U-shaped strap is 1/4″ thick, 2″ wide, and 1′-2″ long.

Activity 24.3—Repeat the instructions in Activity 24.1 for the isometric connection detail on the right in Figure 24.7. The beam is 4″ wide × 12″ deep and rests on a 12″ wide concrete wall. The angle is 4″ × 4″ × 1/4″ × 2″ and the anchor bolts are 3/4″ in diameter.

Activity 24.4—Redraw Figure 24.9 making the following changes: Use three-hinged arches and connect the arch to the footing in a manner similar to the isometric detail on the right in Figure 24.6.

Activity 24.5—Redraw Figure 24.9 making the following changes: Use A-frame arches and connect the arches to the footing in a manner similar to the isometric detail on the left in Figure 24.6. The arches are 6″ wide × 12″ deep and are 24′-0″ high at the crest.

SECTION VI

Civil Engineering Drafting and Piping

Property Maps and Plot Plans

OBJECTIVES

Upon completion of this unit, the student will be able to:

■ Explain the concepts of location, direction, and distance.

■ Calculate/plot azimuths.

■ Calculate/plot bearings.

■ Plot traverses.

■ Develop property maps/plot plans from survey notes showing bearings and distance.

■ Develop property maps/plot plans from metes and bounds descriptions.

■ Develop property maps/plot plans using the rectangular system of legal descriptions.

■ Draw plot plans containing all necessary and required information.

FUNDAMENTALS OF CIVIL ENGINEERING DRAFTING

Civil engineering drafting is the field that deals with plot plans, site preparation, mapping, legal descriptions of property, highways, and roads (Figures 25.1 and Figure 25.2). Civil drafting technicians work in civil engineering, surveying, and land development companies as well as in city, county, state, and federal agencies that deal with land, roads, and bridges. Civil drafting technicians also work in organizations that are using Geographic Information Systems (GIS) (Unit 27). The fundamentals of civil engineering drafting include:

■ Location, direction, and distance

■ Legal descriptions

■ Plot plans

■ Contour lines

■ Highway and road plans

■ Site preparation and earthwork

The first three areas are discussed in this unit, and the last three are discussed in Unit 26.

LOCATION, DIRECTION, AND DISTANCE

Whether developing plot plans, road plans, or maps, civil drafting technicians must understand the fundamentals of location, direction, and distance. Every parcel of land has a specific and unique location on Earth. In order to identify that location and describe it, the concepts of direction and distance are used together. The following concepts should be understood by civil drafting technicians:

■ Direction

■ Azimuths

■ Bearings

■ Plotting traverses

You may think of direction as driving west or turning left at the light; but, in civil engineering terms, direction is the angular relationship between two lines. For example, if due north or due south is considered the zero baseline, then any line radiating from a point will form an angle (Figure 25.3). In this figure, Line A makes a 30-degree angle with the due north or zero baseline. Line B makes a 60-degree angle. Line C makes an angle of 30 degrees with the due south or zero baseline. East and west may also be used as baselines.

Direction (Angular Measurement)

Angles are measured in degrees (°), minutes (′), and seconds (″). There are 360 degrees in a circle, 60 minutes in a degree, and 60 seconds in a minute. Angles are expressed in one of two ways: (1) as degrees, minutes, and seconds; and (2) as decimal degrees:

43° 15′ 12″   (degrees-minutes-seconds form)

68.17°   (decimal form)

It is common practice to use the decimal form for adding and subtracting angles and to use the degrees-minutes-seconds form on drawings. In any case, technicians must know how to:

■ Convert the degree-minutes-seconds form to the decimal form.

■ Convert the decimal form to the degrees-minutes-seconds form.

■ Add, subtract, multiply, and divide angles expressed in degrees, minutes, and seconds.

■ Add, subtract, multiply, and divide angles expressed in decimal form.

Converting Angles to Decimal Form. It is simple to perform mathematical operations on angles if they are expressed in decimal form. Angles express in degrees, minutes, and seconds are converted to decimal form as follows:

$$42° 23′ 12″ = 42.39°$$

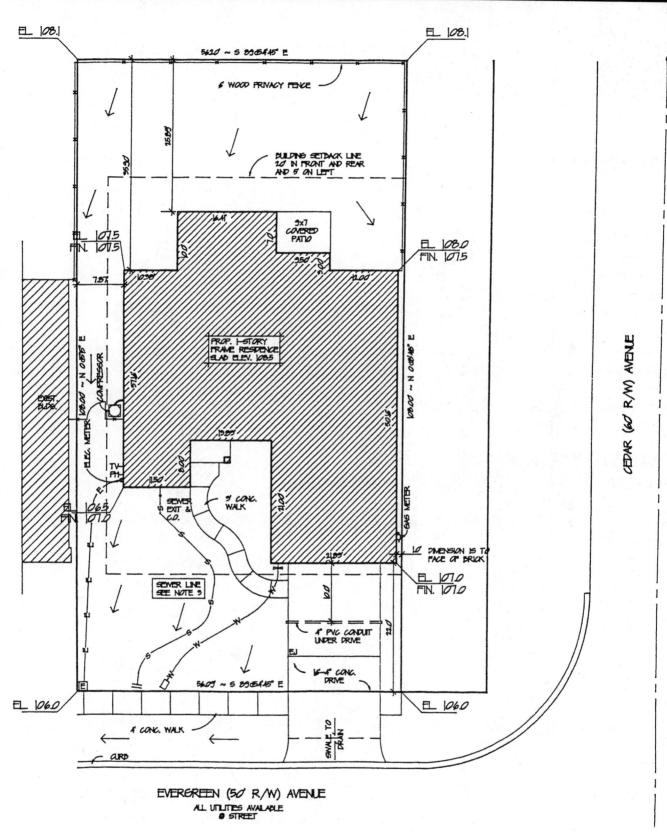

FIGURE 25.1 ■ Plot plan.

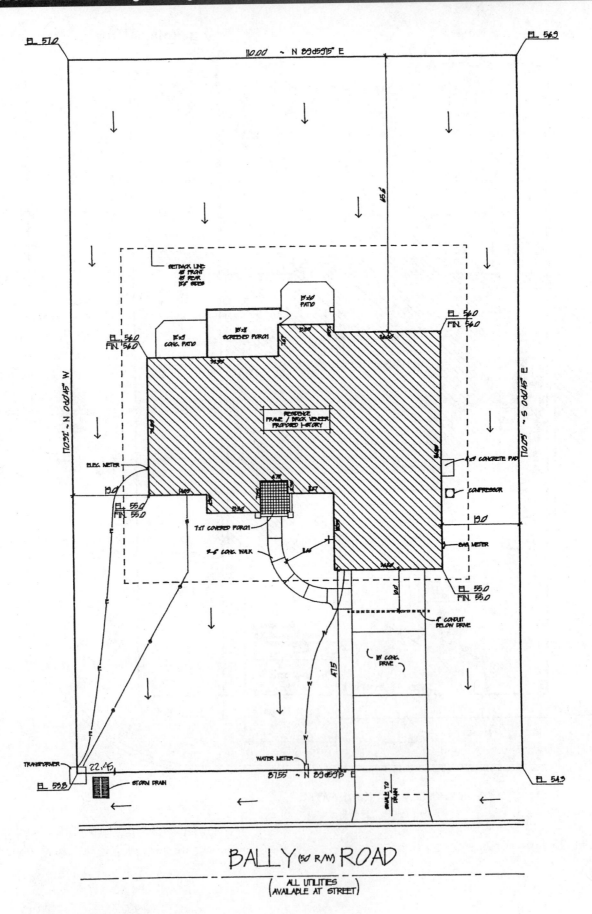

FIGURE 25.2 ■ Plot plan.

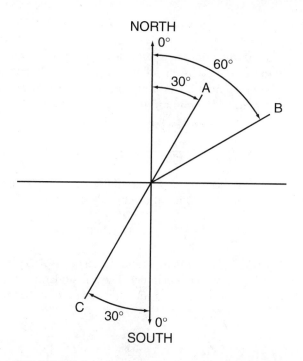

NORTH
0°

60°

30°

A

B

C

30°

0°

SOUTH

FIGURE 25.3 ■ Angles are turned from north or south.

STEP 1 Convert 12 seconds to decimal form by dividing by 60 (there are 60 seconds in a minute).

$$\frac{12''}{60} = 0.2'$$

STEP 2 Add the 0.2 minute to the 23 minutes and convert the sum to decimal form by dividing by 60 (there are 60 minutes in a degree).

$$\frac{23.2'}{60} = 0.3866667°$$

STEP 3 Round off the 0.3866667 degree and add it to the 42 degrees.

42.39°

Converting Decimal Angles to Degrees-Minutes-Seconds. It is common practice to state angles in the degree-minute-seconds format on drawings. Angles expressed in decimal form can be converted to degrees-minutes-seconds as follows:

29.62° = 20° 37′ 12″

STEP 1 Convert the 0.62 degree to minutes by multiplying by 60.

0.62 × 60 = 37.2′

STEP 2 Convert the 0.2 minute to seconds by multiplying by 60.

.02 × 60 = 12″

STEP 3 Rewrite the decimal form as degrees, minutes, and seconds.

29° 37′ 12″

Adding Degrees, Minutes, and Seconds. Degrees, minutes, and seconds can be added using normal addition procedures as follows:

16° 39′ 42″
+26° 44′ 59″
43° 24′ 41″

STEP 1 Begin at the extreme right by adding the seconds.

42″
+ 59″
101″

When, as in this case, the sum is more than 60 seconds, 1 minute (60 seconds) is carried over to the minutes column and the remaining seconds are entered under the seconds column.

42″
+ 59″
41″   (101″ – 60″)

STEP 2 Add the minutes column, including the 1 minute carried over from Step 1.

1′
39′
+ 44′
84′

When, as in this case, the sum is more than 60 minutes, 1 degree (60 minutes) is carried over to the degrees column and the remaining minutes are entered under the minutes column.

1′
39′
+ 44′
24′   (84′ – 60′)

STEP 3 Add the degrees column, including the 1 degree carried over from Step 2.

1°
16°
+ 26°
43°

STEP 4 Rewrite the final sum.

43° 24′ 41″

Subtracting Degrees, Minutes, and Seconds. Degrees, minutes, and seconds are subtracted using normal subtraction procedures as follows:

34° 12′ 14″
−18° 15′ 23″
15° 56′ 51″

STEP 1 Begin at the extreme right by subtracting the seconds. When, as in this case, the larger number is being subtracted, you must borrow. Borrow 1 minute (60 seconds) from the minutes column and perform the transaction.

$$
\begin{array}{r}
14'' + 60'' = 74'' \\
\underline{-23'' \qquad -23''} \\
51''
\end{array}
$$

STEP 2 Subtract the minutes column, remembering that the top number has already been reduced by 1 minute. When, as in this case, the larger number is being subtracted, you must borrow. Borrow 1 degree (60 minutes) from the degrees column and perform the transaction.

$$
\begin{array}{r}
11' + 60' = 71' \\
\underline{-15' \qquad -15'} \\
56'
\end{array}
$$

STEP 3 Remember that the top number in the degrees column has been reduced by 1. Perform the subtraction operation.

$$
\begin{array}{r}
33° \\
\underline{-18°} \\
15°
\end{array}
$$

STEP 4 Rewrite the final result.

$$15° \; 56' \; 51''$$

Multiplying and Dividing Degrees, Minutes, and Seconds. Multiplication is a shortcut for addition. Angles can be multiplied by applying normal multiplication procedures as follows:

$$
\begin{array}{r}
11° \; 42' \; 33'' \\
\underline{\times 3} \\
33° \; 126' \; 99'' \\
\underline{+1 \; -60} \\
33° \; 127' \; 39'' \\
\underline{+2 \; -120} \\
35° \quad 7' \; 39''
\end{array}
$$

STEP 1 Beginning at the extreme right, multiply the seconds, minutes, and degrees times 3 and record the results under the appropriate column. This procedure yields a result of:

$$33° \; 126' \; 99''$$

STEP 2 Because the seconds element is more than 60, it must be reduced. Take as many minutes (multiples of 60) out of the seconds column as possible by subtracting. In this case, you can subtract 1 minute (60 seconds) from the seconds column and add it to the minutes column.

$$
\begin{array}{r}
33° \; 126' \; 99'' \\
\underline{+1 \; -60} \\
33° \; 127' \; 39''
\end{array}
$$

STEP 3 Because the minutes element is more than 60, it must be reduced. Take as many degrees (multiples of 60) out of the minutes column as possible by subtracting. In this case, you can subtract 2 degrees (120 minutes) from the minutes column and add them to the degrees column to produce the final answer.

$$
\begin{array}{r}
33° \; 127' \; 39'' \\
\underline{+2 \; -120} \\
35° \quad 7' \; 39''
\end{array}
$$

Division is a shortcut for subtraction. Angles can be divided by applying normal division procedures as follows:

$$\frac{43° \; 15' \; 19''}{2} = 21° \; 37' \; 40''$$

STEP 1 Divide 43 degrees by 2, and the result is 21 degrees with a 1 degree remainder. The remainder is converted to minutes (60) and added to the minutes element.

$$
\begin{array}{r}
21° \\
2\overline{)43°} \\
\underline{42} \\
1 \quad (60 \text{ minutes}) + 15' = 75'
\end{array}
$$

STEP 2 Divide 75 minutes (15 minutes + 60 minutes remainder from Step 1) by 2, and the result is 36 minutes with a 1-minute remainder. The remainder is converted to seconds (60) and added to the seconds element.

$$
\begin{array}{r}
37' \\
2\overline{)75'} \\
\underline{74} \\
1 \quad (60 \text{ seconds}) + 19'' = 79''
\end{array}
$$

STEP 3 Divide 79 seconds (19 seconds + 60 seconds remainder from Step 2) by 2 and the result is 39.5 seconds. The final answer is:

$$21° \; 37' \; 40''$$

Azimuths

An **azimuth** is a direction measured as a horizontal angle from a zero baseline that runs north and south. The lines shown in Figure 25.3 are azimuths. A *true azimuth* is a horizontal angle measured from a *true north* baseline (a baseline that points to the location of the North Pole). True north is not the same as *magnetic north* or north as shown on a magnetic compass.

Grid azimuths are used with the rectangular surveying system wherein the north-south grid lines become zero baselines. *Magnetic azimuths* use magnetic north as the zero baseline. Magnetic azimuths and true azimuths can differ by several degrees because of a concept called *magnetic declination*.

Magnetic declination is the horizontal angle between true north and magnetic north (Figure 25.4). Magnetic declination

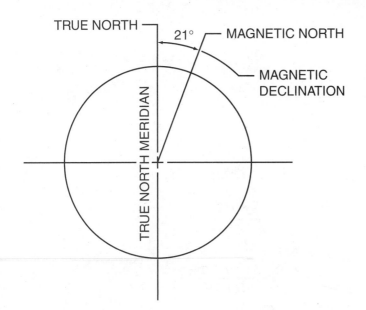

FIGURE 25.4 ■ Magnetic declination.

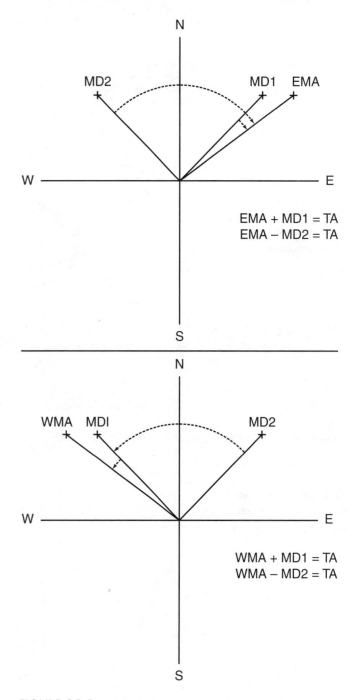

$$EMA + MD1 = TA$$
$$EMA - MD2 = TA$$

$$WMA + MD1 = TA$$
$$WMA - MD2 = TA$$

FIGURE 25.5 ■ Calculating a true azimuth.

can be toward the east as shown in this figure or toward the west. Magnetic declination is a continually changing concept depending on the magnetic conditions at work at any given time at a given location in the world. Consequently, it cannot be said that the magnetic declination is always a certain number of degrees. Rather, it can be said only that at this time in this location the magnetic declination is so many degrees.

A *local attraction* is any factor that can cause the magnetic declination to change. For example, large concentrations of steel, iron, or electricity (e.g., power lines, metal buildings, bridges, towers, or iron ore deposits) can affect the magnetic declination.

Civil drafting technicians must be able to calculate the true azimuth when a magnetic azimuth and the magnetic declination are known. When making this type of calculation, the following abbreviations apply:

Abbreviation	Term
MD	Magnetic declination
EMD	East magnetic declination
WMD	West magnetic declination
EMA	East magnetic azimuth
WMA	West magnetic azimuth
TA	True azimuth

A surveyor's notes will indicate whether an azimuth is true or magnetic. Figure 25.5 shows how to calculate the true azimuth when the magnetic declinations and magnetic azimuths are known.

Bearings

The *bearing* of a line is the direction it takes on a compass. A line's bearing is determined by the quadrant in which it is located and the angle it makes with the north-south baseline, (Fig-

ure 25.6). For example, the bearing in the northeast quadrant of the figure is N45 degrees 00 minutes 00 seconds E. It is actually as follows:

From the north meridian, turn 45 degrees toward the east.

The azimuths that correspond with the bearings in Figure 25.6 are shown below them. Notice that, whereas bearings relate directly to a specific quadrant, azimuths begin at the north meridian and sweep the entire 360 degrees of the compass.

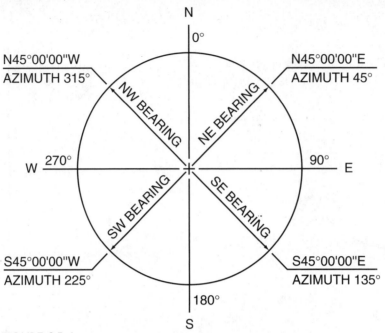

FIGURE 25.6 ■ Compass quadrants with bearings and azimuths.

Figure 25.7 illustrates how to convert an azimuth to a bearing and a bearing to an azimuth. Because Angle C is known (62 degrees 20 minutes 30 seconds), the bearing of the line is known. The angle formed by the south meridian and the line and the bearing are the same. However, in order to calculate the azimuth, one must know Angle B. By subtracting the known angle (Angle C) from 90 degrees (one quadrant), Angle B can be determined. It is 27 degrees 39 minutes 30 seconds. Now the

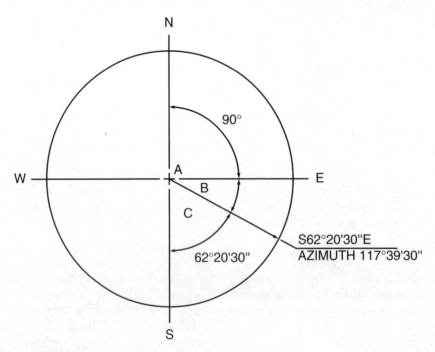

ANGLE A = 90°
ANGLE B = 90°–62°20'30" = 27°39'30"
ANGLE C = 62°20'30"
BEARING OF LINE = 62°20'30"
AZIMUTH = 117°39'30"

FIGURE 25.7 ■ Calculating bearings and azimuths.

azimuth can be calculated by adding Angle B and 90 degrees (27 degrees 39 minutes 30 seconds + 90 degrees 00 minutes 00 seconds = 117 degrees 39 minutes 30 seconds).

Calculating bearings and azimuths will be easier if you remember the following:

■ Bearings will be 90 degrees or less because they occur within one of the four quadrants. The bearing will always be the angle the line in question makes with either the north or the south meridian.

■ Azimuths are continuous angles turned from the zero baseline (usually the north meridian). Azimuths that occur in each quadrant represent the following angles:

NE quadrant	0° to 90°
SE quadrant	>90° to 180°
SW quadrant	>180° to 270°
NW quadrant	>270°

Plotting Traverses

The graphic description of a piece of property consists of lines with bearings and dimensions. Plotting a piece of property in civil drafting is similar to the geometric concept of plotting a traverse. A traverse is a polygon. As such, the following rules relating to polygons apply when plotting a traverse (Figure 25.8):

■ The angles inside of a four-sided polygon must total 360° if the polygon is to close.

■ The total of the angles inside of a polygon with more than four sides can be calculated using the following formula (n = number of angles):

$$n - 2 \times 180° = \text{total degrees}$$

In Figure 25.8, if all of the included angles in the polygon do not total 720 degrees, the traverse does not close. This typically

FOUR-SIDED POLYGON

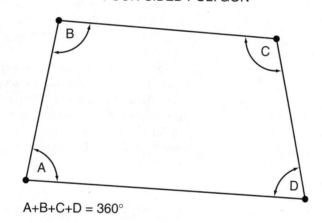

A+B+C+D = 360°

MORE-THAN-FOUR-SIDED POLYGON

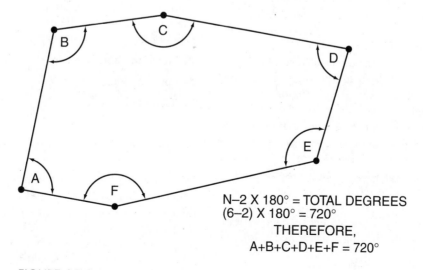

N–2 X 180° = TOTAL DEGREES
(6–2) X 180° = 720°
THEREFORE,
A+B+C+D+E+F = 720°

FIGURE 25.8 ■ Included angles in polygons.

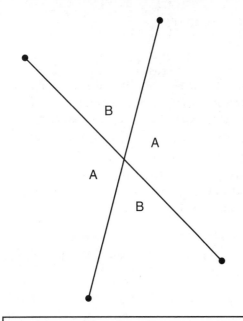

WHEN LINES INTERSECT, OPPOSITE ANGLES ARE EQUAL.

A = A
B = B

FIGURE 25.9 ■ Geometric rule for intersecting lines.

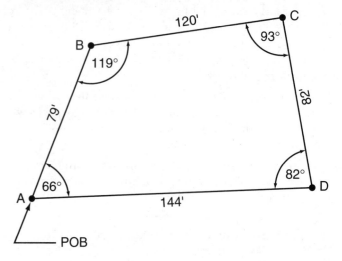

FIGURE 25.10 ■ Converting angles to bearings when plotting a traverse.

means that there is a problem with the survey. One or more of the angles is off. In such cases, the surveyor might have to go back and re-shoot the lines.

The rules of geometry relating to intersecting lines can be helpful when plotting a traverse (Figure 25.9). The pertinent rule is as follows:

When lines intersect, opposite angles are equal.

This rule can be helpful when it is necessary to calculate an unknown angle using other known angles.

Figure 25.10 shows a situation that civil drafters often face. This sketch shows a traverse consisting of lines AB, BC, CD, and DA—each of a specified length. It also shows the included angles of the traverse. Since there are four included angles and they total 360 degrees, we know the traverse closes. But we do not know the bearings of each line, and bearings are required on a plat plan. With the information given in Figure 25.10, the bearings can be calculated.

Calculate the bearing of one line at a time beginning at the point of the beginning (POB) and moving around the traverse in a clockwise manner. Construct north/south and east/west baselines placing Point A at their intersection as shown in Figure 25.11, and draw Line AB in relation to the baselines. The bearing of Line AB is north to east. The north/east quadrant contains 90 degrees. Line AB makes a 66-degree angle with the horizontal baseline. Hence, the bearing is calculated as follows:

$$90° - 66° = 24°$$

Therefore, the bearing of Line AB is N 24 E. If you will remember that each quadrant contains 90 degrees, the bearing of each line in the traverse can be calculated in a similar manner.

Once the bearing for each line is known, plotting the traverse is a simple drafting exercise. With CAD, the process is simplified even further. The coordinate geometry of COGO capabilities of CAD systems allow drafting technicians to simply enter the included angles and distances for the traverse in question. The CAD system performs the necessary calculations, computes the bearings, and plots the traverse.

LEGAL DESCRIPTIONS

Legal descriptions are word pictures that describe parcels of land. Every piece of property has a specific location in the world. A legal description identifies this location and describes the size and shape of the property. The three most widely used types of legal descriptions are:

■ Metes and bounds

■ Lot and block

■ Rectangular system

Metes and Bounds Descriptions

All **metes and bounds** descriptions start at a permanent monument or some other fixed and observable point of beginning. Metes are property lines that are measured in established units (e.g., feet, inches, meters, yards, rods, links). Bounds are property boundaries that are formed by the lines of adjacent properties, rivers, streams, roads, lakes, bays, and so on. Figure 25.12

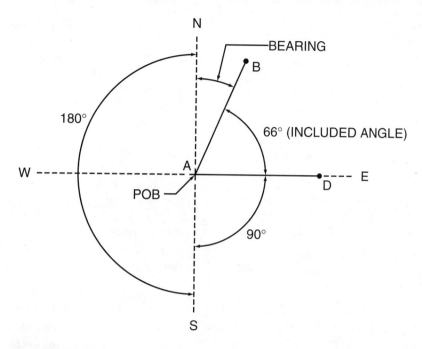

FIGURE 25.11 ■ Calculating the bearing of Line AB.

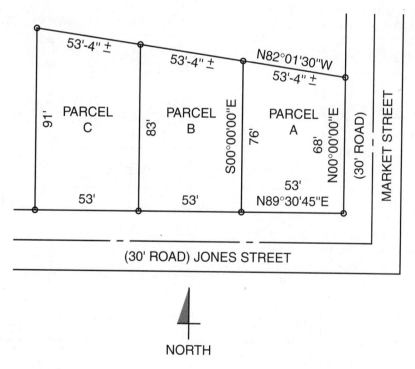

FIGURE 25.12 ■ Three parcels with metes and bounds descriptions.

contains three parcels of land with metes and bounds descriptions. The description for Parcel A reads as follows:

A parcel of land in Okaloosa County, Florida, beginning at a permanent survey marker located at the intersection of the north right-of-way of Jones Street and the west right-of-way of Market Street. From this point, go N 00° 00′ 00″ E 68′ to a metal stake. Then turn to a heading of S 00° 00′ 00″ E and go 76′ to a metal stake located on the north right-of-way of Jones Street. Then turn to a heading of N 89° 30′ 45″ E and go 53′ back to the point of beginning.

With the compass bearings and the distances known, a parcel of property can be drawn. Compass bearings are given in degrees, minutes, and seconds. The compass bearing N 82° 01′ 30″ W is read as follows:

From north, turn 82 degrees, 1 minute, and 30 seconds toward the west.

Compass bearings used in metes and bounds descriptions will always begin with either north or south and turn toward either the west or the east. For example, the bearing N 00° 00′ 00″ E is interpreted as follows:

From north, turn zero degrees, zero minutes, and zero seconds toward the east.

In other words, go due north. The compass bearing N 00° 0′ 00″ W is also due north. When going due north or due south, the east and west components can be used interchangeably.

Lot and Block Descriptions

Property that is subdivided, platted, and filed in the official records of the office of the county clerk may be described using **lot and block** numbers. **Subdivisions** are typically divided into blocks, and the blocks are, in turn, divided into lots (Figure 25.13). This figure shows Block A of a subdivision known as Elmwood Estates. The actual metes and bounds description for the entire subdivision, as well as a subdivision **plat** (property map), are on file in the official records of the county in question. Consequently, lots in the subdivision can be described using lot and block designations (e.g., Lot 3, Block A, and Elmwood Estates Subdivision). Anyone who needs a more definitive description of a parcel in the subdivision can refer to the subdivision plat in the county's official records.

Rectangular System

The U.S. Bureau of Land Management developed a method of describing land called the **rectangular system**. The system is in the following states:

- Alabama
- Alaska
- Arizona
- Arkansas
- California
- Colorado
- Florida
- Idaho
- Illinois
- Indiana
- Iowa
- Kansas

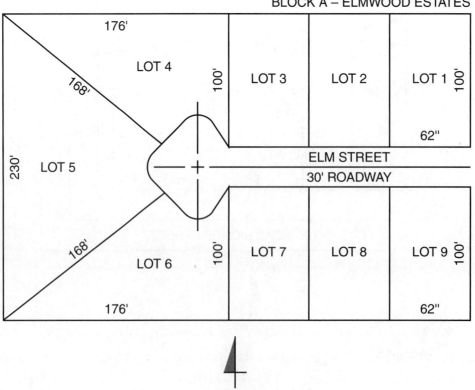

FIGURE 25.13 ■ Nine parcels with lot and block descriptions.

- Louisiana
- Michigan
- Minnesota
- Mississippi
- Missouri
- Montana
- Nebraska
- Nevada
- New Mexico
- North Dakota
- Ohio
- Oklahoma
- Oregon
- South Dakota
- Utah
- Washington
- Wisconsin
- Wyoming

Land in these states is divided into a grid of blocks called **townships** (Figure 25.14). Each major segment of land has both a township and a range number. For example, a designation might be *T24NR36E*. This designation reads as follows:

Township 24 north, Range 36 east

Township designations come from the rows that run east and west (Figure 25.14). Range designations come from the vertical columns that run north and south in this figure. The designations east and west relate to the line marked as the principal meridian. The designations north and south relate to the line marked as the baseline.

Subdividing Townships. A full-sized township is a tract of land 6 miles square. Exceptions to this rule are created by geographical features such as shorelines, natural borders, rivers, and large bodies of water. For example, many of the townships in Florida are irregularly shaped and less than full size. Regardless of their actual shape and size, townships are divided into **sections** as shown in Figure 25.14.

Subdividing Sections. Sections can be subdivided into an almost unlimited number of smaller units. For example, the section shown in Figure 25.15 contains several smaller subunits.

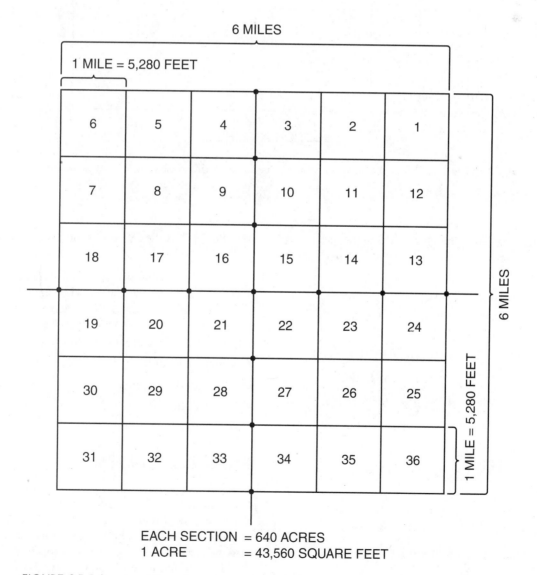

EACH SECTION = 640 ACRES
1 ACRE   = 43,560 SQUARE FEET

FIGURE 25.14 ■ One full-sized township.

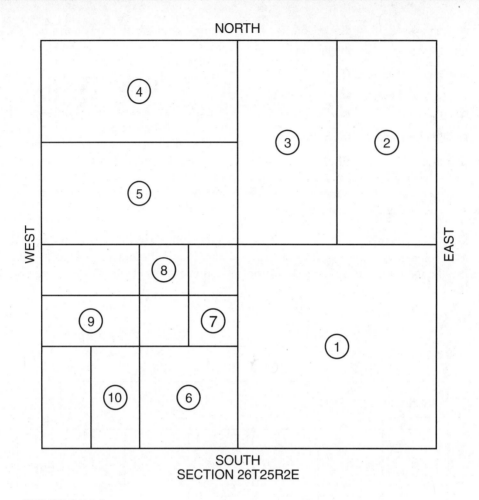

NORTH

WEST

EAST

SOUTH
SECTION 26T25R2E

FIGURE 25.15 ■ A section (640 acres) divided into ten parcels using the rectangular system.

Ten of these parcels have been numbered. Using the rectangular system, the legal descriptions of these numbered parcels are as follows:

1. The SE 1/4 of Section 26T25R2E (Reads as follows: The southeast one-quarter of Section 26 in Township 25, Range 2 east)

2. The E 1/2 of the NE 1/4 of Section 26T25R2E

3. The W 1/2 of the NE 1/4 of Section 26T25R2E

4. The N 1/2 of the NW 1/4 of Section 26T25R2E

5. The S 1/2 of the NW 1/4 of Section 26T25R2E

6. The SE 1/4 of the SW 1/4 of Section 26T25R2E

7. The SE 1/4 of the NE 1/4 of the SW 1/4 of Section 26T25R2E

8. The NW 1/4 of the NE 1/4 of the SW 1/4 of Section 26T25R2E

9. The S 1/2 of the NW 1/4 of the SW 1/4 of Section 26T25R2E

10. The E 1/2 of the SW 1/4 of the SW 1/4 of Section 26T25R2E

The key to reading this type of legal description is to begin at the end and read backward. This simplifies the process by beginning with the large and working toward the small. For example, look at Parcel 7 in this list of legal descriptions. Reading backward, the component parts are:

■ Section 26T25R2E

■ SW 1/4 of the section

■ NE 1/4 of that parcel

■ SE 1/4 of that parcel

PLOT PLANS

A set of plans for a commercial building or a residential dwelling must contain a **plot plan** (Figure 25.16). The plot plan locates the building and other key features on the piece of property in question. The requirements for what information must be contained on a plot plan vary from state to state and even from community to community. However, the following

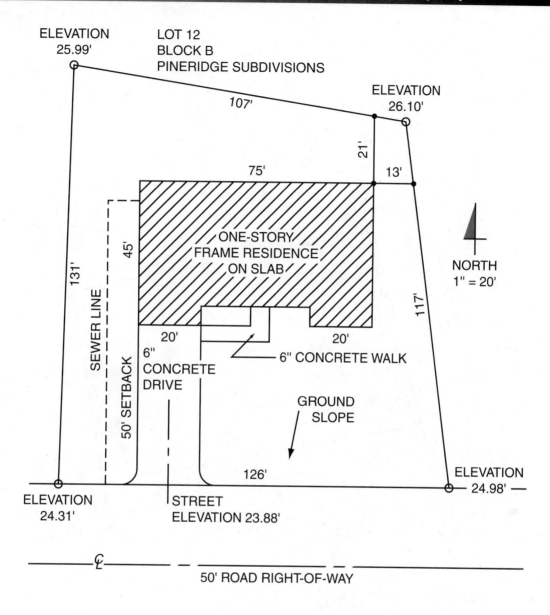

ELEVATION 25.99'

LOT 12
BLOCK B
PINERIDGE SUBDIVISIONS

107'

ELEVATION 26.10'

21'

75'

13'

131'

45'

ONE-STORY
FRAME RESIDENCE
ON SLAB

SEWER LINE

50' SETBACK

NORTH
1" = 20'

117'

20'

6"
CONCRETE
DRIVE

6" CONCRETE WALK

20'

GROUND
SLOPE

126'

ELEVATION 24.98'

ELEVATION 24.31'

STREET
ELEVATION 23.88'

50' ROAD RIGHT-OF-WAY

FIGURE 25.16 ■ Plot plan for a residential dwelling.

list contains the types of information that are typically required:

■ Legal description of the property (using lot and block, metes and bounds, or rectangular systems)

■ North arrow and drawing scale

■ Roads (existing, planned, and proposed)

■ Driveways, walkways, parking areas, patios, and decks

■ Current, planned, and proposed structures on the property

■ Locations (as applicable) of wells, water lines, gas lines, power lines, sewer lines, septic tanks, drain fields, leach

lines, soil test holes, rain drains, footing drains, and drive/walkway drains

■ Elevations of property corners and at the street (measured at the centerline of the driveway)

■ Ground slope arrow

■ Setback dimensions

■ Easements

In addition, land without buildings that is subdivided out of a larger piece must be surveyed. The survey is typically accompanied by a drawing such as the one in Figure 25.17.

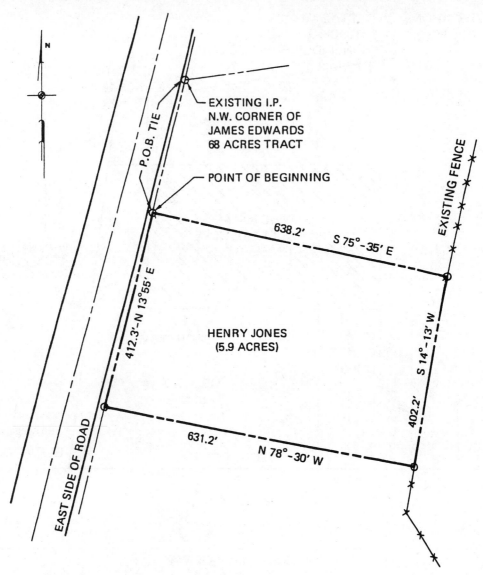

FIGURE 25.17 ■ Plot plan for a vacant piece of property.

SUMMARY

■ The fundamentals of civil engineering drafting include the following: location, direction, and distance; legal descriptions; plot plans; contour lines; highway and road plans; and site preparation and earthwork.

■ Every parcel of land has a specific location on Earth. In order to identify that location, we use direction, azimuths, bearings, and traverses.

■ Angles are measured in degrees, minutes, and seconds. There are 360 degrees in a circle, 60 minutes in a degree, and 60 seconds in a minute.

■ An azimuth is a direction measured as a horizontal angle from a zero baseline that runs north and south. A true azimuth is a horizontal angle measured from true north. True north is not the same as magnetic north. Magnetic azimuths use magnetic north as the zero baseline. Magnetic azimuths and true azimuths can differ by several degrees because of a concept known as magnetic

declination (the horizontal angle between true north and magnetic north).

■ The bearing of a line is the direction it takes on a compass. The bearing N 45 E reads as follows: From the north meridian, turn 45 degrees toward the east.

■ Plotting a piece of property in civil drafting is similar to plotting a traverse in geometry. A traverse is a polygon. Hence, the geometric rules relating to polygons apply to traverses.

■ Legal descriptions are word pictures that describe parcels of land. Every parcel has a specific location in the world. A legal description describes the size, shape, and location of the parcel in question. Three types of legal descriptions are widely used: metes and bounds, lot and block, and rectangular system.

■ A plot plan locates buildings and other features on a piece of property. Plot plans contain the following types of information: legal description of the property, north arrow, drawing scale, roads, driveways, walkways, parking areas, patios, decks, current and planned structures, wells, water lines, gas lines, power lines, sewer lines, septic tanks, drain fields, leach lines, soil test holes, rain drains, footing drains, drive/walkway drains, elevations of property corners, ground slope arrow, setback dimensions, and easements.

REVIEW QUESTIONS

1. What is an azimuth?

2. Distinguish between a true azimuth and a magnetic azimuth.

3. Explain the concept of magnetic declination.

4. What is the bearing of a line? Interpret the following bearing: S 32 W.

5. What is a legal description?

6. List and explain the three most widely used types of legal descriptions.

7. What is a plot plan, and what information should one contain?

CAD ACTIVITIES

GENERAL INSTRUCTIONS

The following activities may be completed on any CAD system. Before reading the *specific instructions* for each activity (below), go through each step in the following planning checklist. The checklist applies to any CAD system and will help ensure the optimum use of your time and resources.

1. Analyze the problem carefully. Decide exactly what you are being asked to do.

2. Determine what resources and references you will need in order to complete the problem and collect them.

3. Decide if any particular standards apply to the project and have those standards available.

4. Determine what types of views will be required and how many of each.

5. Determine what the final plotted scale of the drawing will need to be, and select the appropriate paper size for plotting/printing (make sure the appropriate paper size is available).

6. Plan your drawing sequence. In what order will you develop the drawing (i.e., lines, features, dimension lines, leaders, dimensions, notes, etc.)?

7. Review the various CAD commands you will have to use in order to develop the drawing.

8. Examine your CAD system to ensure that everything is in working order, then begin the project.

SPECIFIC INSTRUCTIONS

Activity 25.1—Convert the following angles to decimal form: 43° 22′ 13″ and 52° 14′ 24″.

Activity 25.2—Convert the following angles to degrees, minutes, and seconds: 32.53 and 75.69.

Activity 25.3—Figure 25.18 contains a drawing of Section 26T25R2E. Parcel 1 has been divided in half by a dotted line.

Using the rectangular system, develop a legal description for the south half of Parcel 1.

Activity 25.4—Figure 25.18 contains a drawing of Section 26T25R2E. Parcel 3 has been divided in half by a dotted line. Using the rectangular system, develop a legal description for the north half of Parcel 3.

Activity 25.5—Figure 25.18 contains a drawing of Section 26T25R2E. Parcel 5 has been divided in half by a dotted line. Using the rectangular system, develop a legal description for the west half of Parcel 5.

Activity 25.6—Figure 25.19 contains the outline of an irregularly shaped piece of property, Redraw the plot plan, and add a north arrow.

Activity 25.7—Figure 25.20 contains the outline of an irregularly shaped piece of property. Redraw the plot plan, add a north arrow, and label each corner with an elevation that would be realistic in your community.

Activity 25.8—Figure 25.21 contains the outline of an irregularly shaped piece of property. Redraw the plot plan, add a north arrow, add fictitious address information, and label each corner with an elevation that would be realistic in your community.

Activity 25.9—Figure 25.22 contains a partially completed plot plan for a vacant lot and a description of the property lines. Redraw that plot plan, and complete it by adding all required bearings to property lines. Add fictitious address information, and label each corner with an elevation that would be realistic in your community.

Activity 25.10—Figure 25.23 contains a partial plot plan for a vacant lot. It also contains the survey notes for the property lines. Use the drawing and the notes to calculate the bearings of each line. Then place a house with a driveway, a walkway, a deck, and a well on the parcel. Indicate that all utilities are available at the street. Write a metes and bounds legal description for the parcel, and write it out in the lower left-hand corner of your plot plan. Use Point A as your point of beginning (POB).

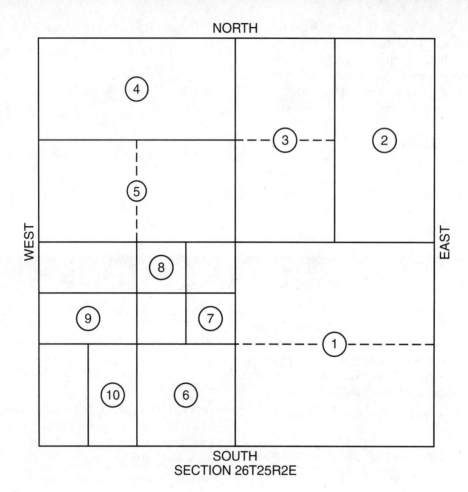

FIGURE 25.18 ■ CAD Activity 25.3 through Activity 25.5.

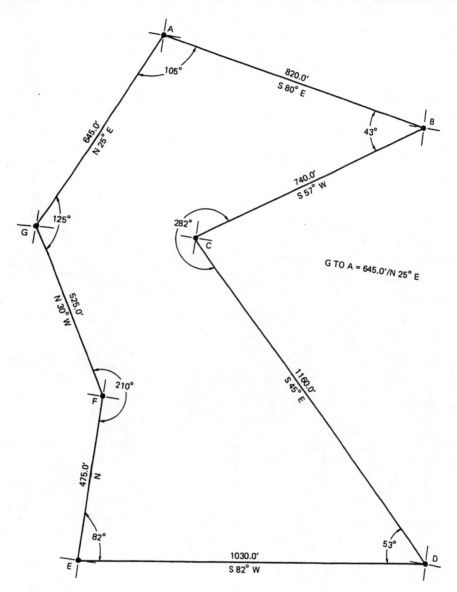

FIGURE 25.19 ■ CAD Activity 25.6.

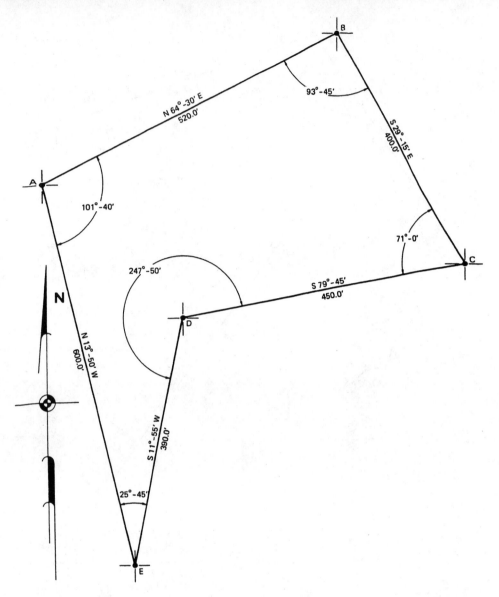

FIGURE 25.20 ■ CAD Activity 25.7.

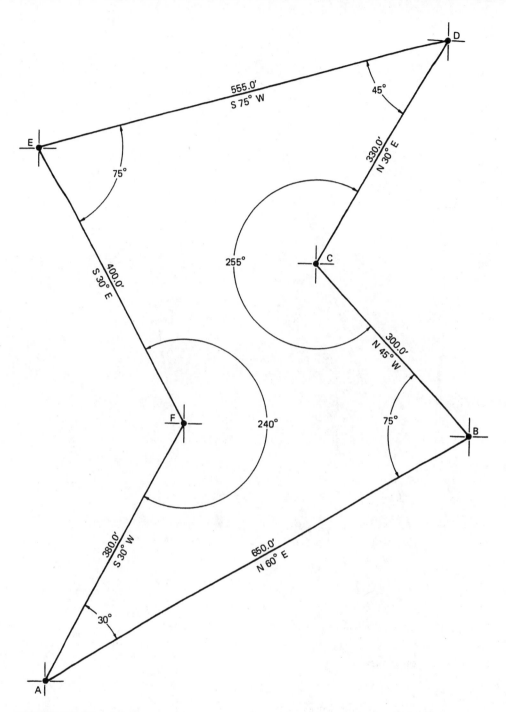

FIGURE 25.21 ▪ CAD Activity 25.8.

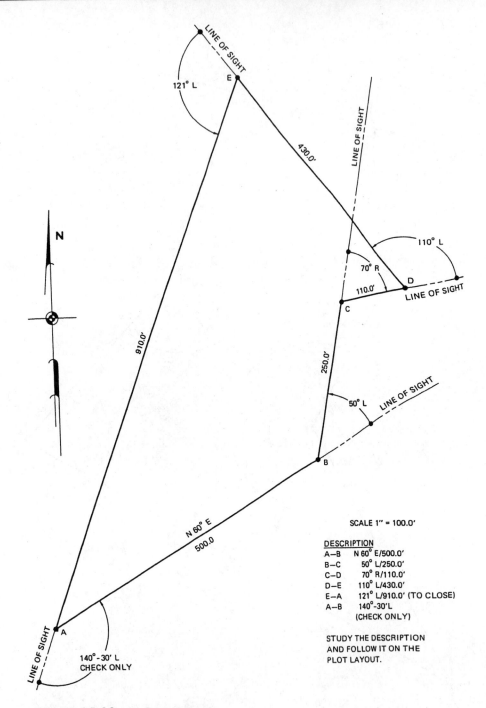

SCALE 1″ = 100.0′

<u>DESCRIPTION</u>
A—B   N 60° E/500.0′
B—C   50° L/250.0′
C—D   70° R/110.0′
D—E   110° L/430.0′
E—A   121° L/910.0′ (TO CLOSE)
A—B   140°-30′L
      (CHECK ONLY)

STUDY THE DESCRIPTION
AND FOLLOW IT ON THE
PLOT LAYOUT.

FIGURE 25.22 ■ CAD Activity 25.9.

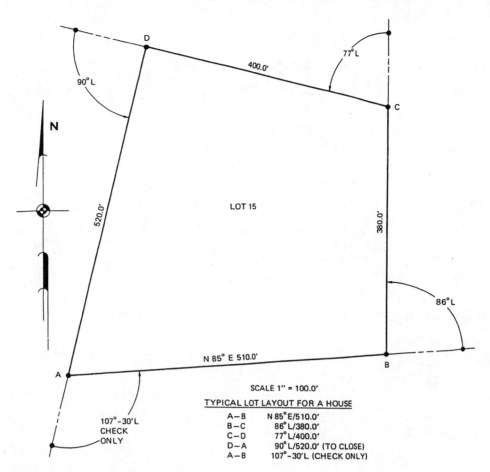

N

LOT 15

90° L

77° L

400.0'

D

C

520.0'

380.0'

86° L

N 85° E 510.0'

A

B

107°-30'L
CHECK
ONLY

SCALE 1″ = 100.0'

<u>TYPICAL LOT LAYOUT FOR A HOUSE</u>

A – B	N 85° E/510.0'
B – C	86° L/380.0'
C – D	77° L/400.0'
D – A	90° L/520.0' (TO CLOSE)
A – B	107°-30'L (CHECK ONLY)

FIGURE 25.23 ∎ CAD Activity 25.10.

26 UNIT

Contour Lines, Profiles, and Roadwork

OBJECTIVES

Upon completion of this unit, the student will be able to:

- Interpret contour lines.
- Plot contour lines from survey notes.
- Construct contour map profiles.
- Construct profile leveling drawings.
- Construct plan and profile/drawings.
- Construct highway layout drawings.
- Construct earthwork (cut-and-fill) drawings.

CONTOUR LINES

Contour lines are used on topographical maps to connect points of elevation. To a person who can interpret their meaning, contour lines tell whether the land in question is flat or hilly, gently sloped, concave, or convex.

In reading contour lines, a key characteristic to consider is *interval*. The greater the interval between contour lines, the more gradual the slope. A shorter interval indicates a steeper slope (Figure 26.1). Streams and ridges are special topographical features that can also be represented using contour lines. Figure 26.2 shows how contour lines appear at a stream. Notice that they form a "V" that points upstream. Figure 26.3 shows how a ridge can be indicated by contour lines.

Three different types of contour lines are shown in Figure 26.4. *Index contour lines* are thicker than the others and are spaced at intervals of five so that every fifth line is an index line. The index line is broken periodically and labeled with the elevation.

Intermediate contour lines are thin, unbroken lines between the index contour lines. *Supplementary contour lines* are dashed and typically represent half the distance between contour lines. Regardless of the relative slope of the land, every fifth contour line is an index line. Consequently, the contour interval value (i.e., 2, 5, 10 . . . feet) can vary.

Plotting Contour Lines from Survey Notes

The best way to accurately describe the shape of land is by ground survey. The survey team records spot elevations at known points. These records are called either *survey points* or *field notes*. These

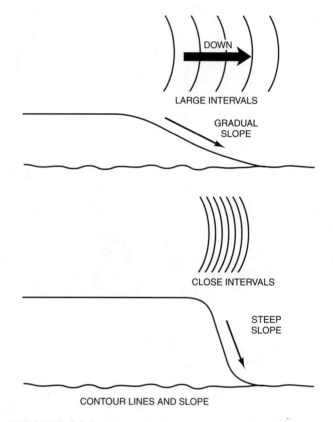

FIGURE 26.1 ■ Contour lines and slopes.

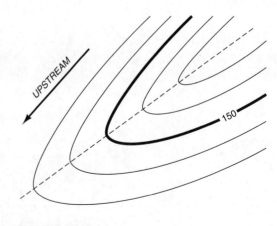

FIGURE 26.2 ■ Contour lines and streams.

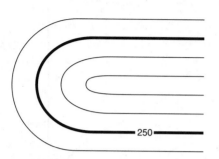

FIGURE 26.3 ■ Contour lines and ridges.

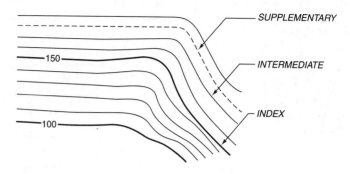

FIGURE 26.4 ■ Types of contour lines.

notes can be used as the basis for plotting a contour map. Figure 26.5 contains a page from a surveyor's field notes. Figure 26.6 is a contour map constructed from these notes.

By examining the various elevations plotted in Figure 26.5, the drafting technician can determine that the ground in question takes the shape of a hill with a summit of 156 feet. Based

on the differences in elevation of the points surveyed, a contour interval of 10 feet would be appropriate. With this decision made, the contour map can be developed.

Notice that most of the points surveyed fall between even intervals of 10 feet. Beginning at the summit of the hill (156 feet), draw the contour lines at even intervals approximating their

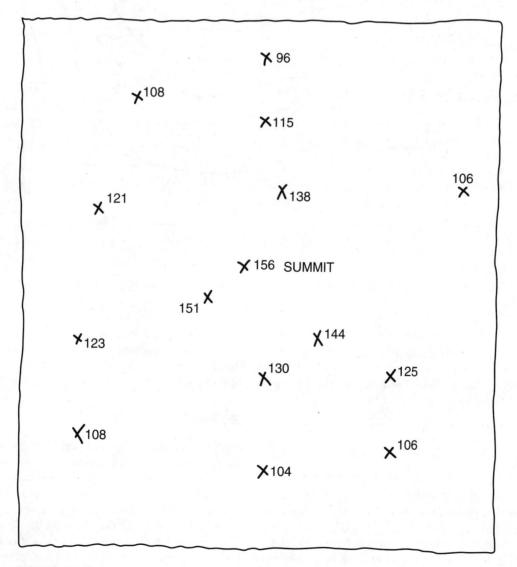

FIGURE 26.5 ■ Elevations recorded in a surveyor's field notes.

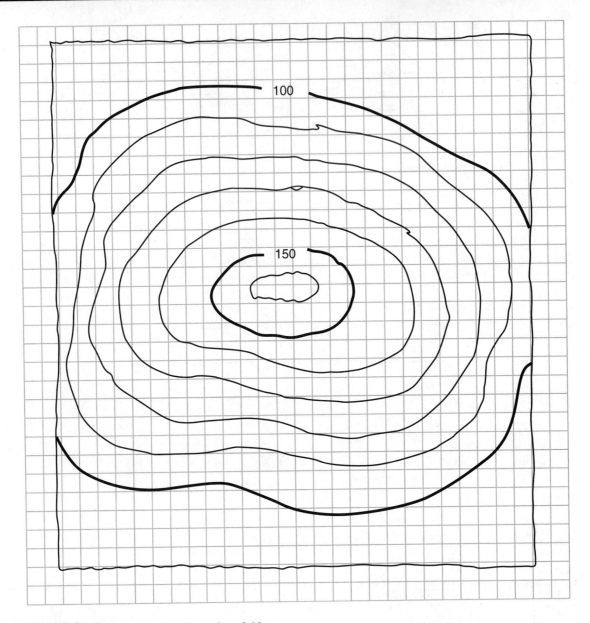

FIGURE 26.6 ■ Contour lines drawn from field notes.

distance from the actual survey points. For example, the point that is at the 104-foot elevation should fall slightly less than halfway between the contour lines drawn to represent 100 and 110 feet, respectively. This method of plotting contour maps is similar to the childhood game of connect the dots, except that the dots do not always connect exactly due to differences in elevations.

Grid Surveys

Surveyors sometimes use **grid surveys** for checking elevations at selected points on a parcel of property. First, the property is divided into a grid like the one in Figure 26.7. The grid interval is determined by the shape of the land. The flatter the land, the larger the grid squares. Stakes are driven into the ground at

each grid intersection. Then the surveyor checks the elevation at each stake, recording the elevation as shown in Figure 26.8. The drafting technician then uses the grid and the elevations from the field notes to plot the contour lines (Figure 26.9).

PROFILES

When a road, ditch, sewer line, gas line, or some other similar entity is to be constructed across a parcel of land, a contour map profile is developed. A profile is similar to a sectional view in structural drafting with the road, ditch, or gas line representing where the section is to be cut.

The first step in developing a profile drawing is to draw a line representing the centerline of the entity in question (road, ditch, sewer line, gas line, etc.) on the contour map. The con-

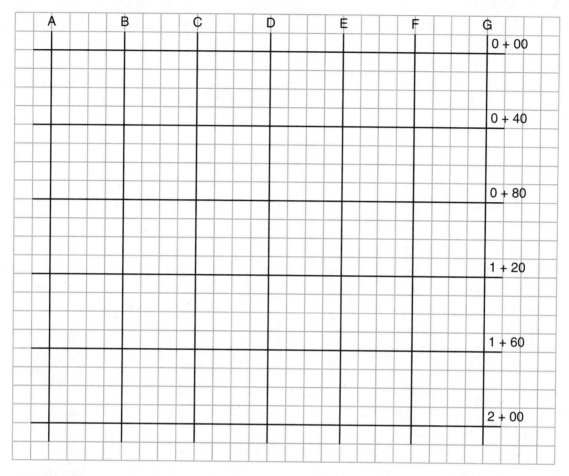

FIGURE 26.7 ■ Elevations are checked at each grid intersection.

tour map is then turned so that the line is horizontal, as shown in Figure 26.10. Line AB in this figure represents the centerline of a gas line to be constructed.

The next step in developing a profile drawing is to develop a vertical grid by projecting vertical lines down from each point where Line AB intersects a contour line to a horizontal baseline (Figure 26.11). The scale for the vertical grid does not necessarily have to match the scale used on the contour map. In fact, a different scale is often chosen to exaggerate the shape of the land in the profile. For example, in Figure 26.11, the contour map was drawn at a scale of 1″ = 100′ while the vertical grid for the profile was established at 1″ = 10′.

The elevation at each intersection of Line AB and the contour lines is plotted on the vertical grid as shown in Figure 26.11. These points are then connected to complete the profile.

Profiles of Curved Lines

The profile layout process illustrated in Figure 26.11 works for straight lines, but not all lines are straight. Roads, ditches, gas lines, sewer lines, and other entities often curve. Laying out a profile of a curved line requires an additional step in the process (Figure 26.12).

In this figure, Line AB represents the centerline of a road. The centerline is parallel to the horizontal from Point 1 through Point 7, then it curves. To develop a profile, the curved portion—Point 8 through Point 13—must be transferred to the horizontal (Line AB′). This is done by measuring the distance between contour lines along the curved portion of Line AB using dividers or an engineer's scale. These distances are plotted one at a time on the horizontal (Line AB′) using Point 7 as the beginning point for measurements.

With Point 8′ transferred to the horizontal (Line AB′), Point 9′ is measured from it and so on through Point 13′. Once Point 8′ through Point 13′ have all been transferred to the horizontal (Line AB′), the remaining steps are the same as those in the previous step.

Profile-Leveling Drawings

Profile-leveling drawings are used to display graphically the elevation characteristics of a given line or feature (road, railroad,

Station	Elevation	Station	Elevation
A 0 + 00	81	E 0 + 00	79
A 0 + 40	85	E 0 + 40	79
A 0 + 80	91	E 0 + 80	81
A 1 + 20	101	E 1 + 20	84
A 1 + 60	101	E 1 + 60	84
A 2 + 00	93	E 2 + 00	83
B 0 + 00	81	F 0 + 00	78
B 0 + 40	84	F 0 + 40	79
B 0 + 80	88	F 0 + 80	79
B 1 + 20	96	F 1 + 20	79
B 1 + 60	91	F 1 + 60	78
B 2 + 00	87	F 2 + 00	76
C 0 + 00	78	G 0 + 00	78
C 0 + 40	83	G 0 + 40	79
C 0 + 80	87	G 0 + 80	79
C 1 + 20	91	G 1 + 20	79
C 1 + 60	92	G 1 + 60	78
C 2 + 00	88	G 2 + 00	77
D 0 + 00	78.5		
D 0 + 40	81		
D 0 + 80	85		
D 1 + 20	88		
D 1 + 60	86		
D 2 + 00	86		

FIGURE 26.8 ■ Elevations recorded in a surveyor's field notes.

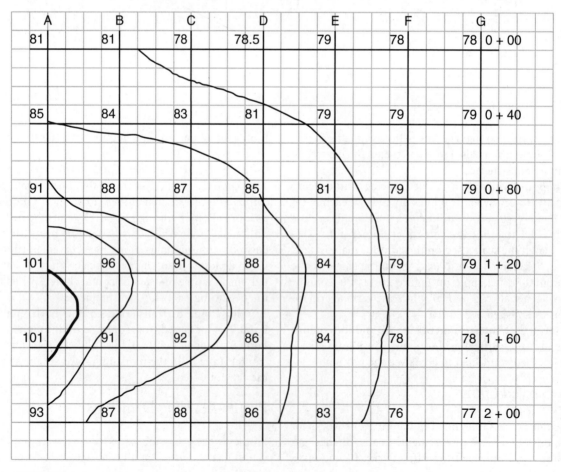

FIGURE 26.9 ■ Elevations recorded at each grid intersection.

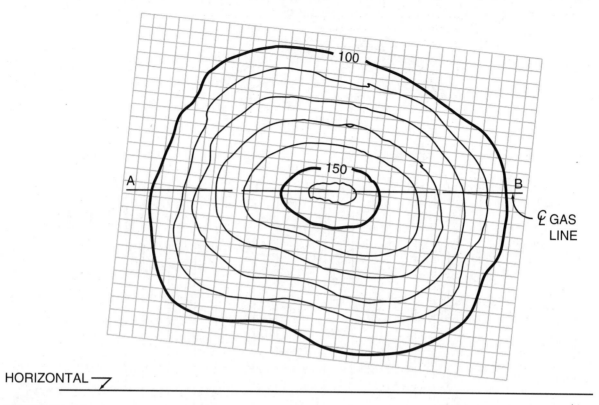

HORIZONTAL

FIGURE 26.10 ■ Line AB must be aligned parallel to the horizontal by rotating the contour map (if necessary).

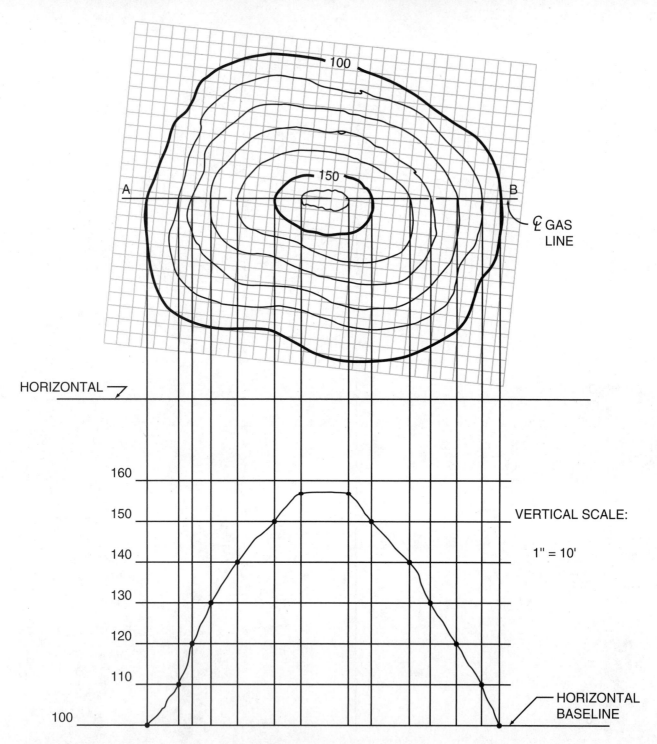

FIGURE 26.11 ■ Development of the profile of a straight line.

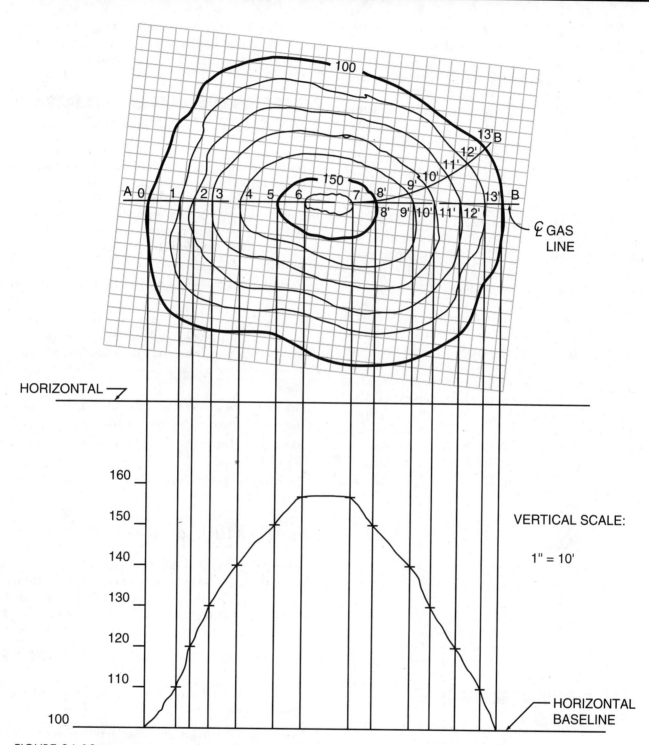

FIGURE 26.12 ■ Development of the profile of a curved line.

sewer line, gas line, etc.) along a given route. CAD technicians develop profile-leveling drawings from field notes provided by surveyors. The survey team marks off a specific portion of the linear feature in question and takes readings along the centerline of the feature at specific intervals (Figure 26.13).

The survey team places its elevation rod at a location where the elevation is known. In Figure 26.13, this point is Station 14 + 05.11 at the center of Washington Street. A backsite (BS) is shot to determine the height of the survey instrument. The elevation rod is then placed at specific intervals measured along the centerline of the linear feature, and elevations are recorded. These shots are called *intermediate foresight readings (IFS)*. It is common practice to set stakes for readings at 50 or 100-foot intervals. In Figure 26.13, the interval is 50 feet. The stations set at 100-foot intervals are called *full stations* and read as follows: 14 + 00, 15 + 00, 16 + 00 . . . . Intermediate readings taken between full stations are called *plus stations* and read as follows: 14 + 50, 15 + 50, 16 + 50 . . . .

In Figure 26.13, all of the points along the centerline of the linear entity (a road) could be shot from the original point where the survey instrument was set up. This will not always be possible. When IFS readings can no longer be shot from the original location, the survey instrument is moved. This is done by placing the survey instrument on a turning point (TP) and taking a foresight (FS) reading to re-establish the instrument height. This process is repeated as often as necessary to complete the readings for the portion of the route in question.

Note that the elevations taken at the stations designated in Figure 26.13 are as follows:

Station	Elevation
14 + 00	12.20
14 + 50	11.57
15 + 00	10.89
15 + 50	10.40
16 + 00	9.77
16 + 50	9.29
17 + 00	8.89
17 + 50	8.96
18 + 00	9.90
18 + 14.94	10.40

PLAN AND PROFILE DRAWINGS

Plan and profile drawings are like the plan and front elevation views used in structural drafting. Plan and profile drawings are just more detailed versions of the profile-leveling drawings explained in the previous section. Figure 26.14 through Figure 26.16 are examples of typical plan and profile drawings.

Notice that while profile-leveling drawings show only the profile of the centerline, plan and profile drawings show the profile of the road's thickness and other features.

HIGHWAY AND ROAD LAYOUT DRAWINGS

Laying out a highway drawing is a matter of converting field notes from a surveying team into lines on a property plot or a contour map. The one skill needed beyond those developed thus far is plotting curve data. If one can lay out the centerline of a road, one can lay out the road. Figure 26.17 illustrates the various items of *curve data* needed to plot curves in roads. The terms contained in this figure have the following meanings:

Point of curve (PC)—Point at which the straight line begins to curve.

Point of tangency (PT)—Point at which the curve ends. At this point, the line becomes straight again or curves again. If it curves again, the point is called the *point of reverse curve,* which is where one curve ends and a new one with its own curve data begins.

Radius of curve (R)—Radius of the curve. It must be measured along a line that is perpendicular at the point of curve.

Delta angle of curve (Δ)—Included angle of the curve between the PC and the PT.

Length of curve (L)—Length of the centerline between the PC and the PT.

Laying Out Highways/Roads

The straight portions of highways and roads are layed out by plotting the bearing and distance of the line up to the point of curve. At this point, there are two methods drafting technicians can use to lay out the curves. The first method involves using the curve data as just described. It proceeds as follows:

1. Draw a line perpendicular to the straight portion of the road line at the PC (Figure 26.17).

2. Measure the R distance along this line and mark the radius center point.

3. Turn the included angle using the PC and radius center point as points of reference.

4. Draw a line from the radius center point to inscribe the included angle (Δ). Then draw the curved line until it intersects this line at the PT.

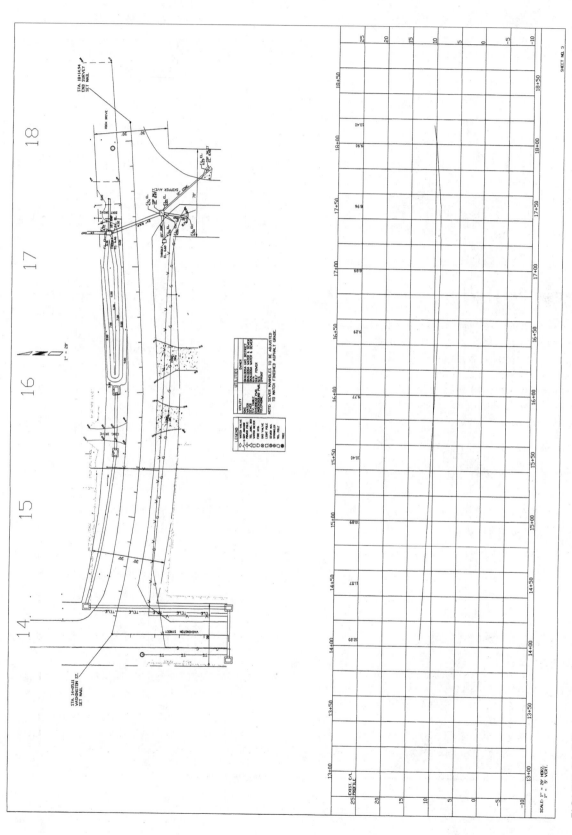

FIGURE 26.13 ■ Simple profile of a road.

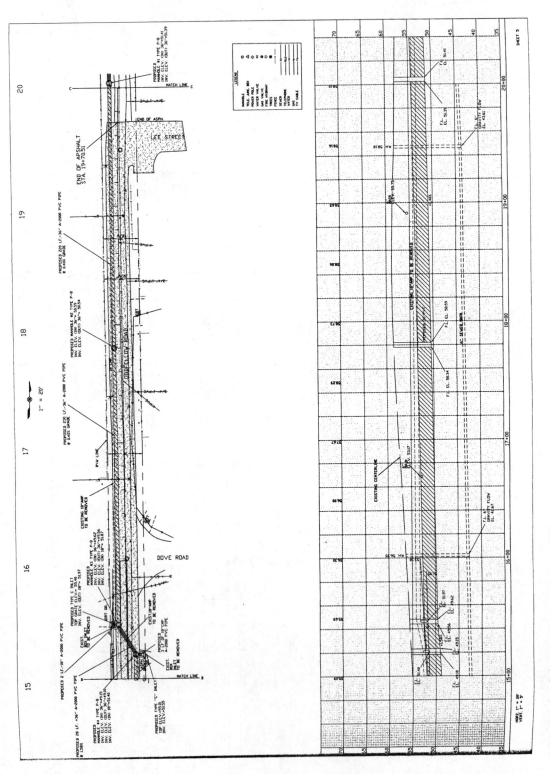

FIGURE 26.14 ■ Plan and profile for a road.

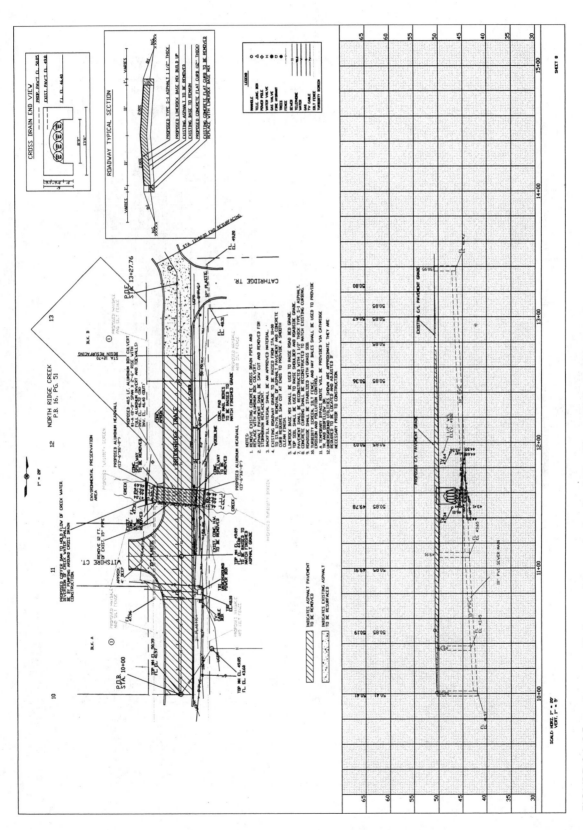

FIGURE 26.15 ■ Plan and profile for a road (notice the detail and other features in the profile).

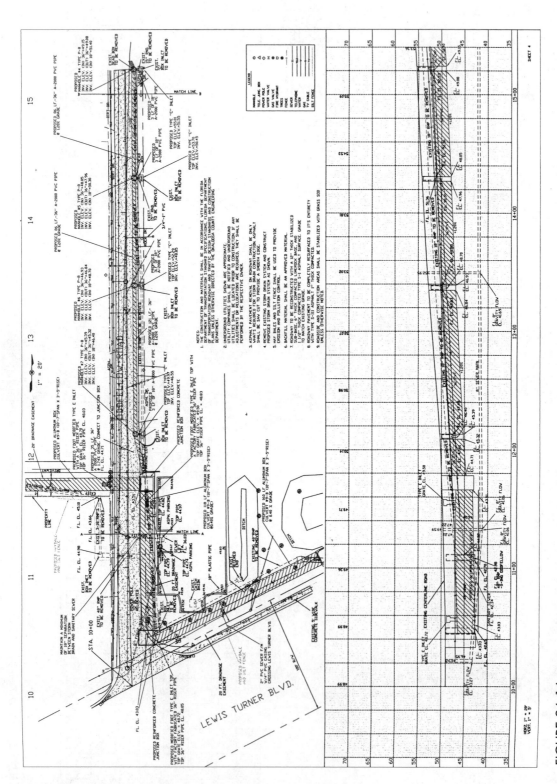

FIGURE 26.16 ■ A complex and detailed plan and profile drawing.

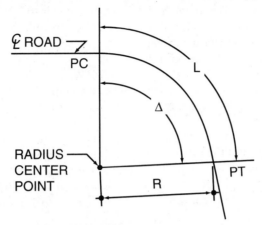

PC = POINT OF CURVE
PT = POINT OF TANGENCY
R = RADIUS OF CURVE
Δ = DELTA ANGLE OF CURVE
L = LENGTH OF CURVE

FIGURE 26.17 ■ Curve data for highway layouts.

The second method used for laying out highways and roads uses bearings and distances to lay out the straight portions of lines (as in the first method) and curve radii for creating the curves (Figure 26.18). This is known as the *points-of-intersection* method.

In Figure 26.18, the surveying team provided two bearings and distances for the centerline of a road (N 90°00′00″ E for 360′ and S 45°00′00″ E for 396′) and a suggested radius (R). The curve is created using the following steps:

1. Lay out the two straight lines using the bearings and distance provided by the surveyor.

2. Create a parallel line to each of the straight lines from Step 1 using the suggested radius (R) as the distance between the two lines.

3. Where the two dashed parallel lines in Figure 26.18 intersect is the radius point for drawing the curve.

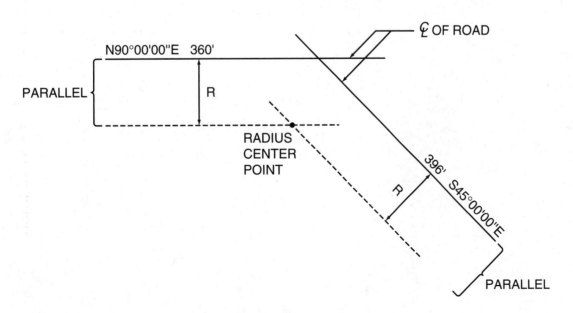

FIGURE 26.18 ■ Points-of-intersection method of laying out curves.

Using bearings, distances, and one of the two curve methods just described, drafting technicians can lay out roads from the simple to the complex. Look closely at Elderberry Lane in Figure 26.19, and you can see the curve data used to lay out this road.

EARTHWORK DRAWINGS

If the earth were perfectly flat, the construction of roads and highways would be simplified significantly. But the earth is not flat, even in areas that have no mountains or valleys. Consequently, any time a highway is to be built, high spots of earth must be cut away and low spots must be filled in. This process is called **cut and fill.**

CAD software has simplified the process of drawing cut-and-fill cross sections so much that the process is almost automatic. However, CAD technicians need to understand what the CAD software does for them and how.

Cut-and-Fill Cross Sections

Cut-and-fill drawings help engineers determine how much earth must be moved and how much land must be purchased when a road is to be built. Cut-and-fill cross sections are developed by applying the following steps:

1. Lay out the road in question on a contour map (Figure 26.20).

2. Develop a profile grid off the end of the road and perpendicular to it (Figure 26.21). The *angle of repose* is provided by the surveyor or engineer for the project. The distance between contour lines is 10 feet in this case and may be drawn at a convenient scale on the profile grid. The angle of repose in this case is 2 to 1, meaning a slope defined by two horizontal units of distance for every one unit of vertical distance.

3. Projectors from each grid point (110, 120, . . . 160 and 60, 70 . . . 100) are drawn parallel to the centerline of the road. Where these projectors intersect, their corresponding contour lines are points that define the outline of the areas to be cut and filled (see the shaded areas in Figure 26.21).

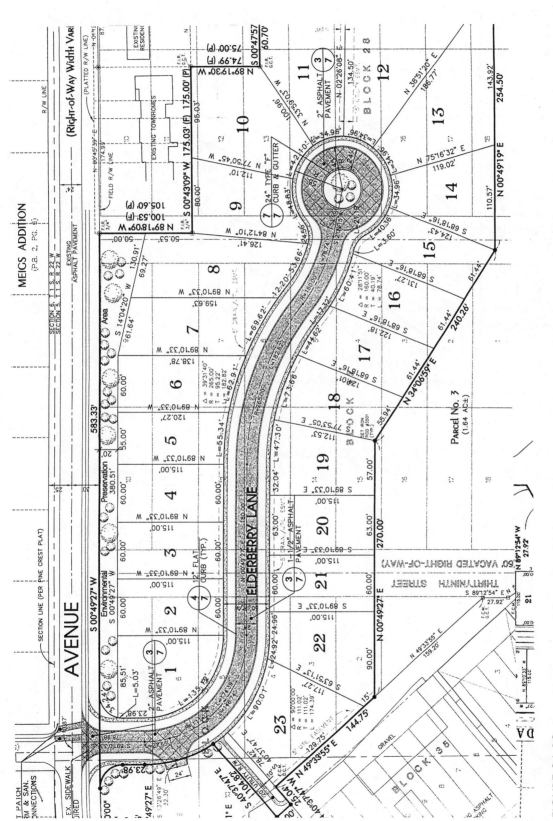

FIGURE 26.19 ■ Road leading to a cul-de-sac in a subdivision.

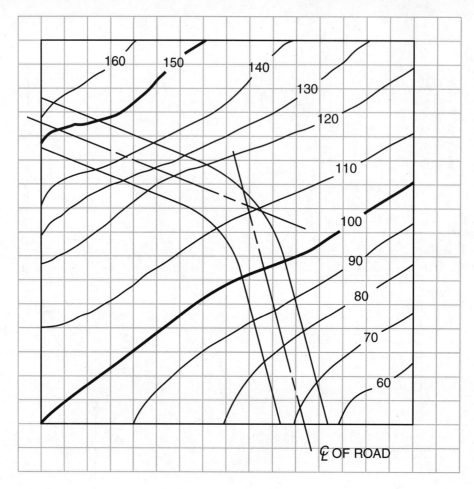

FIGURE 26.20 ■ Plotting the road on a contour map.

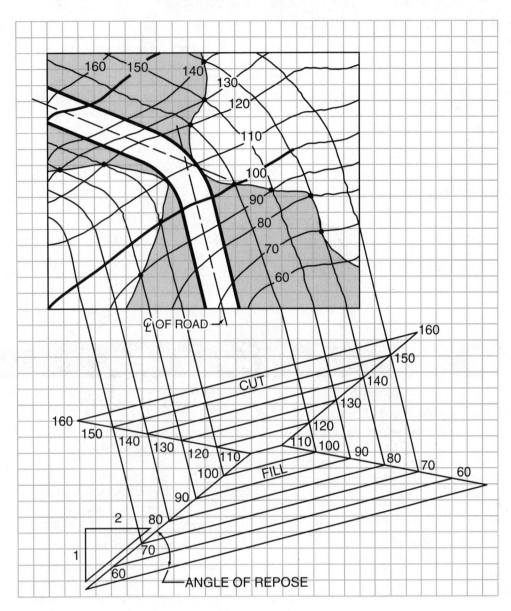

FIGURE 26.21 ■ Plotting the cut-and-fill area on a contour map.

SUMMARY

■ Contour lines are used on topographical maps to connect points of elevation. The greater the interval between contour lines, the more gradual the slope. Index contour lines are thicker than other contour lines and are spaced at intervals of five so that every fifth line is an index line. Intermediate contour lines are thin, unbroken lines between the index contour lines.

■ Contour lines are plotted by CAD technicians working from survey notes (also called field notes). Grid surveys involve establishing a grid over a selected parcel of land and checking the elevation at each grid intersection.

■ A contour map profile is a cross-sectional view cut through the earth to reveal the profile of the ground in a given area.

Contour profiles are developed by establishing parallelism between the centerline of the feature that is being cut through the earth (road, gas line, etc.) and a horizontal line.

■ Profile-leveling drawings are used to display graphically the elevation characteristics of a given linear feature (road, railroad, sewer line, gas line, etc.). Such drawing are developed by CAD technicians from field notes provided by a surveying team.

■ Plan and profile drawings are like the plan and elevation views used in structural drafting. They are just more detailed versions of profile-leveling drawings.

■ In order to lay out roads and highways, CAD technicians need to know how to plot curves from curve data. Key concepts in plotting curves are: point of curve (PC), point of tangency (PT), radius of curve (R), delta angle of curve (Δ), and length of curve (L).

■ Cut-and-fill drawings help engineers determine how much earth must be moved when constructing a road or highway and how much right-of-way must be purchased. When constructing a road through hilly terrain, high spots must be cut away and low spots must be filled. Cut-and-fill drawings show these areas graphically.

REVIEW QUESTIONS

1. What is the purpose of contour lines?

2. What are the different types of contour lines?

3. When plotting contour lines, where do CAD technicians get their information?

4. What is a grid survey and how is it used?

5. What is a contour map profile and how is one used?

6. How are profile-leveling drawings used?

7. What is a plan and profile drawing? How is one different than a profile-leveling drawing?

8. Define the following curve data terms:
   • Point of curve
   • Point of tangency
   • Radius of curve
   • Delta angle of curve
   • Length of curve

9. What is a cut-and-fill drawing and how is one used?

CAD ACTIVITIES

GENERAL INSTRUCTIONS

The following activities may be completed on any CAD system. Before reading the *specific instructions* for each activity (below), go through each step in the following planning checklist. The checklist applies to any CAD system and will help ensure the optimum use of your time and resources.

1. Analyze the problem carefully. Decide exactly what you are being asked to do.

2. Determine what resources and references you will need in order to complete the problem and collect them.

3. Decide if any particular standards apply to the project and have those standards available.

4. Determine what types of views will be required and how many of each.

5. Determine what the final plotted scale of the drawing will need to be, and select the appropriate paper size for plotting/printing (make sure the appropriate paper size is available).

6. Plan your drawing sequence. In what order will you develop the drawing (i.e., lines, features, dimension lines, leaders, dimensions, notes, etc.)?

7. Review the various CAD commands you will have to use in order to develop the drawing.

8. Examine your CAD system to ensure that everything is in working order, then begin the project.

SPECIFIC INSTRUCTIONS

Activity 26.1—Using the surveyor's field notes shown in Figure 26.22, plot the contour lines. Use a contour interval of 10 feet. Select an appropriate scale.

Activity 26.2—Using the surveyor's grid and notes from Figure 26.23, plot the contour lines using intervals of 10 feet.

Activity 26.3—Reconstruct the contour map in Figure 26.24 with centerline AB passing through the center of the map horizontally. Develop a contour map profile for this parcel of land.

Activity 26.4—Using the field notes in Figure 26.25, lay out the centerline for the road.

Activity 26.5—Using the *points-of-intersection* method and the suggested radius of 300 feet, lay out the centerline of the road shown in Figure 26.26.

Activity 26.6—At the scale indicated, reconstruct the contour map in Figure 26.27, then draw the 100-foot-wide road right-of-way that cuts through the parcel (distances and an angle are provided). Using an angle of repose of 1½ to 1, construct a cut-and-fill grid, plot the cut-and-fill areas on the contour map, and shade these areas.

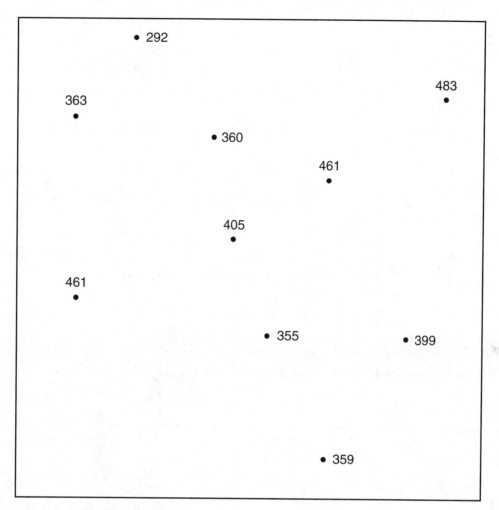

FIGURE 26.22 ■ Field notes for CAD Activity 26.1.

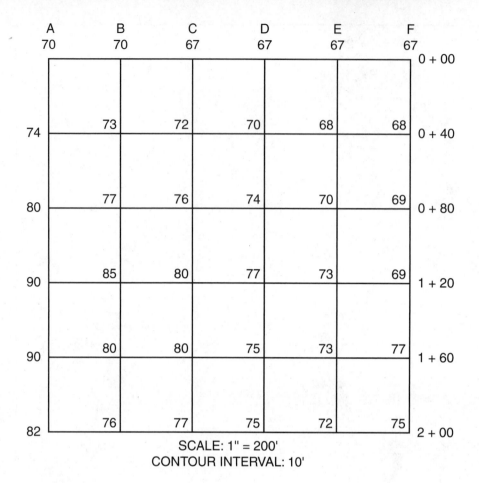

SCALE: 1" = 200'
CONTOUR INTERVAL: 10'

FIGURE 26.23 ■ Surveyor's grid for CAD Activity 26.2.

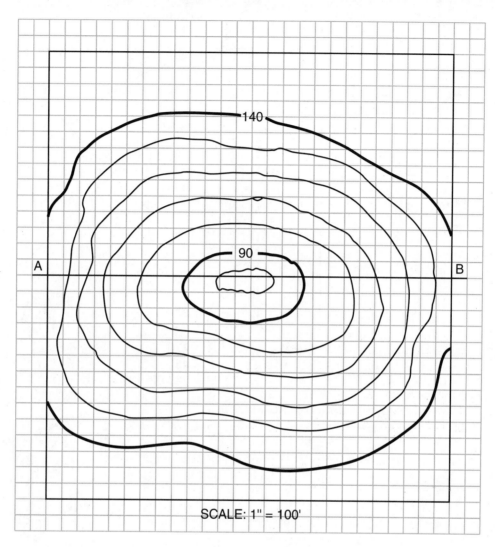

SCALE: 1" = 100'

FIGURE 26.24 ■ Contour lines drawn from field notes for CAD Activity 26.3.

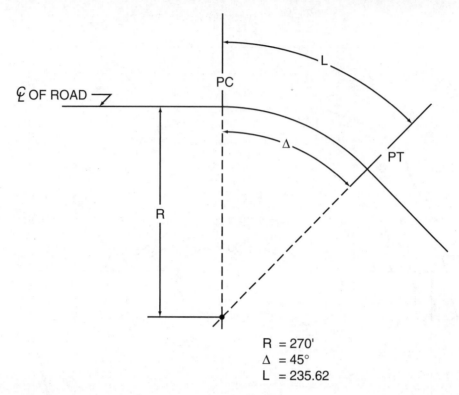

R = 270'
Δ = 45°
L = 235.62

FIGURE 26.25 ■ Field notes for CAD Activity 26.4.

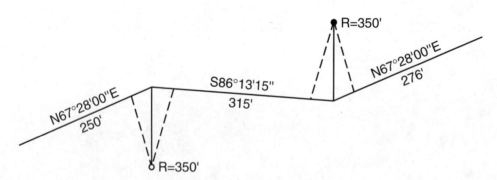

FIGURE 26.26 ■ Field notes for CAD Activity 26.5.

ℓ of 100′ ROAD RIGHT-OF-WAY

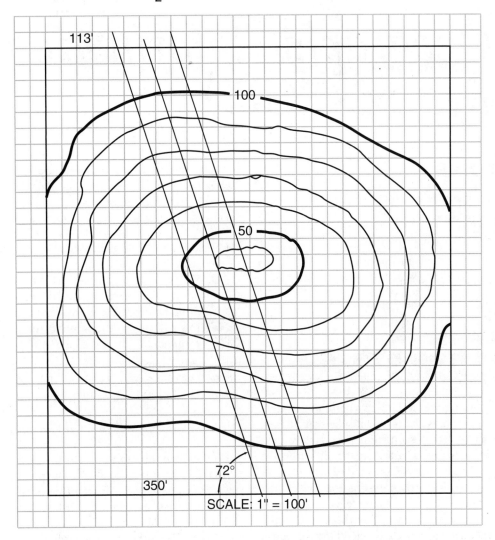

FIGURE 26.27 ■ Field notes and contour map for CAD Activity 26.6.

Geographic Information Systems (GIS)

OBJECTIVES

Upon completion of this unit, the student will be able to:

■ Explain key concepts in the field of GIS.

■ Explain GIS capabilities: Spatial analysis.

■ Explain GIS capabilities: Attribute queries.

■ Explain how GIS and CAD work together.

■ Describe the most common applications of GIS.

GEOGRAPHIC INFORMATION SYSTEM FEATURES

A **Geographic Information System (GIS)** is a computer system that collects, stores, retrieves, analyzes, and displays spatial and attribute data relating to features on, under, and above the ground. Spatial data in this sense are data that describe the location of features on and under the ground. These features are represented by points, lines, or polygons that have their own unique location in the world. For example, consider a parcel of land in a typical community and all of the various types of interrelated data that may be collected concerning the parcel (Figure 27.1). The types of data shown in this figure are just a few examples of the types that can be collected and used with a GIS.

Attribute data can be any type of data about a feature other than its location in the world. For example, if the feature is a building, attribute data may include the following:

■ Square footage

■ Number of floors

■ Principle building material (steel, concrete, wood frame)

■ Square footage by type of room (offices, meeting rooms, break rooms)

■ Any other data that describe or define the building beyond its location

What makes GIS software special is its ability to examine spatial relations and answer *what-if* questions. For example, the following types of questions may concern city or county engineers and planners in a given community:

■ What if we rezone an area to allow for additional construction?
  ■ Will there be any negative effects on the water table?
  ■ Can the area accommodate the new infrastructure that would be needed (e.g., roads, utilities, and drainage)?
  ■ How will the additional solid and liquid waste treatment requirements affect the area?

■ What if drilling for oil is permitted in a given area?
  ■ How will the local river be affected?
  ■ What impact will drilling have on the nearby children's park?
  ■ Will underground utilities have to be rerouted?

A GIS map file can represent the entire world or any part of it to the desired level of detail. The part of the world put in the system can have many different types of attributes (e.g., political boundaries, population density, zip codes, census designations, streets, and bridges). A GIS map file can be created for any type of data that have geographic linkages.

Everything that is under, on, or above the land is interrelated (Figure 27.2). Any action that affects one feature in a spatial system can affect other features. A major benefit of a GIS is that it can help prevent decisions that create spatial conflicts that can have adverse and unintended consequences.

One of the difficulties with making land-use decisions is that they cannot be made in a vacuum. One seemingly isolated and well-intended activity such as constructing a culvert system to prevent flooding may unwittingly cause an even worse problem than flooding. For example, it may cut through underground utility, water, and/or sewer lines that engineers forgot were there or it may transfer toxic runoff into the community's principal water source. A GIS allows planners to consider the potential domino effects one action can have on all features within a spatial system (Figure 27.2).

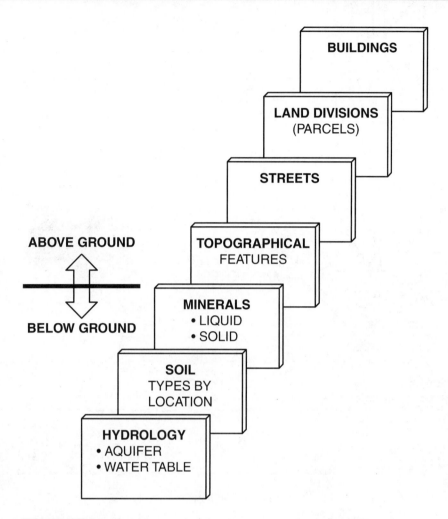

BUILDINGS

LAND DIVISIONS
(PARCELS)

STREETS

ABOVE GROUND

TOPOGRAPHICAL
FEATURES

BELOW GROUND

MINERALS
• LIQUID
• SOLID

SOIL
TYPES BY
LOCATION

HYDROLOGY
• AQUIFER
• WATER TABLE

FIGURE 27.1 ■ Types of data that might be collected and stored in a GIS database.

KEY GIS CONCEPTS

The first concept to understand about GIS is the *georelational data model*. This model serves as the tie between *spatial data* and *attribute data*. Any attribute data that can be linked to geographic locations is linked by the georelational data model. For example, a GIS map may show the geographic locations of the largest electricity or water users in the community.

Spatial data are stored as points, lines, and polygons that occupy a specific location in space. In this way, planners are able to relate various features to each other as well as *attributes* relating to those features. For example, the county's records for solid waste disposal can be linked by the GIS with the locations of residential, commercial, and/or government buildings in the county.

Attribute data can be almost any kind of information relating to what is under, on, and/or above the ground in any specified

location or anything that is inside of a given component. For example, say the feature in question is a tree. What kind of tree is it? How tall is it? When was it last harvested? How old is it? The answers to these questions are attribute data. A GIS uses a type of software called *relational database management systems (RDBMS)* to relate data stored in the database. The relationship between and among files is based on a common element contained in the files. The computer will key on this common element when retrieving data in response to *what-if* scenarios posed by planners, engineers, and technicians.

A final but fundamental GIS concept is *topology*. This concept is a mathematic system that defines relationships among geometric objects/shapes (such as polygons that represent parcels of land, buildings, or other features). These relationships include adjacency, connectivity, contiguity, and proximity. Are several parcels (buildings, features, etc.) adjacent? Are they connected? Are they contiguous? Humans can *see* these spatial

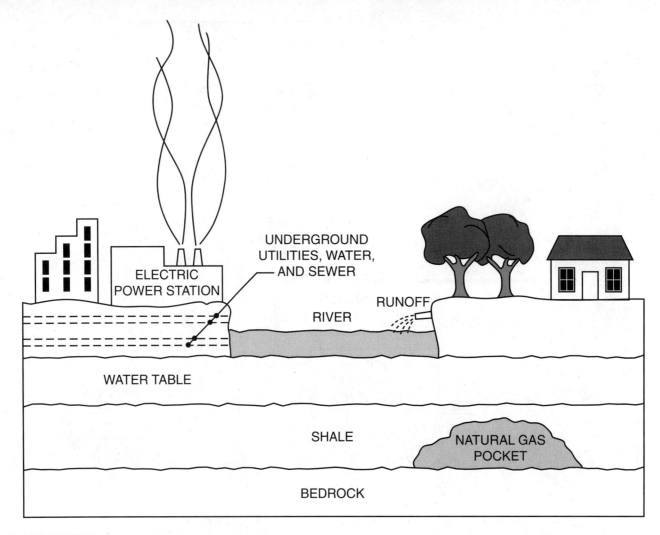

FIGURE 27.2 ■ Related features in a selected location.

relationships, but computers are blind in the human sense. Consequently, they must determine spatial relationships mathematically. Topology gives computers this capability. Figure 27.3 and Figure 27.4 are examples of GIS databases that were developed for specific applications. Figure 27.3 is a database that was developed to support an environmental application. The list of items on the left in the figure shows what the user wanted to be able to do (Desired Capability). The user in this case was a military base.

The base's environmental management office had 15 specific ongoing tasks it wanted to be able to perform. The column on the right-hand side in Figure 27.3 would show the types of data that were collected and stored in the GIS pertaining to each desired capability. For example, locational and attribute data about industry and such sensitive features as schools on, around, or near the base may be put in the GIS. Now, if the base's environmental management office is concerned about how storm runoff, noise from training maneuvers, the disposal of toxic substances, or any other action with potential environmental consequences may affect the surrounding community, it could retrieve and display a map that superimposes the base on the local community. Potential conflicts will be readily apparent and observable. The database in Figure 27.4 was developed to assist in the launch of missiles into space. The desired capabilities on the left are supported by the data on the right.

ENVIRONMENTAL APPLICATION

Desired Capability ⟶ **Relevant Data Stored (to be entered)**

1. Total Restoration/Compliance Software Systems Integration ("Point-and-Click Management")

2. Pollution Prevention Opportunity Assessment

3. Plume Delineation

4. Dispersion Modeling

5. Concentration Gradients

6. Land Utilization Records

7. Resource Assessment

8. Master Planning Site Avoidance

9. Noise Analysis

10. Regional/Community Planning Oversight

11. Hazardous Material Tracking ("Pharmacy Concept")

12. Graphical Permit Support

13. Wetlands Delineation

14. Air Emission Modeling

15. Storm Water Runoff Tracking and Management

FIGURE 27.3 ■ GIS database for an environmental application.

SPACE LAUNCH FACILITIES APPLICATION

Desired Capability ⟶ **Relevant Data Stored**

■ Telemetry Data Display

■ Range Tracking

■ Command Destruct Criteria

■ Down Range Depiction

■ Safety Buffer Zone Analysis (in Real Time)

■ Large Vehicle Routing

■ Security Planning

■ Tracking station locations, 3-D spatial relationships

■ Moving location, dynamic zones of impact probability

■ Graded overall constraints zones

■ Visualization of safety constraints, real-time tracking

■ Dynamic quantity/distance explosive zones

■ Alternative route selections, automated evacuation

■ Security zones, SCIP locations, alarmed areas

FIGURE 27.4 ■ GIS database for a space launch facilities application.

HOW A GIS IS USED

The best way to understand a GIS is to consider an actual example of how one is used. Figure 27.5 contains a map of a college campus. A GIS could be used by the college's Facilities Management Office for maintaining the master plan for maintenance, renovation, remodeling, and numerous other functions.

The map shows the location on the campus of all facilities including buildings, roads, parking lots, and athletic facilities.

The legend shows the letter designation for each building, which, in turn, has a descriptive name based on its function (e.g., Administration, Gymnasium/Wellness Center, and Arts Center). All pertinent locational data relating to buildings and facilities are stored in the GIS database.

In addition to the locational data, there are also attribute data for all pertinent features below and on the ground. For example, consider Building E, the Learning Resources Center, from the campus map in Figure 27.5. Attribute data for this building may include the following:

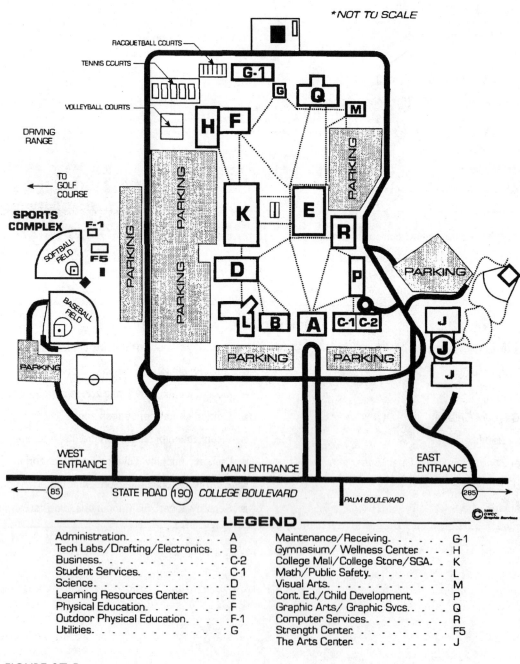

LEGEND

Administration.	A	Maintenance/Receiving.	G-1
Tech Labs/Drafting/Electronics.	B	Gymnasium/ Wellness Center	H
Business.	C-2	College Mall/College Store/SGA.	K
Student Services.	C-1	Math/Public Safety.	L
Science.	D	Visual Arts.	M
Learning Resources Center.	E	Cont. Ed./Child Development.	P
Physical Education.	F	Graphic Arts/ Graphic Svcs.	Q
Outdoor Physical Education.	F-1	Computer Services.	R
Utilities.	G	Strength Center.	F5
		The Arts Center.	J

FIGURE 27.5 ■ GIS map for a college campus.

- *Floor plans for each level*
  - Maintenance data for all rooms on all floors
  - Past, present, and planned renovation data for all rooms
  - Contents of each room (chairs, tables, computers, audiovisual equipment, etc.)
- *Air conditioning and heating plan for the building*
  - Maintenance schedules
  - Equipment specifications
- *Electrical plan for the building*
  - Maintenance schedules
  - Specifications for electrical equipment, lines, and fixtures
- *Data and voice lines plan for the building*
  - Maintenance schedules
  - Specifications for lines and equipment
- *Fire alarm plan for the building*
  - Maintenance schedules
  - Specifications for the system and all components
- *Sprinkler (fire suppression)*
  - Maintenance schedule
  - Specifications for the system

The same information could be stored about all of the buildings on campus. Similar information could be stored about all features on the campus. Having this information—along with all of the locational data—readily available gives the college's facilities management personnel powerful capabilities.

GIS CAPABILITIES: SPATIAL ANALYSIS

Every GIS has certain standard capabilities. This section illustrates those standard capabilities that fall under the heading of *spatial analysis*. *Spatial queries* allow users to pose what-if scenarios concerning features on the map. The GIS bases its responses to spatial queries on points, lines, and polygons. *Point-in-polygon* responses locate specific points that lie within specific polygons. For example, say a company needed to show the local fire department the location of all fire extinguishers in a given building. The polygon would be the building, and each fire extinguisher would show up as a point superimposed on the building's floor plan as displayed on a computer terminal screen or printed in hard form.

Polygon overlay involves the generation of a new polygon by overlaying two or more polygons. For example, say company personnel want to make sure that the installation of fiber optic cables will not interfere with the existing electrical system. The GIS could overlay the existing electrical system (one polygon) over the floor plan of the building in question (a second polygon). It could then overlay the proposed fiber optic system plan over the first two. The newly created polygon when displayed would point out visually any potential wiring conflicts.

Buffering generates new polygons around a set of points, lines, or other polygons. For example, say the college in Fig-

ure 27.5 wants to add batting cages north of its softball field but is concerned about several wells dug to provide water for the college's irrigation system. The wells are represented in the GIS database by points that occupy specific locations on the campus. The new polygons generated by buffering would be the footprints for the proposed batting cages (Figure 27.6). From this figure, it is easy to see that the proposed batting cages will interfere with Well 2.

Nearest neighbor is a GIS capability that identifies the closest features to another feature. This capability is especially valuable in keeping track of features as they relate to other features. For example, say the college in Figure 27.5 wants to add a new business and industry training center in the area between its west entrance and its main entrance. What features are the nearest neighbors on the ground and underground? If one of the nearest neighbors turns out to be a large tract of environmentally sensitive wetlands in this area, the college may decide to put the new training center elsewhere.

Network analysis is a GIS capability that allows users to perform flow analysis, routing determinations, and/or orders stops. The college in Figure 27.5 may use this capability to analyze traffic flow patterns, student routing patterns between classes, or storm drainage flow analysis.

GIS CAPABILITIES: ATTRIBUTE QUERIES

Attribute queries pull information about the attributes of features that are shared in the database. How they are organized and displayed depends on the nature of the specific inquiry in question.

Extracting is a GIS capability that selects a subset of data from a broader set. For example, say the college in Figure 27.5 wants to know how many classrooms it has that have wall-mounted videotape players. This would be the subset. The broad set of data includes *all* classrooms.

Generalizing is a capability that combines polygon features based on common or similar attributes. For example, the college in Figure 27.5 may want to see a list of all rooms on campus that have been recarpeted within the last 6 months. A floor plan could be displayed with the subject rooms highlighted.

GIS AND CAD

The wedding of GIS and CAD makes for a perfect marriage. GIS is, at its heart, a database system. As such, a GIS is best at storing, analyzing, and displaying data. CAD, on the other hand, is graphics based. It displays pictures. Put these two concepts together, and users get the best of two worlds—conveniently organized data that are augmented visually by drawings. For example, the campus map in Figure 27.5 is a product of the marriage of CAD and GIS. The ability to display data in a graphic format simplifies matters when it comes to the user interface. Even nontechnical users can understand a picture.

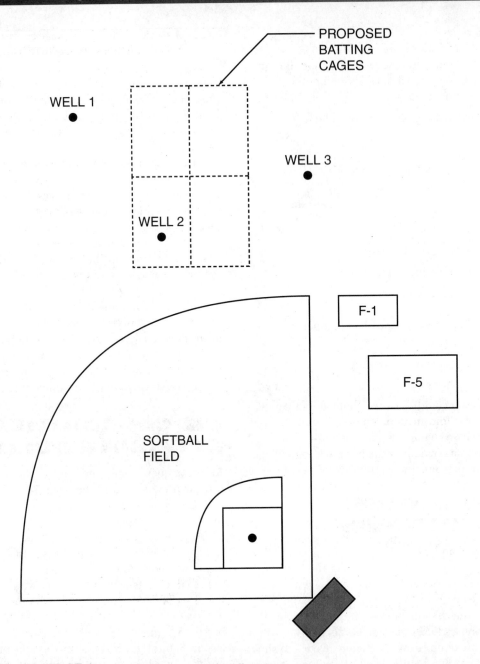

FIGURE 27.6 ■ GIS map of a portion of a college campus.

GIS APPLICATIONS CHECKLIST FOR A COMPANY

- Comprehensive/Master Planning
- Facilities Management
- Facilities Siting
- Utilities Planning
- Water Resource Projects
- Natural Resources Management
- Wetlands Inventory
- Environmental Restoration/Compliance
- Hazardous Material/Waste Management
- Environmental Health Risk Assessment
- Historical/Archaeological Site Protection
- Fire and Emergency Vehicle Dispatch
- Maintenance Team Dispatch
- Security Response and Planning
- Space Utilization
- Pavement Management
- Floodplain Management
- Terrain Analysis
- Communication Network Planning
- Logistics Planning

FIGURE 27.7 ■ Uses of a GIS at a company.

LAYERS OF LOCATIONAL DATA

- Above-Ground Features
  - Buildings
  - Fire Hydrants
  - Parking Lots
  - Fire Extinguishers within Buildings
  - Roads
  - Routes of Ingress and Egress
  - Sidewalks
  - Recreational Facilities
  - Sports Facilities
  - Wetlands Locations
  - High and Low Elevation
- Underground Features
  - Gas Lines
  - Water Lines
  - Sewer Lines
  - Storm Drainage
  - Electrical Lines
  - Data/Communication Lines

FIGURE 27.8 ■ Layers of locational data.

OTHER COMMON APPLICATIONS OF GIS

GIS began as a land-use and planning tool. These are still its most common applications, as you have seen in this chapter. Figure 27.7 shows the various ways a company may use a GIS system. Figure 27.8 shows the layers of locational data found in a GIS. However, the applications of GIS are almost limitless. In addition to community and land planning, GIS is widely used in forestry management, civil engineering, census taking, and demographic analysis. Some actual examples of applications of GIS beyond the traditional uses are:

Marketing research for retail organizations—A major hardware chain uses GIS to analyze data collected by its local retail outlets. These outlets survey their customers to determine where they live and the amount of business they do with the outlet. These data are fed into a GIS system at the corporation's home office. GIS technicians use the data to generate maps that match annual sales with geographic locations. These maps then tell corporate planners where they need to add new outlets or close unproductive outlets.

Franchise application evaluation—A major fitness company uses GIS to evaluate the applications of individuals who want to open new local franchises. The corporation gives each local franchise a specific, contractually protected geographic region. A GIS system in the corporate office displays every local

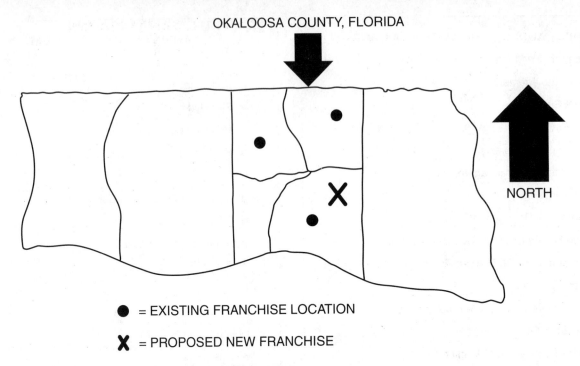

OKALOOSA COUNTY, FLORIDA

NORTH

● = EXISTING FRANCHISE LOCATION

X = PROPOSED NEW FRANCHISE

FIGURE 27.9 ■ GIS map of a portion of Florida's panhandle.

franchise as a point on a map of the United States. The franchise's territory also appears on the map. If a proposed new franchise falls into a protected territory, the GIS will show it immediately (Figure 27.9). In this map, the proposed new franchise clearly falls in the protected territory of an existing franchise. However, there is a territory to the west that is open. Based on this information, the corporation would deny the application for a new franchise in the requested location. However, it may suggest that the applicant consider the open territory to the west.

These are just two applications of GIS beyond land planning and civil engineering. GIS is also being used in the following ways:

- *Newspapers*
  - Demographic analysis
  - Mapping of carrier routes
  - Identifying new subscribers
- *Banks*
  - Building geographic models of the customer base
  - Analyzing the market potential of geographic regions
- *Health care companies*
  - Tracking the demand for specific types of treatment by geographic regions
  - Selecting sites for new facilities
- *Real estate*
  - Showing property on-line
  - Locator maps
  - Property management

The applications of GIS continue to increase in both quantity and variety. This means that the job market for GIS/CAD technicians is wide open and varied. In addition to working in land-planning agencies and civil engineering companies, GIS/CAD technicians may work for large retailers, franchisers, newspapers, banks, health care companies, travel agencies, real estate firms, government agencies (city, county, state, and federal), the military, or any other type of organization that can benefit from the effective management of geographic, locational, and attribute data.

SUMMARY

- A Geographic Information System (GIS) is a computer system that collects, stores, retrieves, analyzes, and displays spatial and attribute data relating to features on and under ground.

- Spatial data describe the location of features on or under the ground. Spatial data are represented by points, lines, and planes that have specific locations on earth.

- The strength of GIS is in its ability to answer *what-if* questions relating to the spatial data of features in a database. It allows planners to consider the domino effect one action can have on all features in a spatial system.

- A GIS map can represent the entire world or any part of it to the desired level. The portion of the world put in the system can have many different types of attributes (e.g., zip codes, political boundaries, and census designations).

- The georelational data model in the GIS serves as the link between spatial data and attribute data. Any attribute data that can be linked to geographic locations is linked by the georelational data model.

■ Spatial data are points, lines, and polygons that occupy a specific location in space. Attribute data can be almost any kind of data about a feature. For example, say the feature is a tree. What kind of tree is it? How tall is it? When was it last fertilized? When was it last harvested? The answers to these questions are attribute data.

■ Topology is a mathematical system that defines spatial relations in a GIS among geometric objects or shapes. These relationships include adjacency, connectivity, contiguity, and proximity.

■ A GIS hardware configuration looks like any other modern configuration and much like a CAD configuration. In addition to a powerful personal computer, the configuration may also contain scanners, digitizers, and plotters.

■ When making spatial queries of a GIS, the responses fall into one of the following categories: point in polygon, polygon overlay, buffering, nearest neighbor, and network analysis.

■ When making attribute queries of a GIS, the responses fall into one of the following categories: extracting and generalizing.

■ Common applications of GIS include the following: land planning, community planning, forestry management, civil engineering, census taking, demographic analysis, market research, geographic demographic modeling, and property management.

REVIEW QUESTIONS

1. What is a Geographic Information System (GIS)?

2. Explain the most important capability of GIS software.

3. Define the following GIS concepts:
   a. Georelational data model
   b. Spatial data
   c. Attribute data
   d. Topology

4. Describe the components that may be included in a typical GIS hardware configuration.

5. Explain the following spatial-analysis terms:
   a. Spatial queries
   b. Point in polygon
   c. Polygon overlay
   d. Buffering
   e. Nearest neighbor
   f. Network analysis

6. Explain the following concepts relating to attribute queries:
   a. Extracting
   b. Generalizing

7. List five different applications of GIS.

CAD ACTIVITIES

GENERAL INSTRUCTIONS

The following activities may be completed on any CAD system. Before reading the *specific instructions* for each activity (below), go through each step in the following planning checklist. The checklist applies to any CAD system and will help ensure the optimum use of your time and resources.

1. Analyze the problem carefully. Decide exactly what you are being asked to do.

2. Determine what resources and references you will need in order to complete the problem and collect them.

3. Decide if any particular standards apply to the project and have those standards available.

4. Determine what types of views will be required and how many of each.

5. Determine what the final plotted scale of the drawing will need to be, and select the appropriate paper size for plotting/printing (make sure the appropriate paper size is available).

6. Plan your drawing sequence. In what order will you develop the drawing (i.e., lines, features, dimension lines, leaders, dimensions, notes, etc.)?

7. Review the various CAD commands you will have to use in order to develop the drawing.

8. Examine your CAD system to ensure that everything is in working order, then begin the project.

SPECIFIC INSTRUCTIONS

Activity 27.1—Write a report entitled *Careers in GIS*. This report should answer the following questions at a minimum:
   a. What types of organizations hire GIS technicians?
   b. What types of skills are these organizations looking for in entry-level GIS technicians?
   c. Do CAD skills give applicants for GIS jobs an advantage?
   d. How do starting wages for GIS technicians in your region compare to starting wages for drafting and design technicians?

Activity 27.2—Visit an organization in your region that is using GIS. What application(s) did you find? Compile a list of every GIS application you can identify.

28 UNIT
Pipe Drafting

OBJECTIVES

Upon completion of this unit, the student will be able to:

■ List and describe the most widely used kinds of pipe.

■ List and describe the three broad classifications of pipe joints/fittings.

■ List and describe the most commonly used kinds of pipe valves.

■ Explain the uses of the following pipe drawing conventions: crossings, connections, fittings, and machines/devices.

■ Demonstrate proficiency in developing single- and double-line pipe drawings.

■ Demonstrate proficiency in developing pipe drawings that include pumps, tanks, and vessels.

PIPE DRAFTING COMPONENTS

Pipe drafting involves designing and drawing the plans for systems that move and store liquids and gases for industrial operations. Most pipe used in such systems is larger than 4 inches in diameter and might be as large as 48 inches or even more. Piping systems must be drawn in accordance with a variety of standards and codes. The components of these systems include pipe, fittings, valves, pumps, tanks, vessels, and related equipment.

TYPES OF PIPE

The wide variety of fluids used by modern society requires a number of different types of pipe. The most widely used of these are:

■ Steel pipe

■ Cast iron pipe

■ Brass and copper pipe

■ Copper tubing

■ Plastic pipe

Steel Pipe

Steel pipe is well suited for high-pressure and high-temperature applications. It is frequently used in piping systems that trans-

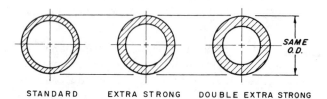

FIGURE 28.1 ■ Schedule of pipe. (*Reprinted from David Goetsch and William Chalk,* Technical Drawing, *4th ed. Delmar Publishers, 1994, Figure 21–1*)

port water, oil, petroleum, and steam. Steel pipe comes in several cross-sectional configurations. The most widely used of these are the standard, extra-strong, and double-extra-strong configurations (Figure 28.1). There are actually ten different cross-sectional configurations for steel pipe.

Steel pipe is specified by a nominal diameter callout. The actual diameter will vary slightly from the nominal. The American National Standard Institute (ANSI) specifies pipe in ten different schedules. Each schedule corresponds to a wall thickness. For example, standard and extra-strong pipe are Schedule 40 and Schedule 80, respectively. The nominal diameter for pipe up to 12 inches refers to the inside diameter. Callouts are given in inches or millimeters. The nominal diameter for pipe over 12 inches refers to the outside diameter.

Cast Iron Pipe

Cast iron pipe is used most frequently for underground applications—to transport water, gas, and sewage. It is well suited for low-pressure steam connections also.

Brass and Copper Pipe

Brass and copper pipe are used in applications where corrosion will be a problem. Because brass and copper are able to withstand corrosion, the expected lifetime of a piping system in a high-corrosive setting will be longer if brass or copper pipe is used.

Copper Tubing

Copper tubing is used extensively in applications where vibration and misalignment are important factors. It is used in hydraulic and pneumatic applications, such as industrial settings involving robots, and in automotive settings.

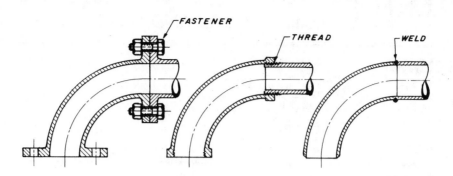

FIGURE 28.2 ■ Pipe connections. (*Reprinted from David Goetsch and William Chalk, Technical Drawing, 4th ed. Delmar Publishers, 1994, Figure 21–2*)

Plastic Pipe

Plastic pipe is used extensively in modern piping settings. It is highly resistant to corrosion and chemical degradation. Plastic pipe is flexible and can be easily installed. However, it is not used where heat or pressure is a factor. Plastic pipe does not have the chemical makeup to withstand heat or the wall strength to withstand high pressures.

TYPES OF JOINTS AND FITTINGS

Each kind of pipe explained in the previous section comes in straight lengths. However, piping systems require turns and branches and changes of size. In every instance in which a pipe must change directions or size, there is a joint. Joints are accomplished using fittings. The three broad classifications of fittings are:

Screwed

Flanged

Welded

Figure 28.2 illustrates these three types of fittings.

Engineers and CAD technicians need to be able to specify pipe fittings. A given pipe fitting is specified by stating the nominal pipe size, the name of the fitting, and the material out of which the fitting is made. Any fitting that is used to connect different sizes of pipes is referred to as a *reduction fitting*. When specifying reduction fittings, you must state the nominal pipe sizes for both the large and small ends of the fitting. The large size is stated first. There are enough differences among screwed, flanged, and welded fittings that each must be studied separately.

Screwed Fittings

Screwed fittings are generally used in applications requiring small-diameter pipe of 2.5 inches or less. The threaded end of the pipe and the internal threads on the fitting are usually coated with a special lubricant to seal the joint and to ease the connection process. Figure 28.3, Figure 28.4, Figure 28.5, and Figure 28.6 contain information on the most commonly used threaded-type fittings.

Figure 28.3 contains information on the 90-degree elbow, the 90-degree street elbow, the 45-degree elbow, the 45-degree street elbow, and the 90-degree reducing elbow. The table accompanying each illustration gives the type of sizes with which that particular fitting can be used, the various dimensions in which it is available, and the weight of the fitting itself. Figure 28.4 contains information for blind-flange, screwed-flange, and flat-band-cap fittings, as well as bushings. Figure 28.5 contains information on union, coupling, cross, tee, and reducing-tee fittings. Figure 28.6 contains information on plug, return-bend, and lock-nut fittings, as well as reducers and nipples.

AVAILABLE STYLES AND SIZES

90° ELBOW

REFERENCE	PIPE SIZE, INCHES														
	1/8	1/4	3/8	1/2	3/4	1	1 1/4	1 1/2	2	2 1/2	3	3 1/2	4	5	6
A	11/16	13/16	15/16	1 1/8	1 5/16	1 7/16	1 3/4	1 15/16	2 1/4	2 11/16	3 1/8	3 7/16	3 3/4	4 1/2	5 1/8
Weight	.055	.060	.095	.145	.210	.355	.705	.790	1.180	1.670	2.590	3.250	4.065	6.900	9.800

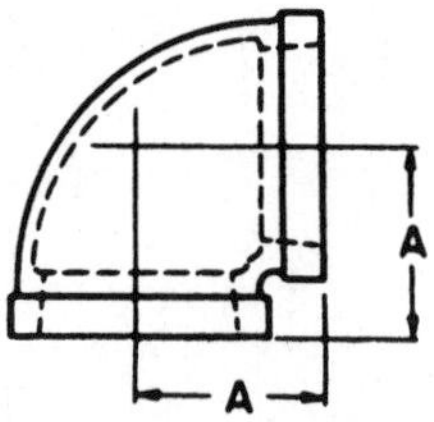

90° STREET ELBOW

REFERENCE	PIPE SIZE, INCHES										
	1/8	1/4	3/8	1/2	3/4	1	1 1/4	1 1/2	2	2 1/2	3
A	11/16	13/16	15/16	1 1/8	1 5/16	1 7/16	1 3/4	1 15/16	2 1/4	2 11/16	3 1/8
B	1 1/8	1 5/16	1 7/16	1 5/8	1 7/8	2 1/8	2 1/2	2 11/16	3 3/16	3 13/16	4 1/2
Weight	.025	.045	.085	.140	.180	.205	.495	.750	1.250	1.850	2.900

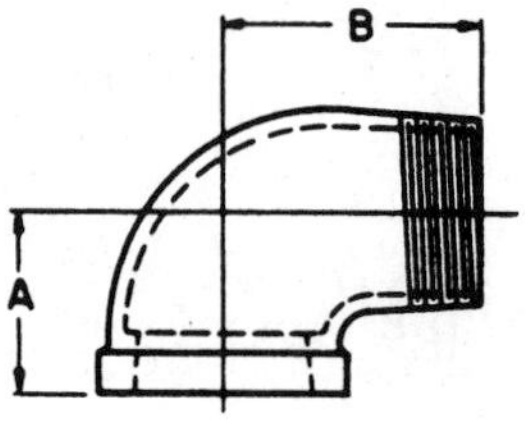

45° ELBOW

REFERENCE	PIPE SIZE, INCHES														
	1/8	1/4	3/8	1/2	3/4	1	1 1/4	1 1/2	2	2 1/2	3	3 1/2	4	5	6
A	11/16	3/4	13/16	7/8	1	1 1/8	1 5/16	1 7/16	1 11/16	1 15/16	2 3/16	2 3/8	2 5/8	3 1/16	3 15/32
Weight	.040	.040	.075	.115	.200	.260	.455	.605	.970	1.420	1.925	2.530	3.335	5.650	8.700

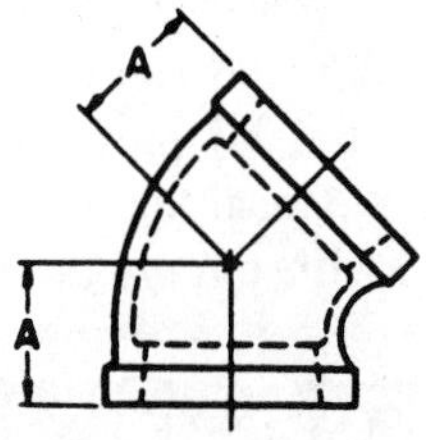

45° STREET ELBOW

REFERENCE	PIPE SIZE, INCHES									
	1/8	1/4	3/8	1/2	3/4	1	1 1/4	1 1/2	2	2 1/2
A	—	—	—	1 1/16	1 3/16	1 3/8	1 17/32	1 5/16	1 5/8	2 1/16
B	—	—	—	1 3/16	1 3/8	1 17/32	1 23/32	2 1/4	2 3/8	2 1/2
Weight	—	—	—	.140	.180	.205	.495	.700	1.000	1.625

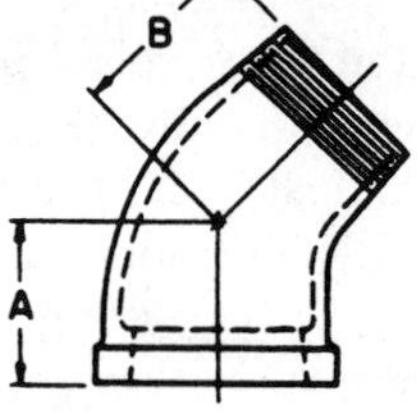

90° REDUCING ELBOW

Size	Weight
1/2" x 1/4"	.115
1/2" x 3/8"	.120
3/4" x 1/2"	.190
1" x 1/2"	.260
1" x 3/4"	.300
1 1/4" x 3/4"	.405
1 1/4" x 1"	.470

Size	Weight
1 1/2" x 1"	.560
1 1/2" x 1 1/4"	.720
2" x 1 1/4"	1.000
2" x 1 1/2"	1.030
2 1/2" x 2"	1.745
3" x 2"	2.205
3" x 2 1/2"	2.315

Size	Weight
4" x 3"	3.065

LATERALS — 1. 45° Y

a. Size Range - 1/2'' through 3''

b. Dimensional Standard - F-52618-C (Revision)

NOTE: Any reducing fittings not specifically listed can be produced on a Special Order basis and will be subject to a Special Order charge of 25% of the listed retail price.

FIGURE 28.3 ■ Pipe fittings. (*Reprinted from David Goetsch and William Chalk,* Technical Drawing, *4th ed. Delmar Publishers, 1994, Figure 21–3*)

AVAILABLE STYLES AND SIZES

FLANGE - Blind

REFERENCE	PIPE SIZE, INCHES										
	½	¾	1	1¼	1½	2	2½	3	3½	4	6
A	3½	3⅞	4¼	4⅝	5	6	7	7½	8½	9	11
B	7/16	7/16	7/16	½	9/16	⅝	11/16	¾	13/16	15/16	1
C	2⅜	2¾	3⅛	3½	3⅞	4¾	5½	6	7	7½	9½
D	⅝	⅝	⅝	⅝	⅝	¾	¾	¾	¾	¾	⅞
E	4	4	4	4	4	4	4	4	8	8	8
Weight	.275	.385	.560	.765	1.015	1.640	2.440	3.075	4.730	5.285	9.250

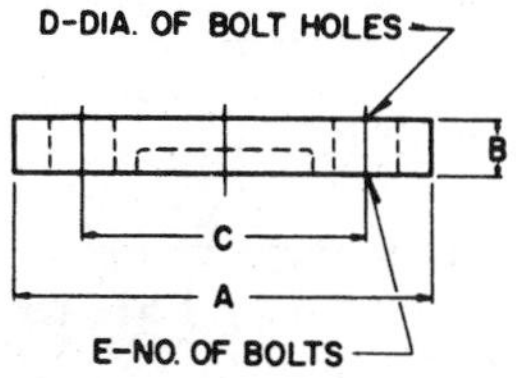
D-DIA. OF BOLT HOLES
B
C
A
E-NO. OF BOLTS

FLANGE - Screwed

FOR REDUCING, SLIP-ON, FLOOR AND WELDING NECK FLANGES

REFERENCE	PIPE SIZE, INCHES											
	½	¾	1	1¼	1½	2	2½	3	3½	4	5	6
A	3½	3⅞	4¼	4⅝	5	6	7	7½	8½	9	10	11
B	⅝	⅝	11/16	13/16	⅞	1	1 3/16	1¼	1¼	1 5/16	1 7/16	1 9/16
C	2⅜	2¾	3⅛	3½	3⅞	4¾	5½	6	7	7½	8½	9½
D	⅝	⅝	⅝	⅝	⅝	¾	¾	¾	¾	¾	⅞	⅞
E	4	4	4	4	4	4	4	4	8	8	8	8
Weight	.255	.375	.550	.740	.970	1.530	2.385	2.835	4.250	4.565	5.650	6.850

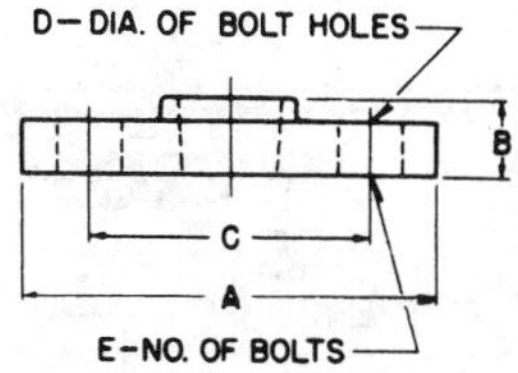
D-DIA. OF BOLT HOLES
B
C
A
E-NO. OF BOLTS

CAP - Flat Band

REFERENCE	PIPE SIZE, INCHES														
	⅛	¼	⅜	½	¾	1	1¼	1½	2	2½	3	3½	4	5	6
A	19/32	25/32	27/32	1 1/16	1 5/32	1¼	1 5/16	1 15/32	1 9/16	2 1/32	2 1/16	2 3/32	2 3/16	2⅜	2⅝
Weight	.015	.025	.050	.075	.100	.150	.290	.350	.450	.760	1.435	1.825	3.080	4.950	6.500

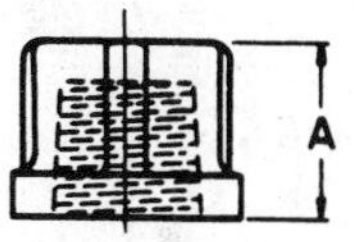
A

BUSHINGS

Size	Weight	Size	Weight	Size	Weight	Size	Weight
¼" x ⅛"	.010	1" x ¾"	.070	2" x 1¼"	.325	3½" x 3"	.750
⅜" x ⅛"	.020	1¼" x ⅜"	.170	2" x 1½"	.285	4" x 1½"	1.860
⅜" x ¼"	.015	1¼" x ½"	.165	2½" x 1"	.785	4" x 2"	1.720
½" x ⅛"	.045	1¼" x ¾"	.145	2½" x 1¼"	.755	4" x 2½"	1.535
½" x ¼"	.030	1¼" x 1"	.130	2½" x 1½"	.715	4" x 3"	1.255
½" x ⅜"	.030	1½" x ½"	.215	2½" x 2"	.605	5" x 2"	2.600
¾" x ¼"	.055	1½" x ¾"	.210	3" x 1¼"	1.095	5" x 3"	2.350
¾" x ⅜"	.050	1½" x 1"	.205	3" x 1½"	1.015	5" x 4"	2.000
¾" x ½"	.045	1½" x 1¼"	.125	3" x 2"	.895	6" x 3"	3.950
1" x ¼"	.110	2" x ½"	.370	3" x 2½"	.705	6" x 4"	3.600
1" x ⅜"	.105	2" x ¾"	.370	3½" x 2"	1.120	6" x 5"	3.250
1" x ½"	.090	2" x 1"	.365	3½" x 2½"	.920		

FIGURE 28.4 ■ Pipe fittings. (*Reprinted from David Goetsch and William Chalk,* Technical Drawing, *4th ed. Delmar Publishers, 1994, Figure 21–4*)

AVAILABLE STYLES AND SIZES

UNION

REFERENCE	PIPE SIZE, INCHES												
	1/8	1/4	3/8	1/2	3/4	1	1 1/4	1 1/2	2	2 1/2	3	3 1/2	4
A	1 9/16	1 13/16	1 15/16	2	2 1/8	2 3/4	3	3	3 1/4	3 5/8	4 1/8	4 7/8	5
Weight	.145	.130	.170	.255	.330	.505	.790	.815	1.250	1.745	2.865	5.000	5.800

COUPLING FOR HALF COUPLINGS AND HEAVY WALL COUPLINGS

REFERENCE	PIPE SIZE, INCHES														
	1/8	1/4	3/8	1/2	3/4	1	1 1/4	1 1/2	2	2 1/2	3	3 1/2	4	5	6
A	19/32	3/4	29/32	1 1/16	1 11/32	1 5/8	1 31/32	2 15/64	2 23/32	3 5/8	3 15/16	4 7/16	4 15/16	6 1/16	7 3/16
B	15/16	1 1/32	15/32	15/16	1 9/16	1 13/16	2 1/16	2 5/16	2 9/16	2 7/8	3 1/16	3 7/16	3 7/16	4 1/8	4 1/8
Weight	.020	.025	.035	.060	.100	.135	.230	.310	.500	.730	1.015	1.740	2.040	3.300	4.500

CROSS

REFERENCE	PIPE SIZE, INCHES												
	1/8	1/4	3/8	1/2	3/4	1	1 1/4	1 1/2	2	2 1/2	3	3 1/2	4
A	11/16	13/16	15/16	1 1/8	1 5/16	1 7/16	1 3/4	1 15/16	2 1/4	2 11/16	3 1/8	3 7/16	3 3/4
Weight	.050	.085	.155	.215	.340	.565	.825	1.115	2.115	2.730	4.000	5.200	6.210

TEE FOR LATERALS

REFERENCE	PIPE SIZE, INCHES														
	1/8	1/4	3/8	1/2	3/4	1	1 1/4	1 1/2	2	2 1/2	3	3 1/2	4	5	6
A	11/16	13/16	15/16	1 1/8	1 5/16	1 7/16	1 3/4	1 15/16	2 1/4	2 11/16	3 1/8	3 7/16	3 3/4	4 1/2	5 1/8
Weight	.060	.070	.130	.180	.285	.435	.745	1.060	1.760	2.250	3.035	4.135	5.580	9.510	13.000

REDUCING TEE

NOTE: In the listing of reducing tees, the last dimension shown is the size of the branch.

Size	Wt.
1/2" x 1/2" x 1/4"	.195
1/2" x 1/2" x 3/8"	.190
1/2" x 1/2" x 3/4"	.320
3/4" x 1/2" x 1/2"	.320
3/4" x 1/2" x 3/4"	.305
3/4" x 3/4" x 3/8"	.315
3/4" x 3/4" x 1/2"	.305
3/4" x 3/4" x 1"	.465
1" x 1/2" x 1/2"	.190
1" x 3/4" x 1/2"	.515
1" x 3/4" x 3/4"	.495
1" x 3/4" x 1"	.465
1" x 1" x 3/8"	.495

Size	Wt.
1" x 1" x 1/2"	.485
1" x 1" x 3/4"	.465
1" x 1" x 1 1/4"	.850
1" x 1" x 1 1/2"	1.300
1 1/4" x 3/4" x 1 1/4"	.830
1 1/4" x 1" x 3/4"	.885
1 1/4" x 1" x 1"	.855
1 1/4" x 1 1/4" x 1/2"	.850
1 1/4" x 1 1/4" x 3/4"	.850
1 1/4" x 1 1/4" x 1"	.800
1 1/4" x 1 1/4" x 1 1/2"	1.170
1 1/2" x 3/4" x 1 1/2"	1.220
1 1/2" x 1" x 1"	1.300

Size	Wt.
1 1/4" x 1" x 1 1/2"	1.180
1 1/2" x 1 1/4" x 1"	1.230
1 1/2" x 1 1/4" x 1 1/4"	1.170
1 1/2" x 1 1/2" x 1/2"	1.245
1 1/2" x 1 1/2" x 3/4"	1.220
1 1/2" x 1 1/2" x 1"	1.180
1 1/2" x 1 1/2" x 1 1/4"	1.115
1 1/2" x 1 1/2" x 2"	1.280
2" x 1 1/2" x 1 1/2"	1.315
2" x 1 1/2" x 2"	1.755
2" x 2" x 1/2"	0.995
2" x 2" x 3/4"	1.290
2" x 2" x 1"	1.300

Size	Wt.
2" x 2" x 1 1/4"	1.325
2" x 2" x 1 1/2"	1.470
2" x 2" x 2 1/2"	1.910
2 1/2" x 2" x 2"	1.925
2 1/2" x 2 1/2" x 2"	2.095
3" x 3" x 2"	2.690
4" x 4" x 2"	4.015
4" x 4" x 3"	5.095
6" x 6" x 4"	10.850

FIGURE 28.5 ■ Pipe fittings. (*Reprinted from David Goetsch and William Chalk,* Technical Drawing, *4th ed. Delmar Publishers, 1994, Figure 21–5*)

AVAILABLE STYLES AND SIZES

PLUG

COUNTERSUNK PLUGS

REFERENCE	PIPE SIZE, INCHES														
	1/8	1/4	3/8	1/2	3/4	1	1 1/4	1 1/2	2	2 1/2	3	3 1/2	4	5	6
A	5/8	11/16	13/16	7/8	1	1 3/16	1 7/16	1 3/8	1 9/16	1 11/16	1 13/16	2	2 3/16	2 3/8	2 9/16
Weight	.010	.015	.020	.030	.055	.090	.150	.215	.315	.560	.755	.910	1.500	2.150	3.500

COUNTERSUNK PLUGS - Square Head   Size Range 3/8'' through 4''
Dimension Standard - ANSI   B16.14-1949

RETURN BEND - Close

REFERENCE	PIPE SIZE, INCHES								
	1/8	1/4	3/8	1/2	3/4	1	1 1/4	1 1/2	2
A	–	–	–	1 1/4	1 1/2	1 3/4	2 1/4	2 1/2	3 1/4
B	–	–	–	2 3/8	3 5/32	3 29/32	4 21/32	5 7/32	6 19/32
Weight	–	–	–	.205	.275	.440	.780	1.000	1.840

LOCK NUT

REFERENCE	PIPE SIZE, iNCHES										
	1/8	1/4	3/8	1/2	3/4	1	1 1/4	1 1/2	2	2 1/2	3
A	–	5/32	3/16	1/4	5/16	3/8	7/16	1/2	5/8	3/4	15/16
Weight	–	.015	.020	.025	.045	.080	.105	.175	.345	.425	.740

REDUCERS

Size	Weight
1/4" x 1/8"	.040
3/8" x 1/8"	.060
3/8" x 1/4"	.060
1/2" x 1/4"	.100
1/2" x 3/8"	.095
3/4" x 1/4"	.150
3/4" x 3/8"	.130
3/4" x 1/2"	.160
1" x 1/2"	.215

Size	Weight
1" x 3/4"	.240
1 1/4" x 3/4"	.330
1 1/4" x 1"	.340
1 1/2" x 3/4"	.480
1 1/2" x 1"	.490
1 1/2" x 1 1/4"	.560
2" x 1"	.710
2" x 1 1/4"	.745

Size	Weight
2" x 1 1/2"	.825
2 1/2" x 2"	1.180
2 1/2" x 1 1/2"	1.200
3" x 2"	1.790
3" x 2 1/2"	1.750
4" x 3"	3.375
5" x 4"	4.75
6" x 4"	5.00
6" x 5"	5.00

NIPPLES - Close   ADDITIONAL NIPPLES *

Pipe Size	Length	Weight	Pipe Size	Length	Weight	Pipe Size	Length	Weight	Pipe Size	Length	Weight
1/8"	3/4"	.005	3/4"	1 3/8"	.045	1 1/2"	1 3/4"	.135	3"	2 5/8"	.570
1/4"	7/8"	.010	1"	1 1/2"	.075	2"	2"	.210	3 1/2"	2 3/4"	.720
3/8"	1"	.015	1 1/4"	1 5/8"	.105	2 1/2"	2 1/2"	.415	4"	2 7/8"	.895
1/2"	1 1/8"	.030									

* NIPPLES 1. Long Nipples a. Size Range — 1/8" through 4"
b. Length — through 12"

FIGURE 28.6 ■ Pipe fittings. (*Reprinted from David Goetsch and William Chalk,* Technical Drawing, *4th ed. Delmar Publishers, 1994, Figure 21–6)*

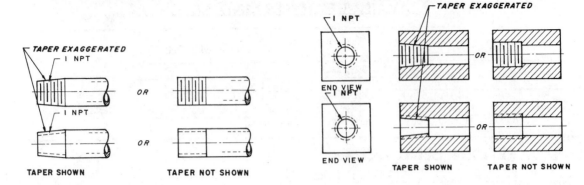

FIGURE 28.7 ■ Pipe thread conventions. (*Reprinted from David Goetsch and William Chalk,* Technical Drawing, *4th ed. Delmar Publishers, 1994, Figure 21–7*)

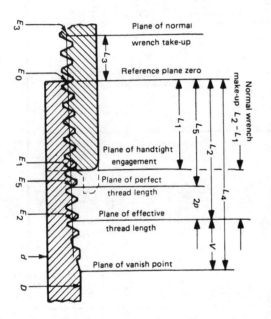

FIGURE 28.8 ■ Pipe thread notation. (*Reprinted from David Goetsch and William Chalk,* Technical Drawing, *4th ed. Delmar Publishers, 1994, Figure 21–8*)

There are two types of American Standard pipe threads—tapered and straight. Tapered threads are more common. Straight threads are usually used only for special applications. The ANSI Standard for Pipe Threads is ANSI/ASME B1.20.1—1983 (R1992). Figure 28.7 illustrates the conventions used for drawing pipe threads. Figure 28.8 illustrates the American National Standard taper pipe thread notation methods. Both types of threads have the same number of threads per inch.

Flanged Fittings and Joints

Some piping applications require that the piping systems occasionally be disassembled. When this is the case, flanged fittings are appropriate. Flanged fittings and adjoining pipes are normally bolted or glued together. Figure 28.9 contains size, weight, and dimensional data for the most common types of flanged fittings: 90-degree elbows, 45-degree elbows, tees, and reducers.

Welded Fittings and Joints

Some piping applications, such as high-pressure and high-temperature systems, require permanent fittings and joints. When this is the case, welded fittings are used. To accommodate the welding process, the connection ends of welded fittings as well as the ends of adjoining pipe are usually beveled. Welded fittings usually weigh less than flanged or screwed fittings and are easier to insulate.

AVAILABLE STYLES AND SIZES

90° ELBOW

REFERENCE	PIPE SIZE, INCHES					
	1½	2	2½	3	4	6
A	4	4½	5	5½	6½	8
Weight	5	6¼	8	8½	10	20

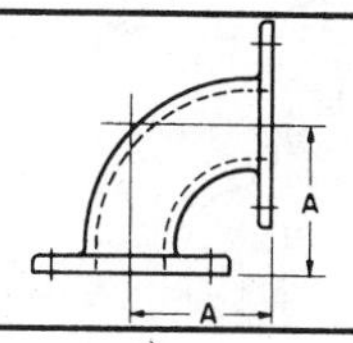

45° ELBOW

REFERENCE	PIPE SIZE, INCHES					
	1½	2	2½	3	4	6
A	2¼	2½	3	3	4	5
Weight	3	4	10	6	6½	16

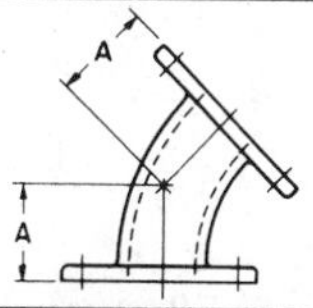

TEE

REFERENCE	PIPE SIZE, INCHES					
	1½	2	2½	3	4	6
A	4	4½	5	5½	6½	8
Weight	5	9	13	13½	20	30

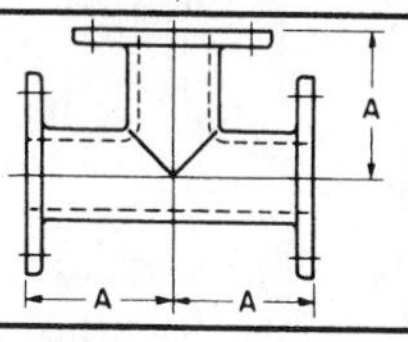

REDUCER

REFERENCE	PIPE SIZE, INCHES					
	2 x 1½	2½ x 1½ 2½ x 2	3 x 2 3 x 2½	4 x 3	6 x 3	6 x 4
A	5	5½	6	7•	9•	9•
Weight	4	6	8	9	12¼	14

356-F

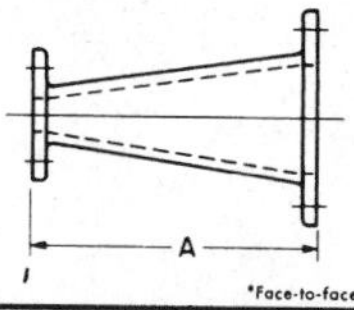

*Face-to-face dimension

All 3'' fittings have a 7 1/2'' flange diameter, a 6'' bolt circle diameter, 4 bolt holes and 3/4'' diameter bolt holes.

All 1 1/2'' fittings have a 5'' flange diameter, a 3 7/8'' bolt circle diameter, 4 bolt holes and 5/8'' diameter bolt holes.

All 4'' fittings have a 9'' flange diameter, a 7 1/2'' bolt circle diameter, 8 bolt holes and 3/4'' diameter bolt holes.

All 2'' fittings have a 6'' flange diameter, a 4 3/4'' bolt circle diameter, 4 bolt holes and 3/4'' diameter bolt holes.

All 6'' fittings have an 11'' flange diameter, a 9 1/2'' bolt circle diameter, 8 bolt holes and 7/8'' diameter bolt holes.

All 2 1/2'' fittings have a 7'' flange diameter, a 5 1/2'' bolt circle diameter, 4 bolt holes and 3/4'' diameter bolt holes.

FIGURE 28.9 ■ Pipe fittings. (*Reprinted from David Goetsch and William Chalk,* Technical Drawing, *4th ed. Delmar Publishers, 1994, Figure 21–9*)

TYPES OF VALVES

Fluids and gases do not just flow freely through piping systems. They must be regulated and, at certain points, stopped. There are a number of different types of valves, including gate, globe, jet, swivelled joint, ball, check, and butterfly. The most frequently used of these are gate, globe, and check valves. These commonly used types of valves are illustrated in Figure 28.10, Figure 28.11, and 28.12.

Gate Valves

Gate valves are used to turn the flow of liquids through a pipe on or off without restricting the flow of the liquids through the valve or with as little restriction as possible. Gate valves are not meant to be used to regulate the degree of flow. Figure 28.13 and Figure 28.14 contain dimensional data for two commonly used gate valve configurations.

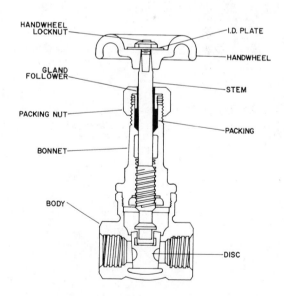

FIGURE 28.10 ■ Gate valve. (*Reprinted from David Goetsch and William Chalk,* Technical Drawing, *4th ed. Delmar Publishers, 1994, Figure 21–10*)

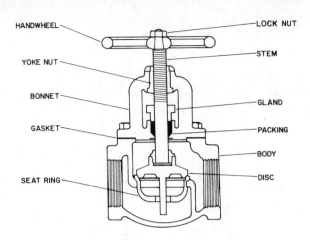

FIGURE 28.11 ■ Globe valve. (*Reprinted from David Goetsch and William Chalk,* Technical Drawing, *4th ed. Delmar Publishers, 1994, Figure 21–11*)

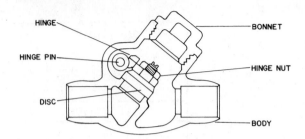

FIGURE 28.12 ■ Check valve. (*Reprinted from David Goetsch and William Chalk,* Technical Drawing, *4th ed. Delmar Publishers, 1994, Figure 21–12*)

Globe Valves

Globe valves are used to turn the flow of liquids through a pipe on and off, and they are used also to regulate the flow of fluids through the valve to the desired level. Figure 28.15 and Figure 28.16 contain dimensional data for two commonly used configurations of globe valves.

Check Valves

Check valves are used to restrict flow of liquids through a pipe in only one direction. A backward flow is checked by the valve, which is activated by any change in the direction of the flow. Figure 28.17 and Figure 28.18 contain dimensional data for two commonly used configurations of check valves.

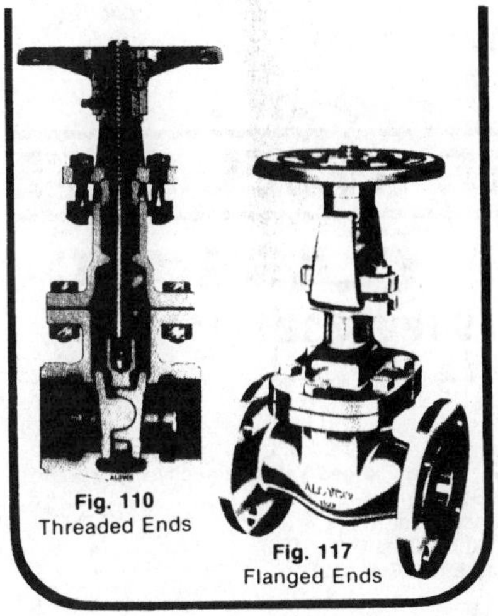
Fig. 110
Threaded Ends
Fig. 117
Flanged Ends

STAINLESS STEEL GATE

CLASS 150

FIGURES: 110 (½″ to 2″), 114 (½″ to 2″), 116 (1½″ to 12″), 117 (½″ to 12″)

DESIGN DESCRIPTION:

Outside Screw and Yoke	Socket Weld Ends (Fig. 114)
Bolted Bonnet	Buttwelding Ends (Fig. 116)
Rising Stem	Flanged Ends (Fig. 117)
Non-Rising Handwheel	Male-Female Bonnet Joint Sizes ½″ to 1″ Incl.
Integral Seats	Double Disc Ball-and-Socket Type Wedge
Threaded Ends (Fig 110)	Flat Faced Flanged Valves, Specify Fig. 117-FF

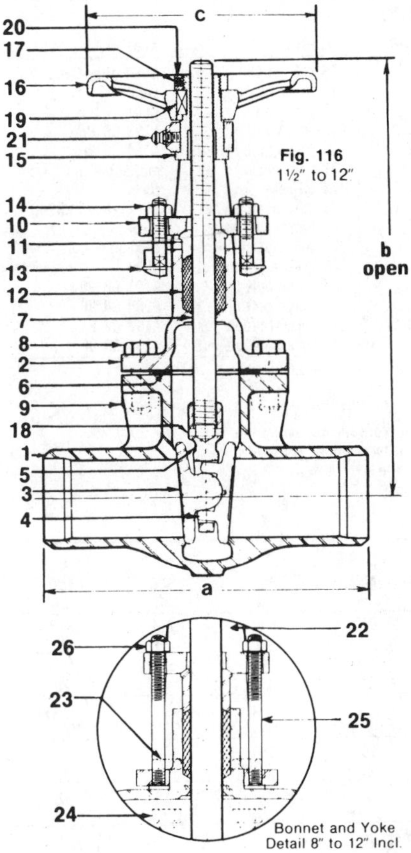
c
20
17
16
19
21
15
14
10
11
13
12
7
8
2
6
9
18
1
5
3
4
Fig. 116
1½″ to 12″
b
open
a
26
22
23
25
24
Bonnet and Yoke
Detail 8″ to 12″ Incl.

PARTS AND MATERIAL LIST:

DESCRIPTION	ALOYCO 18-8 SMO	DESCRIPTION	ALOYCO 18-8 SMO
1 body*	A351 GR CF8M	15 yoke bushing†	wrought B16 or cast B584-C83600
2 bonnet	A351 GR CF8M	16 handwheel	malleable iron
3 male disc	{wrought-type 316 or	17 yoke nut set screw	type 303
4 female disc	{cast A351 GR CF8M	18 disc arm pin	type 316
5 disc arm	A351 GR CF8M	19 handwheel key	steel
6 gasket†	comp. asbestos or teflon	20 yoke bushing nut†	wrought B16 or cast B584-C83600
7 stem	type 316	21 grease fitting	steel
8 bonnet bolt	A193 GR B8	22 yoke	CF8
9 bonnet bolt nut	A194 GR 8F	23 yoke bolt	A193 GR B8
10 gland plate	A351 GR CF8M	24 yoke bolt nut	A194 GR 8F
11 gland follower	wrought-type 316 or cast A351 GR CF8M	25 gland stud	A193 GR B8
12 packing†	braided asbestos or teflon	26 gland stud nut	A194 GR 8F
13 gland bolt	A193 GR B8		
14 gland bolt nut	A194 GR 8F		

Remarks: *Body material on threaded and welding end valves is ELC grade. Other Alloys available. (Refer to Introduction)

†Unless otherwise specified, these valves may be supplied at the manufacturer's option with: PTFE Gasket and Packing; Type 303 yoke bushing and nut. Such valves are limited to a temperature of 550 F and are so tagged.

DIMENSIONS IN MM AND INCHES:

SIZE	MM	15	20	25	32	40	50	65	80	100	150	200	250	300
	INCH	½	¾	1	1¼	1½	2	2½	3	4	6	8	10	12
a (110, 114)	mm	70	80	89	108	114	121	—	—	—	—	—	—	—
	inch	2¾	3	3½	4¼	4½	4¾							
a (116)	mm	—	—	—	—	165	216	241	283	305	403	419	457	502
	inch					6½	8½	9½	11⅞	12	15⅞	16½	18	19¾
a (117)	mm	108	118	127	—	165	178	191	203	229	267	292	330	356
	inch	4¼	4⅝	5		6½	7	7½	8	9	10½	11½	13	14
b (110, 114)	mm	210	210	238	277	286	337	—	—	—	—	—	—	—
	inch	8¼	8¼	9⅜	10⅞	11¼	13¼							
b (116, 117)	mm	210	210	235	—	291	337	394	432	527	781	1016	1219	1524
	inch	8¼	8¼	9¼	—	11 7/16	13¼	15½	17	20¾	30¾	40	48	60
c	mm	89	100	100	127	127	150	178	191	250	305	356	406	457
	inch	3½	4	4	5	5	6	7	7½	10	12	14	16	18
d (114)	mm	13	13	13	13	13	16	—	—	—	—	—	—	—
	inch	½	½	½	½	½	⅝							
d (117)	mm	11	11	11	—	14	16	18	19	24	25	29	30	32
	inch	7/16	7/16	7/16	—	9/16	⅝	11/16	¾	15/16	1	1⅛	1 3/16	1¼

WEIGHTS IN KG AND POUNDS:

110, 114	kg	2.8	2.9	4.0	5.9	7.0	9.0	—	—	—	—	—	—	—
	lb	6.3	6.5	9	13	15.5	20							
116	kg	—	—	—	—	7.0	10	14	16	26	56	91	151	223
	lb					15.5	21	30	35	63	124	201	332	492
117	kg	3.6	4.3	5.4	—	9.9	14	19	24	40	72	124	206	305
	lb	8	9.5	12		22	30	42	51	88	158	275	455	675

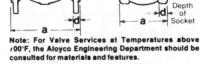

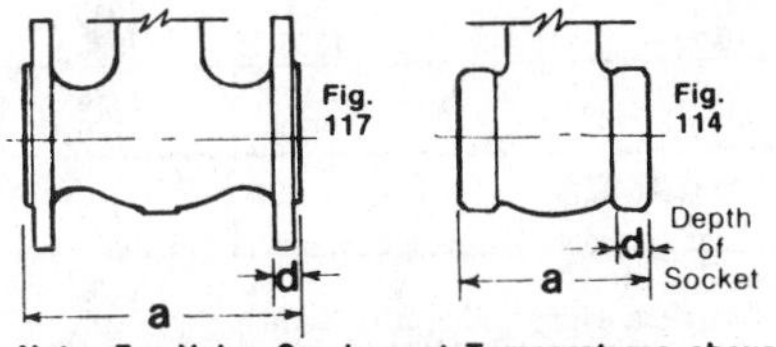
Fig. 117
Fig. 114
Depth of Socket
a
d
a
d

APPLICABLE CLASS 150 STANDARDS:

End Flanges, ANSI B16.5	SW Ends (Bore and Depth), ANSI B16.11
Pipe Threads, ANSI B2.1	BW Ends (Schedule 40), ANSI B16.25 and B16.10
Design, API 603	Face-to-Face, ANSI B16.10
Wall Section, ANSI B16.34	Pressure-Temperature Ratings, ANSI B16.34-1977

Note: For Valve Services at Temperatures above ſ00°F, the Aloyco Engineering Department should be consulted for materials and features.

For Engineering Specifications and Data, see Engineering Section of Catalog.

FIGURE 28.13 ■ Gate valve data. (*Reprinted from David Goetsch and William Chalk,* Technical Drawing, *4th ed. Delmar Publishers, 1994, Figure 21–13*)

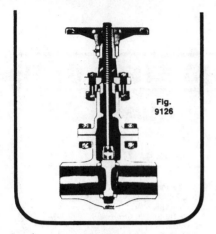

Fig. 9126

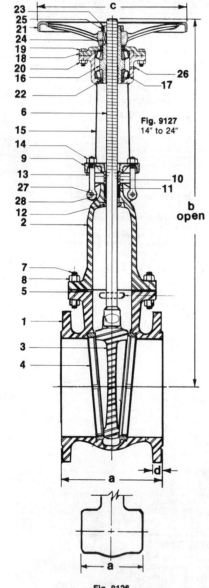

Fig. 9127
14" to 24"

b
open

b

a

d

a

Fig. 9126

STAINLESS STEEL GATE
CLASS 150
FIGURES: 9126, 9127 (14" to 24")

DESIGN DESCRIPTION:

Outside Screw and Yoke	For Flat Faced Flanged Valves, Specify Fig. 9127-FF
Bolted Bonnet	Roller Bearing Yoke
Rising Stem	Renewable Seat Rings
Non-rising Handwheel	Buttwelding Ends (Fig. 9126)
Solid Wedge	Flanged Ends (Fig. 9127)

PARTS AND MATERIAL LIST:

DESCRIPTION	ALOYCO 18-8 SMO	DESCRIPTION	ALOYCO 18-8 SMO
1 body*	A351 GR CF8M	16 yoke bushing	B584 alloy C86200
2 bonnet	A351 GR CF8M	17 roller bearing	steel
3 wedge	A351 GR CF8M	18 yoke cap	CF8
4 seat ring	A351 GR CF8M	19 yoke cap bolt	A193 GR B8
5 gasket	comp. asbestos	20 yoke cap bolt nut	A194 GR 8F
6 stem	type 316	21 handwheel	malleable iron
7 bonnet studbolt	A193 GR B7	22 oil seal	flax
8 bonnet studbolt nut	A194 2H	23 yoke bushing nut	B584 alloy C86200
9 gland plate	A351 GR CF8M	24 handwheel key	steel
10 gland follower	A351 GR CF8M	25 yoke nut set screw	type 303
11 packing	braided asbestos	26 grease fitting	steel
12 stem hole bushing	A351 GR CF8M	27 hinge bolt	A193 GR B8
13 gland eyebolt	A193 GR B8	28 hinge bolt nut	A294 GR 8F
14 gland eyebolt nut	A194 GR 8F	29 yoke studbolt**	A193 GR B8
15 yoke	CF8	30 yoke studbolt nut**	A194 GR 8F

Remarks: *Body material on welding end valves to be ELC grade.
**Not shown.
Other Alloys Available. Refer to Introduction.

**Note: For Valve Services at Temperatures above
700°F, the Aloyco Engineering Department should be
consulted for materials and features.**

DIMENSIONS IN MM AND INCHES:

SIZE	MM INCH	350 14	400 16	450 18	500 20	600 24
a(9126)	mm	571	610	660	711	813
	inch	22½	24	26	28	32
a(9127)	mm	381	406	432	457	508
	inch	15	16	17	18	20
b	mm	1765	1870	2140	2369	2753
	inch	69½	73⅜	84¼	93¼	108⅜
c	mm	508	559	610	660	762
	inch	20	22	24	26	30
d(9127) thickness of flange	mm	35	37	40	43	48
	inch	1⅜	1⁷/₁₆	1⁹/₁₆	1¹¹/₁₆	1⅞

WEIGHTS IN KG AND POUNDS:

9126	kg	342	474	535	708	1043
	lb	754	1046	1180	1560	2300
9127	kg	419	559	644	848	1234
	lb	924	1232	1420	1870	2720

APPLICABLE CLASS 150 STANDARDS

End Flanges, ANSI B16.5	Buttwelding Ends (Schedule 40), ANSI B16.25 and B16.10
Design, ANSI B16.34	Face-to-Face, ANSI B16.10
Wall Section, API 600	Pressure-Temperature Ratings, ANSI B16.34-1977

FIGURE 28.14 ■ Gate valve data. (*Reprinted from David Goetsch and William Chalk,* Technical Drawing, *4th ed. Delmar Publishers, 1994, Figure 21–14*)

STAINLESS STEEL GLOBE
CLASS 150
FIGURES: 422 (½″ to 2″), 427 (½″ to 6″)

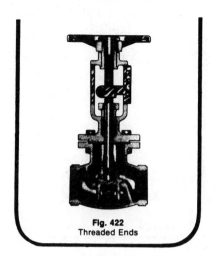

Fig. 422
Threaded Ends

DESIGN DESCRIPTION:

"V" Port Throttling
Bolted Bonnet
Outside Screw and Yoke
Position Indicator
Rising Stem

Non-rising Handwheel
Threaded Ends (Fig. 422)
Flanged Ends (Fig. 427)
When Flat Face End Flanges are
    Required Order as 427-FF

PARTS AND MATERIAL LIST:

DESCRIPTION	ALOYCO 18-8 SMO	DESCRIPTION	ALOYCO 18-8 SMO
1 body	A351 GR CF8M	13 yoke bushing	cast B584 alloy 836 or wrought B16 half hard
2 bonnet	A351 GR CF8M	14 handwheel	malleable iron
3 disc	A351 GR CF8M	15 indicator	CF-16F
4 stem	type 316	16 indicator bolt	A193 GR B8
5 gasket	PTFE	17 escutcheon pins	type 304
6 bonnet bolts	A193 GR B8	18 indicating plate	type 304
7 bonnet bolt nuts	A194 GR 8F	19 yoke bushing nut	cast B584 alloy 836 or wrought B16 half hard
8 gland plate	CF8M	20 handwheel key	steel
9 gland follower	type 316	21 disc pin	type 316
10 packing	PTFE	22 yoke nut set screw	type 303
11 gland bolts	A193 GR B8		
12 gland bolt nuts	A194 GR 8F		

Other Alloys available. (Refer to Introduction)

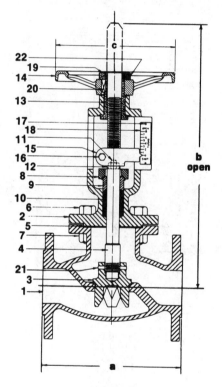

Fig. 427
Sizes ½″ to 6″

Fig. 422 (½″ to 2″)

DIMENSIONS IN MM AND INCHES:

SIZE	MM INCH	15 ½	20 ¾	25 1	40 1½	50 2	65 2½	80 3	100 4	150 6
a	mm	86	95	108	140	165	—	—	—	—
	inch	3⅜	3¾	4¼	5½	6½				
a (F)	mm	108	117	127	165	203	216	241	292	406
	inch	4¼	4⅝	5	6½	8	8½	9½	11½	16
b	mm	180	235	259	283	308	330	368	419	584
	inch	7¹¹/₁₆	9¼	10³/₁₆	11⅛	12⅛	13	14½	16½	23
c	mm	102	127	127	152	178	191	254	305	356
	inch	4	5	5	6	7	7½	10	12	14

WEIGHTS IN KG AND POUNDS:

422	kg	1.8	2.7	4.5	7.2	9.5	—	—	—	—
	lb	4	6	10	16	21				
427	kg	2.7	4	5.9	9.5	13	20	26	43	78
	lb	6	9	13	21	29	43	58	95	173

APPLICABLE CLASS 150 STANDARDS:

End Flanges, ANSI B16.5
Wall Section, ANSI B16.34
Face-to-Face, ANSI B16.10
Pressure-Temperature Ratings, ANSI B16.34-1977
Pipe Threads, ANSI B2.1

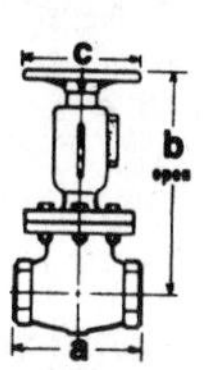

FIGURE 28.15 ■ Gate valve data. (*Reprinted from David Goetsch and William Chalk,* Technical Drawing, *4th ed. Delmar Publishers, 1994, Figure 21–15*)

STAINLESS STEEL GLOBE
CLASS 150
FIGURES: 502 (½″ to 2″), 504 (½″ to 2″), 507 (½″ to 10″)

DESIGN DESCRIPTION:

Outside Screw and Yoke
Bolted Bonnet
Renewable PTFE Disc
Retained Gasket
Rising Stem and Handwheel
For Flat Faced Flanged Valve
  Specify Fig. 507 FF

Integral Seat
Threaded Ends (Fig. 502)
Socket Weld Ends (Fig. 504)
Flanged Ends (Fig. 507)
Valves in Sizes 8″ and 10″ Have
  Disc Guide Below Seat

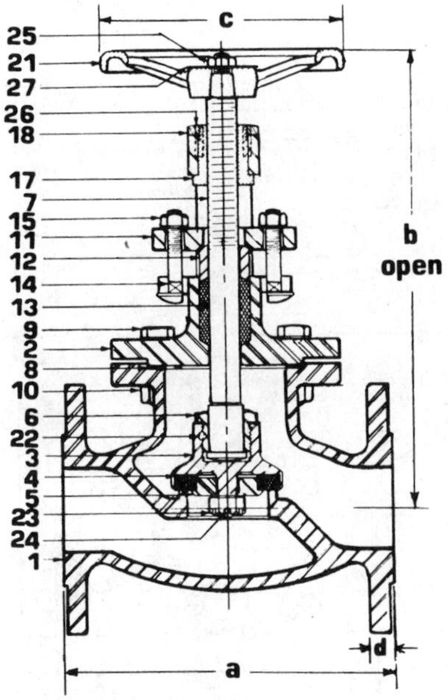

PARTS AND MATERIAL LIST:

DESCRIPTION	ALOYCO 18-8 SMO	DESCRIPTION	ALOYCO 18-8 SMO
1 body*	A351 GR CF8M	14 gland bolt	A193 GR B8
2 bonnet	A351 GR CF8M	15 gland bolt nut	A194 GR 8F
3 disc holder	cast-CF8M, wrought-type 316	16 yoke ***	CF8
4 disc	PTFE	17 yoke bushing	B16 half hard
5 disc holder plate	cast-CF8M,	18 yoke bushing nut	B16 half hard
6 swivel nut	wrought-type 316	19 yoke bolt ***	A193 GR B8
7 stem	type 316	20 yoke bolt nut ***	A194 GR 8F
8 gasket	PTFE	21 handwheel	malleable
9 bonnet bolt	A193 GR B8	22 swivel nut pin	type 316
10 bonnet bolt nut	A194 GR 8F	23 disc holder plate nut**	type 316
11 gland plate	CF8M	24 disc holder nut pin	type 316
12 gland follower	cast-CF8M, wrought-type 316	25 handwheel nut	type 303
		26 yoke nut pin	type 303
13 packing	PTFE	27 I.D. plate	type 304

Remarks: *Body material on threaded and welding end valves is ELC grade.
     **Sizes ½″ to 2″ incl.
   ***Sizes 8″ and 10″ only.
     Other Alloys Available. Refer to Introduction.
   Note: For Valve Services at Temperatures above 700°F, the Aloyco Engineering Department should be consulted for materials and features.

DIMENSIONS IN MM AND INCHES:

SIZE	MM	15	20	25	40	50	65	80	100	150	200	250
	INCH	½	¾	1	1½	2	2½	3	4	6	8	10
a(502,504)	mm	86	95	108	140	165	—	—	—	—	—	—
	inch	3⅜	3¾	4¼	5½	6½						
a(507)	mm	108	117	127	165	203	216	241	292	406	495	622
	inch	4¼	4⅝	5	6½	8	8½	9½	11½	16	19½	24½
b(502,504)	mm	165	210	248	283	305	—	—	—	—	—	—
	inch	6½	8¼	9¾	11⅛	12						
b(507)	mm	165	210	251	283	305	327	359	425	533	622	822
	inch	6½	8¼	9⅞	11⅛	12	12⅞	14⅛	16¾	21	24½	32⅜
c	mm	67	76	102	127	152	178	203	254	305	406	457
	inch	2⅝	3	4	5	6	7	8	10	12	16	18
d(504) Depth of Socket	mm	10	13	13	13	16	—	—	—	—	—	—
	inch	⅜	½	½	½	⅝						
d(507) Flange Thickness	mm	11	11	11	14	16	17	19	24	25	29	30
	inch	⁷/₁₆	⁷/₁₆	⁷/₁₆	⁹/₁₆	⅝	¹¹/₁₆	¾	¹⁵/₁₆	1	1⅛	1³/₁₆

WEIGHTS IN KG AND POUNDS:

502,504	kg	2.7	5	5	7.3	11	—	—	—	—	—	—
	lb	6	10	10	16	24						
507	kg	5	6.4	6.4	10	15	20.8	26	43	78	128.8	188.7
	lb	10	14	14	22	34	46	58	95	173	284	416

APPLICABLE CLASS 150 STANDARDS:

End Flanges, ANSI B16.5
Wall Section, ANSI B16.34
Face-to-Face, ANSI B16.10
Pipe Threads, ANSI B2.1

Socket Weld Ends (Bore and Depth), ANSI B16.11
Pressure-Temp. Ratings, ANSI B16.34-1977
  Maximum Service Temperature, 450°F

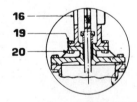

Separable Yoke Detail
Sizes 8″ and 10″ only

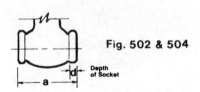

Fig. 502 & 504

FIGURE 28.16 ■ Globe valve data. (*Reprinted from David Goetsch and William Chalk,* Technical Drawing, *4th ed. Delmar Publishers, 1994, Figure 21–16*)

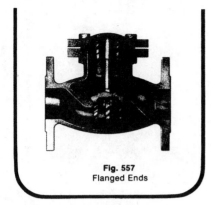

Fig. 557
Flanged Ends

STAINLESS STEEL
LIFT CHECK VALVES
CLASS 150
**FIGURES: 550 (¼″ to 2″), 554 (¼″ to 2″),
556 (½″ to 6″), 557 (½″ to 6″)***

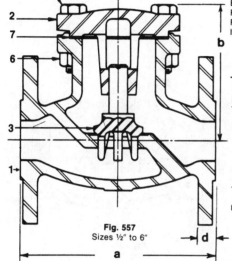

Fig. 557
Sizes ½″ to 6″

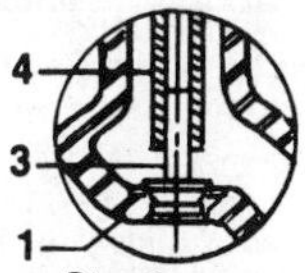

Disc Detail
Sizes ¼″ to ¾″ incl.

DESIGN DESCRIPTION:

Horizontal Type
Bolted Cover
Retained Gasket
Regrinding Disc
Integral Seat

Threaded Ends (Fig. 550)
Socket Weld Ends (Fig. 554)
Buttwelding Ends (Fig. 556)
Flanged Ends (Fig. 557)

PARTS AND MATERIAL LIST:

DESCRIPTION	ALOYCO 18-8 SMO
1 body*	A351 GR CF8M
2 cover	A351 GR CF8M
3 disc	cast CF8M or wrought type 316
4 disc guide**	type 316
5 cover bolt	A193 GR B8
6 cover bolt nut	A194 GR 8F
7 gasket	comp. asbestos

Remarks: *Body material on threaded and welding end valves is ELC grade.
Other Alloys available. (Refer to Introduction)
**Sizes ¼″ to ¾″ incl.
†Unless otherwise specified, these valves may be supplied at the manu-
facturer's option with: PTFE Gasket.

Note: For Valve Services at Temperatures above
700°F, the Aloyco Engineering Department should be
consulted for materials and features.

DIMENSIONS IN MM AND INCHES:

SIZE	MM	6	10	15	20	25	40	50	65	80	100	150
	INCH	¼	⅜	½	¾	1	1½	2	2½	3	4	6
a (550,554)	mm	79	79	86	95	108	140	165	—	—	—	—
	inch	3⅛	3⅛	3⅜	3¾	4¼	5½	6½				
a (556,557)	mm	—	—	108	117	127	165	203	216	241	292	406
	inch			4¼	4⅝	5	6½	8	8½	9½	11½	16
b	mm	67	67	67	76	86	102	116	133	146	168	200
	inch	2⅝	2⅝	2⅝	3	3⅜	4	4⁹/₁₆	5¼	5¾	6⅝	7⅞
d (554)	mm	10	10	10	13	13	13	16	—	—	—	—
	inch	⅜	⅜	⅜	½	½	½	⅝				
d (557)	mm	—	—	11	11	11	14	16	17	19	24	25
	inch			⁷/₁₆	⁷/₁₆	⁷/₁₆	⁹/₁₆	⅝	¹¹/₁₆	¾	¹⁵/₁₆	1

WEIGHTS IN KG AND POUNDS:

		6	10	15	20	25	40	50	65	80	100	150
550,554	kg	1.4	1.4	1.8	2.3	2.9	5.9	7.3	—	—	—	—
	lb	3	3	4	5	6.3	13	16				
556	kg	—	—	2	2.7	3.4	6.8	10	12	18.6	28	51
	lb			4.5	6	7.5	15	22	26	41	61	112
557	kg	—	—	2.3	3	4.1	8	12.7	14	22	34	68
	lb			5	7	9	17.5	28	31	48	75	150

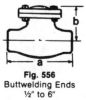

Fig. 556
Buttwelding Ends
½″ to 6″

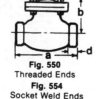

Fig. 550
Threaded Ends
Fig. 554
Socket Weld Ends

APPLICABLE CLASS 150 STANDARDS:

End Flanges, ANSI B16.5
Wall Section, ANSI B16.34
Face-to-Face, ANSI B16.10
Pipe Threads, ANSI B2.1

SW Ends (Bore and Depth), ANSI B16.11
BW Ends (Sch. 40), ANSI B16.25 and B16.10
Pressure-Temp. Ratings, ANSI B16.34-1977

FIGURE 28.17 ■ Check valve data. (*Reprinted from David Goetsch and William Chalk,* Technical Drawing, *4th ed. Delmar Publishers, 1994, Figure 21–17*)

STAINLESS STEEL LIFT CHECK VALVES

CLASS 600

FIGURES: 4550-A (½″ to 2″)
4554-A (½″ to 2″)
4557-A (½″ to 2″)

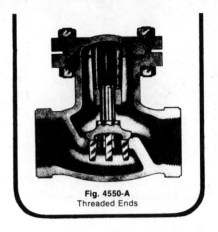

Fig. 4550-A
Threaded Ends

DESIGN DESCRIPTION:

Bolted Cover
Integral Seat
Horizontal Type
Threaded Ends (Fig. 4550-A)
Socket Weld Ends (Fig. 4554-A)
Buttwelding Ends (Fig. 4557-A)

PARTS AND MATERIAL LIST:

DESCRIPTION	ALOYCO 18-8 SMO
1 body	A351 GR CF8M
2 cover	A351 GR CF8M
3 disc	type 316
4 disc guide (½″ and ¾″)	type 316
5 cover studbolt	A193 GR B8
6 cover studbolt nut	A194 GR 8F
7 gasket	comp. asbestos

Other Alloys Available. Refer to Introduction.

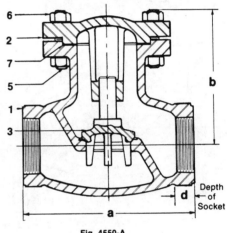

Fig. 4550-A
Sizes ½″ to 2″

DIMENSIONS IN MM AND INCHES:

SIZE	MM INCH	15 ½	20 ¾	25 1	40 1½	50 2
a (4550-A, 4554-A)	mm inch	95 3¾	108 4¼	127 5	165 6½	191 7½
a (4557-A)	mm inch	165 6½	191 7½	216 8½	241 9½	387 15¼
b	mm inch	94 3¹¹/₁₆	94 3¹¹/₁₆	97 3¹³/₁₆	148 5¹³/₁₆	165 6½
d (4554-A)	mm inch	10 ⅜	13 ½	13 ½	13 ½	16 ⅝
d (4557-A)	mm inch	14 ⁹/₁₆	16 ⅝	17 ¹¹/₁₆	22 ⅞	25 1

WEIGHTS IN KG AND POUNDS:

4550-A, 4554-A	kg lb	2.3 5	3 7	5 11	10 22	14 30
4557-A	kg lb	5 11	6.3 14	10 22	15 32	22 48

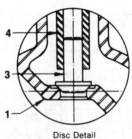

Disc Detail
Sizes ½″ and ¾″

APPLICABLE CLASS 600 STANDARDS:

End Flanges, ANSI B16.5
Wall Section, ANSI B16.34
Face-to-Face, ANSI B16.10
Buttwelding Ends (Schedule 40),
   ANSI B16.25 and B16.10

Pipe Threads, ANSI B2.1
Pressure-Temperature Rating, ANSI B16.34-1977
Socket Weld Ends (Bore and Depth), ANSI B16.11

Note: For Valve Services at Temperatures above 700°F, the Aloyco Engineering Department should be consulted for materials and features.

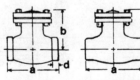

Fig. 4554-A
Socket Weld Ends
½″ to 2″

Fig. 4557-A
Buttwelding Ends
½″ to 2″

FIGURE 28.18 ■ Check valve data. (*Reprinted from David Goetsch and William Chalk,* Technical Drawing, *4th ed. Delmar Publishers, 1994, Figure 21–18*)

PIPE DRAWINGS

Pipe drawings are used, like any other design drawing, to convey the intentions of the designer and engineer to the trades people who will actually construct the piping system. There are two types of pipe drawings: single line and double line. Figure 28.19 is an example of a double-line drawing. Figure 28.20 is an example of the single-line version of the same drawing. Single-line drawings, since they are schematic, can be drawn much faster. However, drafters and engineers should be familiar both with double-line and single-line versions of pipe drawings, since both have applications on the job.

Single-Line Drawings

Single-line pipe drawings consist of schematic, symbolic representations of valves, fittings, and so on connected by a single line that represents the centerline of the pipe. Figure 28.21 contains all of the various schematic symbols used for representing pipe fittings and valves on single-line pipe drawings. Figure 28.22 contains the graphic symbols used for representing piping on single-line pipe drawings.

Like mechanical drawings, pipe drawings may be drawn using orthographic or isometric projection. Single-line orthographic projection is the recommended form of representation

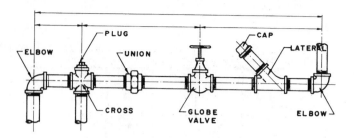

FIGURE 28.19 ■ Double-line pipe drawing. (*Reprinted from David Goetsch and William Chalk,* Technical Drawing, *4th ed. Delmar Publishers, 1994, Figure 21–19*)

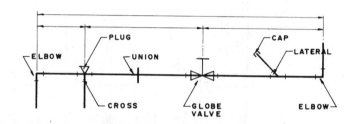

FIGURE 28.20 ■ Single-line pipe drawing. (*Reprinted from David Goetsch and William Chalk,* Technical Drawing, *4th ed. Delmar Publishers, 1994, Figure 21–20*)

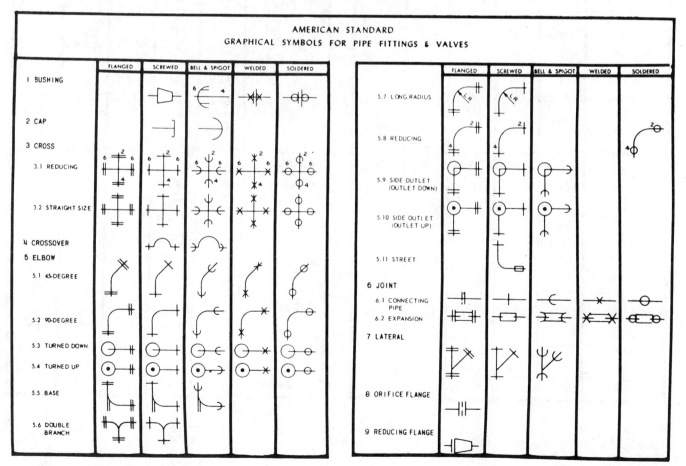

FIGURE 28.21 ■ Symbols for pipe fittings and valves. (*Reprinted from David Goetsch and William Chalk,* Technical Drawing, *4th ed. Delmar Publishers, 1994, Figure 21–21*)

Continued

AMERICAN STANDARD
GRAPHICAL SYMBOLS FOR PIPE FITTINGS & VALVES

	FLANGED	SCREWED	BELL & SPIGOT	WELDED	SOLDERED
10 PLUGS					
10.1 BULL PLUG					
10.2 PIPE PLUG					
11 REDUCER					
11.1 CONCENTRIC					
11.2 ECCENTRIC					
12 SLEEVE					
13 TEE					
13.1 (STRAIGHT SIZE)					
13.2 (OUTLET UP)					
13.3 (OUTLET DOWN)					
13.4 DOUBLE SWEEP					
13.5 REDUCING					
13.6 SINGLE SWEEP					
13.7 SIDE OUTLET (OUTLET DOWN)					
13.8 SIDE OUTLET (OUTLET UP)					
14 UNION					
15 ANGLE VALVE					
15.1 CHECK					
15.2 GATE (ELEVATION)					
15.3 GATE (PLAN)					
15.4 GLOBE (ELEVATION)					
15.5 GLOBE (PLAN)					
15.6 HOSE ANGLE	SAME AS	SYMBOL	23.1		
16 AUTOMATIC VALVE					
16.1 BY-PASS					

	FLANGED	SCREWED	BELL & SPIGOT	WELDED	SOLDERED
16.2 GOVERNOR-OPERATED					
16.3 REDUCING					
17 CHECK VALVE					
17.1 ANGLE CHECK	SAME AS	SYMBOL	15.1		
17.2 (STRAIGHT WAY)					
18 COCK					
19 DIAPHRAGM VALVE					
20 FLOAT VALVE					
21 GATE VALVE					
*21.1					
21.2 ANGLE GATE	SAME AS	SYMBOLS	15.2 & 15.3		
21.3 HOSE GATE	SAME AS	SYMBOL	23.2		
21.4 MOTOR-OPERATED					
22 GLOBE VALVE					
22.1					
22.2 ANGLE GLOBE	SAME AS	SYMBOLS	15.4 & 15.5		
22.3 HOSE GLOBE	SAME AS	SYMBOL	23.3		
22.4 MOTOR-OPERATED					
23 HOSE VALVE					
23.1 ANGLE					
23.2 GATE					
23.3 GLOBE					
24 LOCKSHIELD VALVE					
25 QUICK OPENING VALVE					
26 SAFETY VALVE					
27 STOP VALVE	SAME AS	SYMBOL	21.1		

*ALSO USED FOR GENERAL **STOP VALVE** SYMBOL WHEN AMPLIFIED BY SPECIFICATION

FIGURE 28.21 ■ Continued

AMERICAN STANDARD
GRAPHICAL SYMBOLS FOR PIPING

AIR CONDITIONING

28	BRINE RETURN	— — — BR — — —
29	BRINE SUPPLY	——— B ———
30	CIRCULATING CHILLED OR HOT WATER FLOW	——— CH ———
31	CIRCULATING CHILLED OR HOT WATER RETURN	— — CHR — — —
32	CONDENSER WATER FLOW	——— C ———
33	CONDENSER WATER RETURN	— — — CR — — —
34	DRAIN	——— D ———
35	HUMIDIFICATION LINE	— · — H — · —
36	MAKE-UP WATER	— · — · — · —
37	REFRIGERANT DISCHARGE	——— RD ———
38	REFRIGERANT LIQUID	——— RL ———
39	REFRIGERANT SUCTION	— — — RS — — —

HEATING

40	AIR-RELIEF LINE	— — — — —
41	BOILER BLOW OFF	— — — —
42	COMPRESSED AIR	——— A ———
43	CONDENSATE OR VACUUM PUMP DISCHARGE	— O — O — O —
44	FEED WATER PUMP DISCHARGE	— OO — OO — OO —
45	FUEL-OIL FLOW	——— FOF ———
46	FUEL-OIL RETURN	— — — FOR — — —
47	FUEL-OIL TANK VENT	— — — FOV — — —
48	HIGH-PRESSURE RETURN	— ⫫ — ⫫ — ⫫ —
49	HIGH-PRESSURE STEAM	— ⫫ — ⫫ — ⫫ —
50	HOT-WATER HEATING RETURN	— — — — — —
51	HOT-WATER HEATING SUPPLY	———————

52	LOW PRESSURE RETURN	— — — — — —
53	LOW PRESSURE STEAM	———————
54	MAKE-UP WATER	— · — · — · —
55	MEDIUM PRESSURE RETURN	—⧸—⧸—⧸—
56	MEDIUM PRESSURE STEAM	⧸ — ⧸ — ⧸

PLUMBING

57	ACID WASTE	———— ACID ————
58	COLD WATER	———————
59	COMPRESSED AIR	——— A ———
60	DRINKING-WATER FLOW	— · — · — · —
61	DRINKING-WATER RETURN	— — — — — —
62	FIRE LINE	— F ——— F —
63	GAS	— G ——— G —
64	HOT WATER	— ·· — ·· — ·· —
65	HOT WATER RETURN	— ··· — ··· — ··· —
66	SOIL, WASTE OR LEADER (ABOVE GRADE)	———————
67	SOIL, WASTE OR LEADER (BELOW GRADE)	— — — — — —
68	VACUUM CLEANING	— V ——— V —
69	VENT	— — — — — — —

PNEUMATIC TUBES

70	TUBE RUNS	═══════

SPRINKLERS

71	BRANCH AND HEAD	—— O ——— O ——
72	DRAIN	— S ——— S —
73	MAIN SUPPLIES	——— S ———

FIGURE 28.22 ■ Symbols for piping. (*Reprinted from David Goetsch and William Chalk*, Technical Drawing, *4th ed. Delmar Publishers, 1994, Figure 21–22*)

in most cases. Figure 28.23 is an example of single-line orthographic projection of a pipe drawing. Single-line isometric projection, as illustrated in Figure 28.24, is often used for assembly and layout work because the drawing is easier for trades people to understand. Regardless of whether you are using orthographic or isometric projection, there are certain drawing conventions with which you should be familiar. The most important of these are crossings, connections, fittings, machines, and devices.

Crossings. When lines representing pipe on a drawing cross but do not intersect or make connections, they are usually drawn without breaks. However, on occasion it is necessary to show breaks so that trades people using the plans will be absolutely sure as to which pipe is nearest to them and which is farthest away. When such a need arises, breaks can be used as illustrated in Figure 28.25. When using a break, it should be wide enough to be easily seen but not so wide as to create an inordinate gap. A widely used rule of thumb is to make the break from five to ten times the thickness of the lines representing the pipe.

Connections. All of the various means of making connections at joints and fittings fall into two categories: permanent and detachable. On single-line pipe drawings, permanent connections are indicated by using a heavy dot. Detachable connections are indicated by a single thick line. Both of these methods are illustrated in Figure 28.26.

Fittings. When drawing fittings on single-line drawings, it is sometimes necessary to be able to indicate whether a pipe is coming toward the viewer or going away from the viewer. It is also necessary at times to indicate whether the viewer is looking at a front view of the fitting or a rear view. All of the various types of single-line-drawing situations that you may confront when drawing fittings are covered in Figure 28.21. Refer to Figure 28.22 when drawing any type of single-line fitting symbol.

Machines and Devices. Most piping systems will tie into machines, devices, and other types of apparatus (Figure 28.27). When this is the case, the machines, devices, and other types of apparatus can be drawn using thin phantom lines.

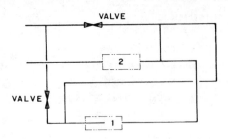

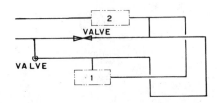

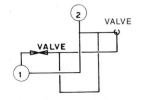

FIGURE 28.23 ■ Orthographic single-line pipe drawing. (*Reprinted from David Goetsch and William Chalk,* Technical Drawing, *4th ed. Delmar Publishers, 1994, Figure 21–23*)

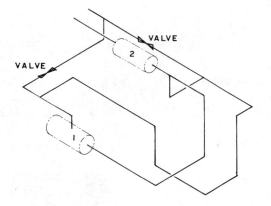

FIGURE 28.24 ■ Isometric single-line pipe drawing. (*Reprinted from David Goetsch and William Chalk,* Technical Drawing, *4th ed. Delmar Publishers, 1994, Figure 21–24*)

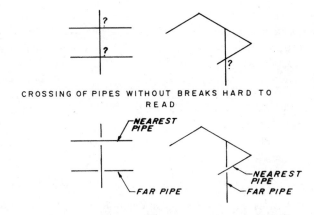

CROSSING OF PIPES WITHOUT BREAKS HARD TO READ

CROSSING OF PIPES WITH BREAK INDICATES NEAREST PIPE

FIGURE 28.25 ■ Drawing crossing pipe. (*Reprinted from David Goetsch and William Chalk,* Technical Drawing, *4th ed. Delmar Publishers, 1994, Figure 21–25*)

Dimensioning Pipe Drawings

There are several dimensioning rules that apply specifically to pipe drawings with which you should be familiar. The most important of these are:

■ Dimensions for pipes and pipe fittings should be shown from center to center of pipe and to the outside face of the pipe end or flange.

■ The length of pipe is not shown on the drawing except in rare cases.

■ Pipe sizes are shown on the drawing using leader lines or as a general note.

■ Fitting sizes are shown on the drawing using leader lines or as a general note.

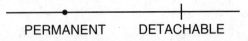

PERMANENT   DETACHABLE

FIGURE 28.26 ■ Drawing pipe connections. (*Reprinted from David Goetsch and William Chalk,* Technical Drawing, *4th ed. Delmar Publishers, 1994, Figure 21–26*)

■ Pipes with bends should be dimensioned from vertex to vertex.

■ The radii of bent pipe should be indicated using a leader line. Supplementary angles of bent pipes should be dimensioned in the normal manner.

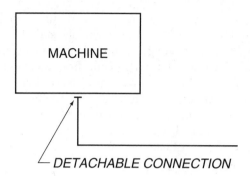

FIGURE 28.27 ■ Drawing adjoining machinery. (*Reprinted from David Goetsch and William Chalk*, Technical Drawing, *4th ed. Delmar Publishers, 1994, Figure 21–27*)

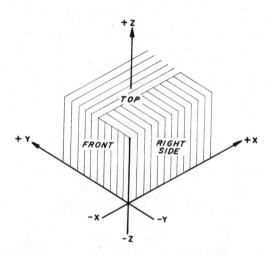

FIGURE 28.28 ■ Isometric axes. (*Reprinted from David Goetsch and William Chalk*, Technical Drawing, *4th ed. Delmar Publishers, 1994, Figure 21–28*)

- Outside diameters and wall thicknesses of pipe may be indicated on the drawing using a leader line or may be made part of a general note.
- A bill of material should be developed and placed directly on the pipe drawing or attached to it.

Single-Line Isometric Piping Drawings

Isometric piping drawings are prepared in a manner similar to mechanical piping drawings. The lines representing pipes are drawn parallel to either the X, the Y, or the Z axis (Figure 28.28). Flanges are represented using short strokes of equal thickness (Figure 28.29). Notice that flanges for horizontal pipe are drawn vertically and flanges for vertical pipe are drawn at the appropriate 30-degree angle to the horizontal.

Figure 28.30 illustrates the various methods used for drawing valves. The unidirectional dimensioning illustrated in Figure 28.31 should be used in isometric piping drawings.

FIGURE 28.29 ■ Drawing flanges. (*Reprinted from David Goetsch and William Chalk*, Technical Drawing, *4th ed. Delmar Publishers, 1994, Figure 21–29*)

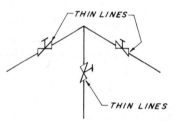

VALVES WITH THREADED CONNECTIONS

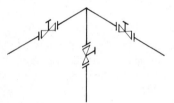

VALVES WITH FLANGE CONNECTIONS

FIGURE 28.30 ■ Drawing valves. (*Reprinted from David Goetsch and William Chalk*, Technical Drawing, *4th ed. Delmar Publishers, 1994, Figure 21–30*)

PUMPS, TANKS, AND VESSELS

Pipes transport liquids and gases from one destination to another. The receiving points at these destinations are typically pieces of equipment that, together with the piping, make up a piping system. The most commonly found equipment in a process piping system consists of pumps, tanks, and vessels.

Pumps

A *pump* is a mechanical device that causes liquid to flow through pipes. It pulls liquid into one end (suction) and pushes it out the other (pressure) to the next destination. Pumps are one of the most common components in a process pipe system, and the centrifugal pump is the most commonly used kind of pump. Figure 28.32 is a schematic drawing that shows the basic elements of a centrifugal pump.

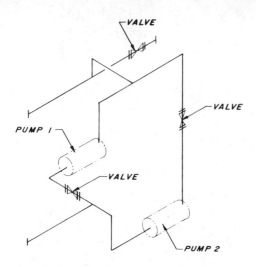

FIGURE 28.31 ■ Isometric pipe drawing.
(*Reprinted from David Goetsch and William Chalk,
Technical Drawing, 4th ed. Delmar Publishers,
1994, Figure 21–31*)

Drawing Pumps

Pumps are drawn in both plan view and elevations. A *general-arrangement drawing* is a plan view that locates pumps on the floor of the building or within the vessel that will house the piping system (Figure 28.33). In general-arrangement drawings, pumps may appear in single-line representation. Most pipe drawings consist of single-line pipes with equipment rep-

resented by symbols. Commonly used pump symbols are shown in Figure 28.34. Gasket symbols and the preferred methods for dimensioning them are shown in Figure 28.35. The actual dimensions for pumps come from catalogs produced by pump vendors and are usually available in both hard copy and on-line versions. When making pipe drawings that include pumps, clearances around the pump, access to the pump, and the arrangement of related piping so as to avoid flow inhibition are all important considerations. What follows are some general rules to follow when designing and drawing piping systems that include pumps.

Access and Clearance

1. Ensure that there is a clear area around all pumps that is at least 3 feet wide. This area should be free of piping, wiring, other equipment, and miscellaneous materials of any kind. Remember to think in three dimensions (length, width, and height). At the motor or drive end of the pump, this clearance distance should be increased to at least 5 feet 6 inches. At the suction end, create a design that puts as few pipes right in front of the pump as possible. Above the pump, design the system so that no pipes run directly over the pump.

2. When positioning pumps in a building or any other kind of structure, position them as near as possible to an access point (large door that opens onto a driveway, etc.).

3. Hand wheels for valves should point toward the access area for the pump (away from the pump).

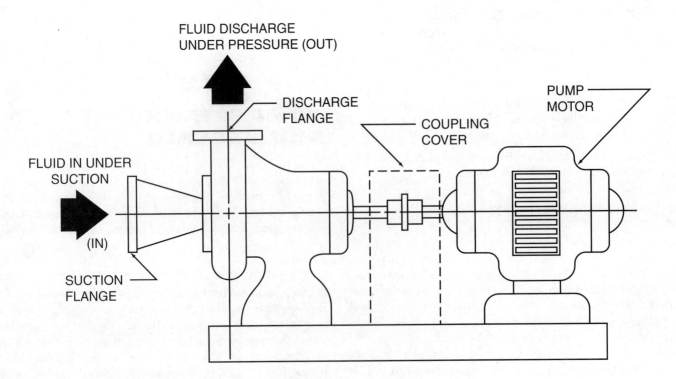

FIGURE 28.32 ■ Pump schematic.

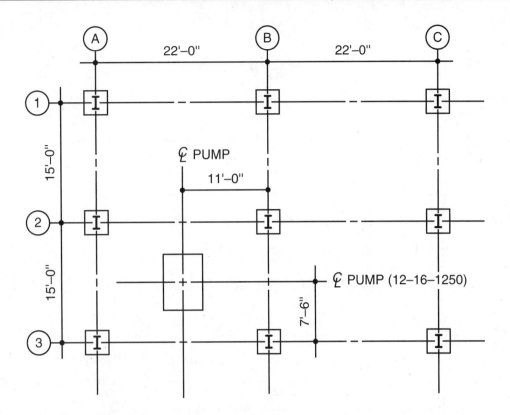

FIGURE 28.33 ■ Pumps are located in the plan view.

FIGURE 28.34 ■ Examples of pump symbols.

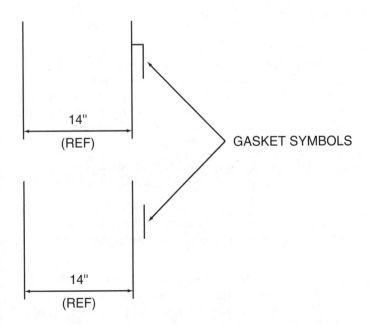

FIGURE 28.35 ■ These symbols show that the gaskets are *not* included in the dimensions.

Pipe, Fittings, and Valves. It is important to design the piping in a system that includes pumps so that the best possible flow is allowed. When designing the piping that goes into and out of a pump, apply the following rules:

1. The ideal suction line is straight and short. If a pipe must be bent at all, make sure to use the longest radius elbow possible and place the elbow in the vertical plane.

2. If vertical suction pipe is necessary, put the flat side of eccentric reducers on the bottom to avoid air pockets in the line (Figure 28.36). Put the flat side on the bottom with horizontal suction piping (Figure 28.36). Air or gas pockets can form if eccentric reducers are not used on the suction side of the pump. When this is the case, be sure to provide vents to release the air or gas.

3. The discharge side of a pump should always be fitted with a gate valve. In addition, there should be a check valve between the gate valve and the discharge flange to prevent backflow.

4. Pipe lines running into the suction end of the pump should slope down slightly toward the pump.

5. When the pipe in the system is 8 inches in diameter or larger, a strainer should be placed between two flanges on the suction side of the pump to prevent the introduction of debris into the pump.

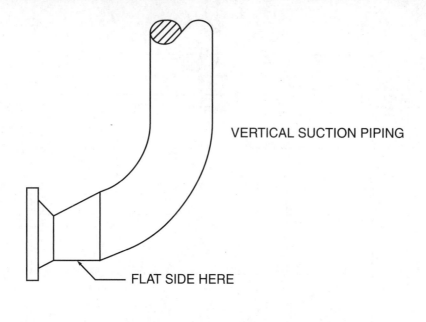

VERTICAL SUCTION PIPING

FLAT SIDE HERE

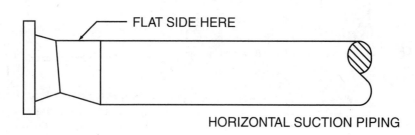

FLAT SIDE HERE

HORIZONTAL SUCTION PIPING

FIGURE 28.36 ■ Orientation of eccentric reducers is important to prevent air pockets.

Tanks and Vessels

Tanks and vessels are frequently part of process piping systems. A *tank* is any container in a piping system that is used to store liquids. Tanks come in numerous shapes and sizes. They can be square, cylindrical, or spherical. They can be horizontal, vertical, or even installed at an angle. A *vessel* is a tank that is specially designed to withstand high operating pressure. In addition, a vessel might be fitted with special controls and components related to a given process. Tanks just store liquids. Vessels may be used to process liquids. Figure 28.37 shows the various standard components of tanks and vessels.

The *shell* is the casing that is the largest component of the tank or vessel and gives it its shape. Shells are typically made of steel but can, depending on the use of the tank or vessel, be made of plastic, fiberglass, or other materials. The *end* and *head* are the caps attached to enclose the shell. The *saddle* consists of the vertical supports or "legs" of the tank or vessel. The saddle is typically made of either steel or concrete. The *nozzle* is a short flanged pipe to which other piping in the system can be attached. The *access way* is an opening that allows the tank or vessel to be maintained. Figure 28.38 shows the various symbols used to depict tanks and vessels on flow diagrams.

Tank and Vessel Drawings

CAD technicians may be called upon to prepare two different types of drawings relating to tanks and vessels. The first type is the *piping layout drawing*. Layout drawings follow the same rules set forth in this chapter for piping systems and use the symbols shown in Figure 28.38 to depict tanks and vessels in the system. In addition, the following general rules apply:

1. Provide sufficient space in the facility above and around where the tank/vessel will be located to allow installation by crane or davit.

2. When routing pipe to a tank/vessel, route it from the highest point first. Also route the largest diameter first.

3. Arrange all ladders, accessways, platforms, and other miscellaneous equipment on one side of the tank/vessel so that the other side has free access for piping connections.

4. Design piping that runs along the outside of a vertical tank/vessel so that it goes outside or alongside platforms but not through them.

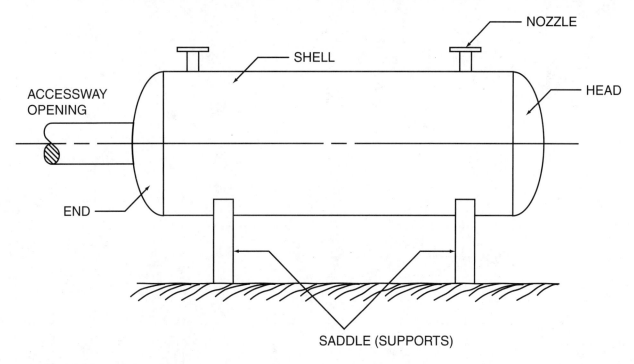

FIGURE 28.37 ■ Typical components of a tank or vessel.

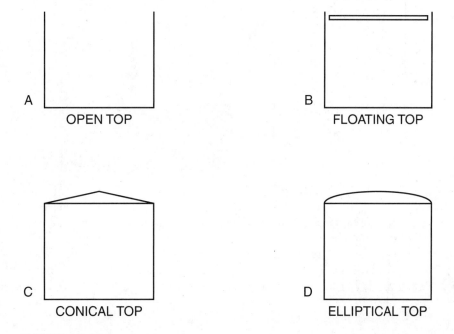

FIGURE 28.38 ■ Tank/vessel symbols for flow diagrams.

5. Ensure that all tanks/vessels have an accessway either at the top or bottom (in the case of vertical equipment) or at one end (in the case of horizontal equipment).

6. Place inlets at the top and outlets at the bottom of vertical tanks/vessels. Place inlets and outlets at opposite ends of horizontal tanks/vessels (inlets on top of the equipment and outlets on the bottom).

7. Ensure that tanks/vessels have the appropriate vents, drains, and steam clean-outs.

8. Ensure that vessels have the appropriate relief valves.

The second type of tank or vessel-related drawing is the shop or *fabrication drawing*. These drawings are used by a shop to actually fabricate the tank or vessel in question. These types of drawings are regular orthographic drawings that require plan and elevation views, sections, details, and bills of material. Figure 28.39 is an example of a layout drawing of a piping system that contains a pump, a tank, and a vessel.

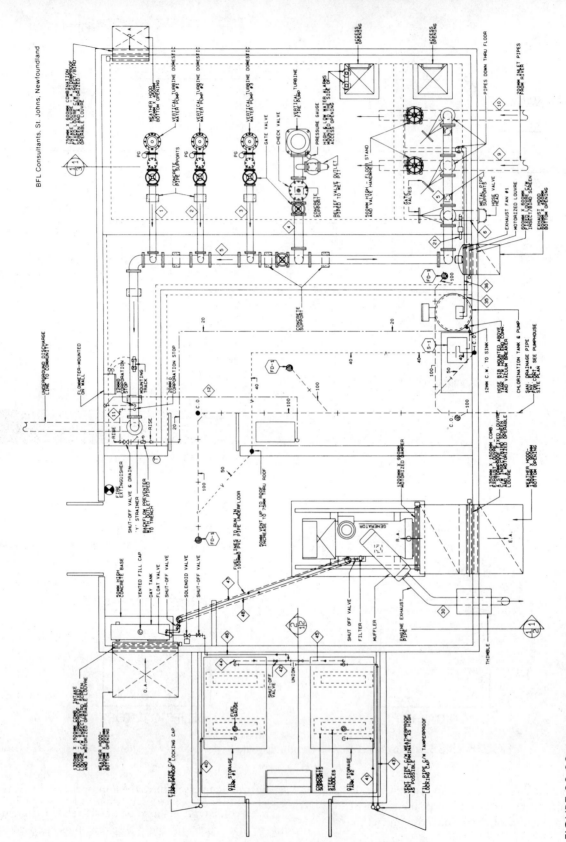

FIGURE 28.39 ■ Double-line representation of a piping system containing pumps, tanks, and vessels. (*Reprinted from David Goetsch and William Chalk, Technical Drawing, 4th ed. Delmar Publishers, 1994, Problem 21–5*)

SUMMARY

■ The most widely used types of pipe are steel, cast iron, brass, copper, copper tubing, and plastic.

■ The three broad classifications of pipe fittings are screwed, flanged, and welded.

■ Fluids and gases flowing through pipe are regulated or stopped at certain points by valves. The most widely used types of valves are gate, globe, jet, swivelled joint, ball, check, and butterfly.

■ Pipe drawings serve the same purpose as other types of technical drawings. They convey the intentions of designers and engineers to the tradespeople who will construct the pipe system. There are two types of pipe drawings: single line and double line.

■ The most common types of equipment found in process piping systems are pumps, tanks, and vessels. A pump is a mechanical device that causes liquid to flow through pipes. A tank is any container in a piping system that is used to store liquids. A vessel is a tank that is specially designed to withstand high operating pressure.

REVIEW QUESTIONS

1. List the most widely used types of pipe.

2. List and describe the three broad classifications of pipe joints/fittings.

3. List and describe the most commonly used kinds of pipe valves.

4. Explain the uses of the following pipe drawing conventions: crossings, connections, fittings, and machines/devices.

5. Define the following terms: pump, tank, and vessel.

6. Explain the term *general-arrangement drawing*.

7. Make a sketch that illustrates the following components of a tank/vessel: shell, end, head, saddle, nozzle, and access way.

CAD ACTIVITIES

GENERAL INSTRUCTIONS

The following activities may be completed on any CAD system. Before reading the *specific instructions* for each activity (below), go through each step in the following planning checklist. The checklist applies to any CAD system and will help ensure the optimum use of your time and resources.

1. Analyze the problem carefully. Decide exactly what you are being asked to do.

2. Determine what resources and references you will need in order to complete the problem and collect them.

3. Decide if any particular standards apply to the project and have those standards available.

4. Determine what types of views will be required and how many of each.

5. Determine what the final plotted scale of the drawing will need to be, and select the appropriate paper size for plotting/printing (make sure the appropriate paper size is available).

6. Plan your drawing sequence. In what order will you develop the drawing (i.e., lines, features, dimension lines, leaders, dimensions, notes, etc.)?

7. Review the various CAD commands you will have to use in order to develop the drawing.

8. Examine your CAD system to ensure that everything is in working order, then begin the project.

SPECIFIC INSTRUCTION

Activity 28.1—Figure 28.40 is a single-line isometric pipe drawing. Convert this drawing into a single-line orthographic drawing showing the following sides: top, front, right, and left.

Activity 28.2—Convert Figure 28.40 into a double-line isometric pipe drawing.

Activity 28.3—Figure 28.41 is a double-line orthographic pipe drawing. Convert it into a single-line orthographic drawing.

Activity 28.4—Convert Figure 28.41 into a double-line isometric pipe drawing.

Activity 28.5—Figure 28.42 is a double-line orthographic pipe drawing. Convert it into a single-line orthographic drawing.

Activity 28.6—Figure 28.43 is a double-line orthographic pipe drawing. Convert it into a single-line orthographic drawing.

Activity 28.7—Locate an existing piping system in your community (water or wastewater pumping station, pump station for the local gas company, process piping system in a local processing plant). Ask for permission from the appropriate authority to make a sketch of a major portion of the system. Then create a single- and double-line drawing from the sketch.

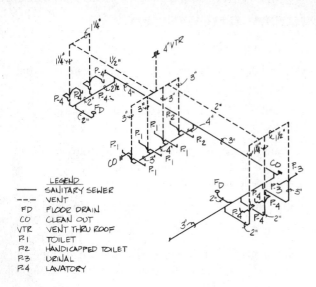

LEGEND
——	SANITARY SEWER
- - -	VENT
FD	FLOOR DRAIN
CO	CLEAN OUT
VTR	VENT THRU ROOF
P.1	TOILET
P.2	HANDICAPPED TOILET
P.3	URINAL
P.4	LAVATORY

SANITARY SEWER RISER DIAGRAM
NTS

FIGURE 28.40 ■ CAD Activity 28.1 and Activity 28.2. (*Reprinted from David Goetsch and William Chalk,* Technical Drawing, *4th ed. Delmar Publishers, 1994, Problem 21–1*)

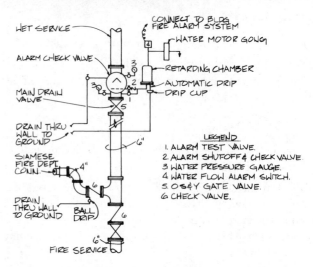

WET PIPE SPRINKLER SERVICE
NTS

LEGEND
1. ALARM TEST VALVE.
2. ALARM SHUTOFF & CHECK VALVE.
3. WATER PRESSURE GAUGE.
4. WATER FLOW ALARM SWITCH.
5. OS&Y GATE VALVE.
6. CHECK VALVE.

FIGURE 28.42 ■ CAD Activity 28.5. (*Reprinted from David Goetsch and William Chalk,* Technical Drawing, *4th ed. Delmar Publishers, 1994, Problem 21–3*)

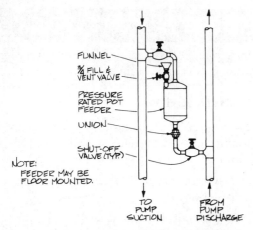

CHEMICAL POT FEEDER DETAIL
N.T.S.

NOTE:
FEEDER MAY BE FLOOR MOUNTED.

FIGURE 28.41 ■ CAD Activity 28.3 and Activity 28.4. (*Reprinted from David Goetsch and William Chalk,* Technical Drawing, *4th ed. Delmar Publishers, 1994, Problem 21–2*)

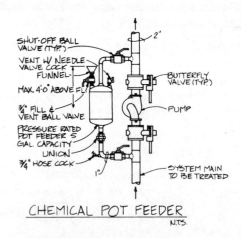

CHEMICAL POT FEEDER
N.T.S.

FIGURE 28.43 ■ CAD Activity 28.6. (*Reprinted from David Goetsch and William Chalk,* Technical Drawing, *4th ed. Delmar Publishers, 1994, Problem 21–4*)

SECTION VII

Employment in Drafting

29 UNIT

Finding a Job

OBJECTIVES

Upon completion of this unit, the student will be able to:

■ List the primary employers of CAD technicians in the heavy construction industry.

■ List the most productive job-finding aids for the CAD technician.

■ Demonstrate how to properly use job-finding aids for identifying potential employers.

EMPLOYMENT IN DRAFTING

Drafting is one of the most important occupations supporting the heavy construction industry. The plans for every apartment building, food store, medical center, educational facility, shopping mall, department store, and numerous other commercial or industrial buildings were prepared by structural CAD technicians.

The heavy construction industry is one of the nation's largest industries, requiring a sizable number of CAD technicians for support. The old adage "it must be drawn before it can be built" is particularly true in heavy construction. This makes drafting an occupation with high employment potential.

Students completing this textbook are prepared to enter the world of work and do the job of a CAD technician in the fields of structural, civil, and piping drafting. However, being able to do the job is only the first step. Students must also learn how to find a job, get a job, and keep a job. This unit deals with finding a job in drafting.

PRIMARY EMPLOYERS OF CAD TECHNICIANS

The first step in finding a job as a CAD technician is learning where to look. Learning where to look for potential employers involves two stages. First, it must be determined who employs CAD technicians and, second, how to locate these employers.

In any city in any state in the United States, the primary employers of CAD technicians for the heavy construction industry are:

■ Consulting engineering firms

■ Steel fabrication companies

■ Surveying firms

■ Precast concrete companies

Consulting Engineering Firms

Consulting engineers work with architects on heavy construction projects. They work on a contract basis and are responsible for preparing the structural, civil, and piping drawings that must accompany the architectural plans. A complete set of architectural plans for a commercial or industrial building includes:

■ The architectural drawings numbered A1, A2, A3, and so on

■ The structural drawings numbered S1, S2, S3, and so on

■ Drawings prepared by electrical and mechanical consulting firms

■ The civil engineering plans numbered C1, C2, C3, and so on

The drawings that go into the complete drawing package are prepared by CAD technicians. Some consulting firms do specialize in a particular type of design and drawing such as steel, precast concrete, poured-in-place concrete, wood, civil, or piping. However, most firms are involved in all of these fields. Therefore, engineering consulting firms usually want to hire those CAD technicians with a broad background encompassing all of these fields.

Consulting engineering firms are engaged only in design and the preparation of plans and do not fabricate products. Therefore, CAD technicians employed in this setting can expect to deal primarily with engineering and layout drawings.

Steel Fabrication Companies

The primary employers for persons wishing to specialize in structural steel drafting consist of the structural steel fabrication companies that employ CAD technicians to prepare shop drawings, advance bills, and shop bills so that the steel members for a job may be ordered, designed, fabricated, and erected.

Precast Concrete Companies

The primary employers for persons wishing to specialize in precast concrete drafting are the precast concrete manufacturing companies. Precast concrete companies employ structural CAD technicians to prepare shop drawings and bills of materials so that the precast members for a job may be designed, cast, shipped, and erected.

SECONDARY EMPLOYERS OF CAD TECHNICIANS

In addition to the primary employers of CAD technicians listed in the previous section, there are also secondary employers that should be familiar to the job-seeking CAD technician. Some of these secondary employers are architects, large construction companies, the federal civil service, and county road departments.

Architects that specialize in the design of large commercial and industrial buildings often maintain a certified engineer on their staff. This engineer requires CAD technicians for support.

In addition to architects, a number of construction companies around the country are large enough and do a volume of work sufficient enough to warrant maintaining a full-time staff of engineers and CAD technicians. The federal civil service also employs CAD technicians. These CAD technicians are involved in projects such as repair and renovations to military and government installations and facilities. The U.S. Army Corps of Engineers is a primary employer of federally employed CAD technicians.

LOCATING POTENTIAL EMPLOYERS

Now that the potential employers of CAD technicians are known, it must be learned how to locate them. There is a systematic approach that can be used to locate potential employers of CAD technicians. Every city in every state in the United States has built-in, job-location aids for aspiring CAD technicians.

The most productive job-finding aids available in any city are:

- The yellow pages in the telephone directory
- The want ads in the Sunday newspaper
- The state employment office
- The *Manufacturer's Directory* of the local chamber of commerce
- The Internet

The Yellow Pages

The yellow pages of the local telephone directory provide an alphabetized listing of potential employers of CAD technicians. The yellow pages give an address and telephone number for each employer of CAD technicians. Used properly, they can be a valuable tool for the job seeker. Architects, construction companies, steel fabricators, and precast concrete manufacturers all appear alphabetically in the yellow pages (Figure 29.1).

The Sunday Want Ads

Many employers in search of qualified CAD technicians solicit in the Sunday edition of the paper if they plan to use the want ads. This is because Sunday newspapers generally have a wider distribution, thereby reaching more readers. Want ads generally are printed in one of two formats: random printing and organized printing alphabetically in occupational categories (Figure 29.2).

In the first category, the job seeker must look carefully through all ads until those scattered throughout pertaining to CAD technicians are located. In an organized format, the job seeker need only turn to the occupational category marked *Professional-Technical*. Under this category, potential jobs are found under C for CAD technicians.

The State Employment Office

Most cities have a branch office of the state employment office. New jobs are posted periodically, usually once each week—on a predetermined day.

The job seeker wishing to investigate the possibility of a job in drafting is given a special code number and directed to a computer terminal. By calling up this code number, the job seeker is shown a listing of jobs available in her area of interest throughout the state.

A word of caution is in order for those aspiring CAD technicians who have completed their training but have no job experience. Most employers that advertise through the state employment office insist that applicants have a certain amount of experience. Unless you have the experience, the state

```
                        YELLOW PAGES

  128 Concrete—Construction

  JONES PRECAST CONCRETE COMPANY
       Southern Drive . . . . . . . . . . . . . . . . . . . . . . 678-6123
  PRESTRESSED PRODUCTS, INC.
       201 Bush Street . . . . . . . . . . . . . . . . . . . . . 678-4040
  RANDALL'S PRECAST COMPANY
       613 Industrial Drive . . . . . . . . . . . . . . . . . . 243-1288
```

FIGURE 29.1 ■ Sample excerpt from the yellow pages.

39 Professional—Technical

STRUCTURAL DRAFTERS!!!!

Immediate, long-term employment opportunities for persons with broad structural background covering steel, precast, poured-in-place, and wood. Good salary and full benefits.

For Appointment, call John Adams
(904) 678-4041

STRUCTURAL STEEL DRAFTER NEEDED

Local steel fabrication company in need of energetic, well-trained individuals for positions as detailers and drafters. Top salaries, full benefits, plenty of overtime, and comfortable working conditions. Inquire in writing and furnish samples of your work:

Drafter
P.O. Box 1416
Niceville, Florida 32578

FIGURE 29.2 ■ Sample excerpt from the Sunday want ads.

PRODUCT SECTION

9622 — Steel Fabricators

ATLANTIC STEEL COMPANY
125 Lektonic Street
Phone: (904) 244-6784
Employees 120M 59F 179T
Personnel Manager — John LaMotte

FIGURE 29.3 ■ Sample excerpt from chamber of commerce *Manufacturer's Directory.*

employment office is unable to send you out on an interview. This is a common problem incurred in using the state employment office. However, on occasion, jobs are available through the state employment office for individuals with no experience but the proper training.

Local Chamber of Commerce

Many local chambers of commerce publish a *Manufacturer's Directory* or guide to industry for their area of responsibility. Every type of manufacturer, fabricator, contractor, and so on in the area is listed either alphabetically by company name or alphabetically in product categories and oftentimes both ways (Figure 29.3).

Categories include *Steel Fabricators, Precast Concrete Manufacturers, Contractors, Civil Engineers, Pipe Fabricators* and many others of interest to the CAD technician seeking a job. Each company listing gives the complete name, address, telephone number, product, and contact person (president, personnel manager) for the company.

The Internet

The advent of the Internet has made the task of finding a job in drafting much easier. All of the other sources listed herein are still valid, but now they and other sources are much easier to locate using the Internet. Some of the more widely used and effective job-search sites of the Internet are:

Monster.com: *http://www.monster.com*

HotJobs.com: *http://www.hotjobs.com*

CareerBuilder: *http://www.CareerBuilder.com*

America's Job Bank: *http://www.AJB.dni.us*

SUMMARY

■ The primary employers of CAD technicians are steel fabrication companies, precast concrete companies, and consulting engineering firms.

■ Secondary employers of CAD technicians are: architects, large construction companies, the federal civil service, and county road departments.

■ Every city has built-in job-location aids for the aspiring CAD technician. Some of the most productive are the yellow pages, the Sunday want ads, the state employment office, the local chamber of commerce, and the Internet.

REVIEW QUESTIONS

1. Name the primary employer of steel CAD technicians.

2. Name the primary employer of precast concrete CAD technicians.

3. List a primary employer of CAD technicians that employs people who have a broad background covering all areas of construction-related drafting.

4. List three secondary employers of CAD technicians.

5. List the five most productive job-finding aids of the CAD technician.

REINFORCEMENT ACTIVITIES

1. Examine the yellow pages of your local telephone directory and make a list of all potential employers of CAD technicians you are able to locate.

2. Examine the Sunday want ads of your local newspaper for several weeks and cut out any ads asking for CAD technicians. If yours is a small town, attempt to secure the newspapers of the nearest large town to your area.

3. Visit the local office of the state employment service in your town and identify the drafting jobs that are available.

4. Visit the local chamber of commerce in your town and examine the *Manufacturer's Directory*. Make a list of the potential employers of CAD technicians you are able to locate.

5. Search the Internet for sites that list job openings in drafting.

30 UNIT

Getting and Keeping a Job

OBJECTIVES

Upon completion of this unit, the student will be able to:

- Write a proper letter of introduction for a job.
- Prepare a well-written resume.
- Develop a portfolio of sample drawings.
- Conduct a positive interview.
- Follow up on an inconclusive interview.

LETTER OF INTRODUCTION

Many times, graduates of technical programs find it advantageous to seek employment out of town. The CAD technician seeking a job in a location other than his hometown will need to know how to write a proper letter of introduction.

The *letter of introduction* is simply a short statement of your interest in working for a particular company. It is used as a cover letter for a resume and should convey the following information:

- You are a CAD technician with specialized training.
- You intend to relocate to the employer's area as soon as you are able to secure a responsible position in your field.
- You would like to meet the employer and discuss present or future drafting openings with his company.
- A resume of your qualifications is enclosed and you look forward to a reply.

A letter of introduction is a selling tool. It should be brief, assertive, and positively stated (Figure 30.1).

RESUME

A well-written resume is one of the CAD technician's best job-seeking aids. It is simply a categorized account of your qualifications relative to drafting. A properly prepared resume sells the job seeker. It amplifies those qualifications possessed by the job seeker that are most important to the job in question.

Mr. John Q. Employer
Jones Steel Company
P.O. Box 229
Houston, Texas 21609

Dear Mr. Employer:

I am a CAD technician with specialized training in all phases of steel and civil engineering drafting. I intend to relocate to Houston as soon as I am able to secure a responsible drafting position. I have examined the Houston job market and find that I would like to work for your company.

I have enclosed a resume of my qualifications and will furnish references and samples of my work upon request. I look forward to hearing from you. Thank you very much.

Sincerely,

Gary Graduate
CAD Technician

GG
Enc: As stated

FIGURE 30.1 ■ Sample letter of introduction.

The resume should begin with personal data that include complete name, mailing address, and telephone number. Such things as marital status, number of children (if any), height, weight, and so on need not be included.

Next, you should give your occupational objective. This is a brief statement of your career goal relative to drafting. It should be broad enough to encompass a wide range of drafting job possibilities but specific enough to show that you have set definite plans for yourself. An example of a well-written occupational objective would be:

"To begin work at a productive level in a drafting position that will allow me to apply my training and to advance at a rate commensurate with my performance on the job."

The next component in a resume should be a chronological record of your education with the highest level of your drafting training listed first. The remaining entrees are presented in reverse order. There is no need to go back any farther than high school.

The next component lists your experience. If you have any experience that relates to drafting, it should be listed first. If not, list the record of your employment in reverse order from the most recent to the earliest. Showing that you have worked, even if it has not been in drafting, is a plus in your favor. If you have held a long list of jobs, select the most important only.

The two final categories are awards and hobbies. These are included to appeal to the human side of the employer and to get your foot in the door for an interview. If one of your awards or hobbies interests the potential employer enough, it might just make the difference in your being granted an interview. Figure 30.2 shows an example of a resume for a student who has completed his training but has no drafting experience. An example of a resume for a student who has actual drafting experience is shown in Figure 30.3.

PORTFOLIO OF DRAWINGS

Once an interview has been granted, the employer will want to see samples of your work. A well-developed portfolio of drawings will impress a potential employer more than anything else in an interview. These drawings show the employer that you can actually do the work that needs to be done. Provided he is well spoken and properly dressed, the CAD technician with a good portfolio usually gets the job.

THE INTERVIEW

The interview is the most critical element in the various phases passed through in the transition from job seeker to employed CAD technician. The wise job seeker will rehearse over and over in his mind the answers to questions that might be asked during the interview. A list of sample interview questions that are commonly asked in drafting interviews is shown in Figure 30.4.

The job seeker should show up for the interview 10 minutes early and appropriately dressed. Dress is important, but do not overdo it. When called in for the interview, be friendly but serious. Your attitude should be positive and assertive. Shake hands with your potential employer firmly but not too tightly. Look the employer right in the eyes and introduce yourself.

While interviewing, answer all questions openly and honestly. Emphasize your positive characteristics and let the employer know you will quickly learn anything that you do not know when you start. If asked to fill out an application, remember, you are a CAD technician. Fill it out in your best drafting lettering.

Make sure that you conclude the interview on a positive note. If you are hired, ask the employer when you are to start. If you are turned down, try to determine exactly why. Tell the employer this information will be very helpful to you in future interviews and that you will appreciate an open, honest appraisal.

If the interview is inconclusive and you are told that you will be contacted later, shake the interviewer's hand, look the interviewer in the eyes and say, "Thank you for your time and consideration. I would like to work for you and I think I can do the job. I hope you will give me the opportunity to prove myself."

INTERVIEW FOLLOW-UP

Interviews are often inconclusive. This does not mean that you are not going to get the job. It usually means that the employer has several people scheduled for interview and wants to see them all before making a decision. Before leaving an inconclusive interview, ask the employer when the decision is expected to be made. Also, tell the employer that you will call back in a specified amount of time if you have not heard about the job. Be assertive but not pushy. If the employer says the decision will be made in a week, wait a week and call back. If a decision has still not been reached, you have at least reminded the employer that you want the job. Avoid becoming a pest, but, on the other hand, do not be afraid to let the employer know that you want the job. Being timid rarely pays off in job interviews.

THE WISE JOB SEEKER

One of the most frustrating experiences of a person's life can be the first job search. After spending long hours training to become a CAD technician, most people are not prepared to spend additional hours trying to find a job. However, being able to do a job is only half of the process. Being able to get a job is equally important.

Hundreds of well-paying jobs are available on any given day throughout the country for well-trained CAD technicians. However, most employers are looking for a person with actual drafting experience. This is the biggest problem facing the new graduate. Fortunately, it is a problem that can be overcome.

Figure 30.5 contains a list of rules for the wise job seeker. These rules can help the well-trained graduate who is lacking in experience to break into the world of work and get that all-important first job. Figure 30.6 contains a list of characteristics that will ensure success in a job once it has been attained.

Resume for

GARY GRADUATE
729 Green Street, Lot No. 1
Fort Walton Beach, Florida 32548
(904) 555-4040

Occupational Objective

To begin work at a productive level in a drafting position that will allow me to apply my training and advance at a rate commensurate with my performance on the job.

Education

ASSOCIATE OF SCIENCE DEGREE. Elmwood Community College, Fort Walton Beach, Florida—May 2004.
My training at Elmwood Community College covered all phases of drafting with specialized, in-depth work in structural and civil engineering drafting.

HIGH SCHOOL GRADUATE. Fort Walton Beach High School—June 2002.

Work Experience

LABORER. Smith Construction Company.
Fort Walton Beach, Florida . . . . .2 years

WAITER. Al's Seafood Restaurant.
Niceville, Florida . . . . .1 year

Awards

Graduated from Elmwood Community College on Dean's List with 3.75 G.P.A.
First Place—10th Annual Elmwood Community College Drafting Contest.

Hobbies

Long distance running, tennis, sailing, and fishing.

FIGURE 30.2 ■ Sample resume for student with no experience.

Resume for

AMANDA GRADUATE
1512 Elm Street
Fort Walton Beach, Florida 32548
(904) 678-1513

Occupational Objective

To continue working in drafting and to advance to a position of checker at a rate commensurate with my performance and development on the job.

Education

ASSOCIATE OF SCIENCE DEGREE. Elmwood Community College, Fort Walton Beach, Florida—May 2004.
My training at Elmwood Community College covered all aspects of drafting with specialized, in-depth study in structural and civil engineering drafting.

HIGH SCHOOL GRADUATE. Niceville High School—June 2002.

Work Experience

STRUCTURAL STEEL DETAILER. Sanders Steel Fabrication Company.
Seminole, Alabama . . . . .1 year

Drafting trainee. Marshall & Thomas, Consulting Engineers.
Seminole, Alabama . . . . .2 years

Awards

Graduated from Elmwood Community College on the Dean's List with a G.P.A. of 3.69.
COMMUNITY SCHOLARSHIP WINNER. 2001 school year.

Hobbies

Running, swimming, reading.

FIGURE 30.3 ■ Sample resume for student with experience.

1. Why did you decide to become a CAD technician?

2. Why do you think you would like to work for us?

3. What do you think a beginning CAD technician will be required to do?

4. Are you willing and eager to learn?

5. Why do you think you will make a good CAD technician?

6. Do you mind starting at the bottom and working your way up?

7. What are your long-term goals in terms of drafting?

8. Do you have any samples of your work?

9. What would you say are your best qualities in terms of drafting?

10. What would you say are your weak points in terms of drafting?

FIGURE 30.4 ■ Sample interview questions.

1. ***Do not be afraid to relocate.*** For some graduates, leaving their hometown is a frightening thought. However, few things in life are more important than one's career. The bottom line in job seeking is: If you want a job, go to where the jobs are.

2. ***Learn to market your skills.*** Study your job-seeking skills as you would your technical skills and develop them. It is almost as important to be a skilled job seeker as it is to be a skilled CAD technician.

3. ***Be positive and assertive in your job search.*** When interviewing, avoid appearing as a timid, unsure person willing to take any job. Go into every interview as a high-trained CAD technician and be assertive enough to say, "I would like to work for your company, I think I can do the job, and I hope you will give me the opportunity to prove myself."

4. ***Do not be afraid to start at the bottom and work your way up.*** Many companies may start you off running blueprints, doing errands, and performing minor corrections and revisions. There is nothing wrong with this. Many successful CAD technicians started their careers this way.

5. ***Do not become frustrated.*** For the inexperienced CAD technician, it might take several interviews to get the first job. However, once the first job hurdle has been crossed and experience is gained, upward mobility will become the rule.

FIGURE 30.5 ■ Rules of the wise job seeker.

*_**Be dependable.**_ Go to work on time and work while you are there. An employer needs to know you can be counted on to get the job done properly and on time.

*_**Be a learner.**_ Do not be afraid to tackle new and unfamiliar assignments. The broader your knowledge and skills become, the more valuable you will be to your employer.

*_**Be a worker.**_ Avoid joining the water cooler clique or the coffee pot gang or being a clock watcher. Everyone needs a break from drawing, and you will too. However, do not allow yourself to become the person the "boss" bumps into everytime he or she passes the coffee pot.

*_**Be self-sufficient.**_ Nothing is wrong with asking questions. In fact, it is to be encouraged. However, before asking a question, try to find the answer yourself. This helps you develop into an independent-thinking problem solver.

*_**Strive to constantly improve.**_ Make note of every new thing that you learn and internalize it. Also work at improving your linework, lettering, accuracy, and speed.

*_**Be personable.**_ Getting along with your fellow employees is important. Drafting tasks are completed by teams of CAD technicians, checkers, and engineers, so developing your abilities to work with people is a must.

FIGURE 30.6 ■ Keeping a job in drafting.

SUMMARY

■ The letter of introduction is a short, written statement introducing the job seeker to a potential employer.

■ The letter of introduction is used as a cover letter for a resume.

■ The letter of introduction is used as a selling tool and should be brief, positive, and assertive.

■ A well-written resume is one of the job seeker's most valuable tools.

■ A resume should contain personal information, an occupational objective, education information, experience, awards, and hobbies.

■ Education and experience are listed in reverse order in one of two ways: from most recent to earliest or from most important in terms of the job in question to least important.

■ Most employers want to see a portfolio of drawings during the interview.

■ The interview is the most critical element in the job-seeking process.

■ The job seeker should show up for an interview 10 minutes early, dressed appropriately.

■ When you are called in for the interview, shake hands firmly but not too tightly, look the employer right in the eyes, and introduce yourself.

■ During the interview, effect a positive, assertive attitude and answer all questions openly and honestly.

■ During the interview, emphasize your positive characteristics and let the employer know you will learn quickly.

■ If asked to fill out an application, fill it out in your best drafting lettering.

■ Before leaving an inconclusive interview, ask the employer when a decision is expected to be made.

■ Avoid being pushy, but let the employer know you want the job.

REVIEW QUESTIONS

1. List the four items that should be conveyed by the letter of introduction.

2. Write an occupational objective for a resume of your qualifications.

3. List the personal information that would go on your resume.

4. List the various parts of a resume in order.

5. What is the portfolio of drawings?

6. When should the job seeker arrive for an interview?

7. How should a CAD technician dress for a job interview?

8. List six characteristics that will ensure success on the job.

REINFORCEMENT ACTIVITIES

1. Write a letter of introduction to the following fictitious company:

   Mr. David H. Lemox
   Chief Engineer
   TechCon Engineering
   Ft. Smith, Arizona 96907

2. Write a complete resume of your qualifications as they will be when you complete your training.

3. Put together a portfolio of drawings from among those prepared for this course.

4. Select a fellow student or friend and conduct several practice interviews using the questions in Figure 30.4.

5. Make two B-size charts:
   a. Five Rules of the Wise Job Seeker
   b. Keeping a Job in Drafting

31 UNIT

Advanced Drafting Projects

OBJECTIVES

Upon completion of this unit, the student should be able to demonstrate proficiency in using CAD techniques to:

■ Prepare prestressed concrete projects that include framing plans, sections, connection details, and fabrication details.

■ Prepare structural steel shop drawings that include fabrication details and bills of material.

■ Prepare structural concrete and structural steel plans using architectural drawings as the original material.

■ Prepare civil engineering drawings based on an engineer's notes.

■ Prepare piping drawings based on an engineer's notes.

OVERVIEW

This unit consists entirely of advanced CAD projects. These projects are designed to give students the opportunity to experience being responsible for the development of a complete project from start to finish. Each project is equivalent to the type of work that is expected of a CAD technician or senior CAD technician in a real-world setting. In addition to the original material provided for each activity, students may have to seek out information from a variety of reference sources, as is often the case when actually employed in this field. Resourcefulness is important in today's competitive workplace.

REINFORCEMENT ACTIVITY 1: BLEACHER ADDITION

Figure 31.1 is the complete drawing package for an addition to a high school's stadium bleachers. Reconstruct the drawing package in its entirety, making the following changes:

■ Shorten the out-to-out dimension by 9″ and adjust all others accordingly.

■ Add one additional row of bleachers on the south side and make all corresponding adjustments.

REINFORCEMENT ACTIVITY 2: PEDESTRIAN BRIDGE

Figure 31.2 is the complete drawing package for a pedestrian bridge. Reconstruct the drawing package in its entirety, making the following changes:

■ Shorten the overall span by 2′-6″ and adjust all other dimensions accordingly.

■ Make the bridge twice as wide by adding a second row of double-tee members that duplicates the first. Change all other components of the drawing package accordingly.

REINFORCEMENT ACTIVITY 3: MIDDLE SCHOOL ADDITIONS

Figure 31.3 is the complete drawing package for two additions to a middle school. The roof members are 6″ cored flat slabs that are 4′ wide. Reconstruct the drawing package in its entirety, making the following changes:

■ Convert the Mark Number 1 cored slabs to 24″ double-tee members and make all corresponding adjustments to the drawing.

■ Change Mark Numbers 2, 2A, 3, 4A, and 5 to 8″ cored flat slabs and revise the drawing accordingly.

REINFORCEMENT ACTIVITY 4: RADAR BUILDING

Figure 31.4 is the complete drawing package for a radar building. The roof is framed in 24″ double tees and two cored flat slabs (Mark Numbers 6 and 6A). Reconstruct the drawing package in its entirety, making the following changes:

■ Add 9′ to the north end of the building and adjust the drawing accordingly.

■ Remove that portion of the roof (and building) covered by Mark Numbers 6 and 6A and adjust the drawing accordingly.

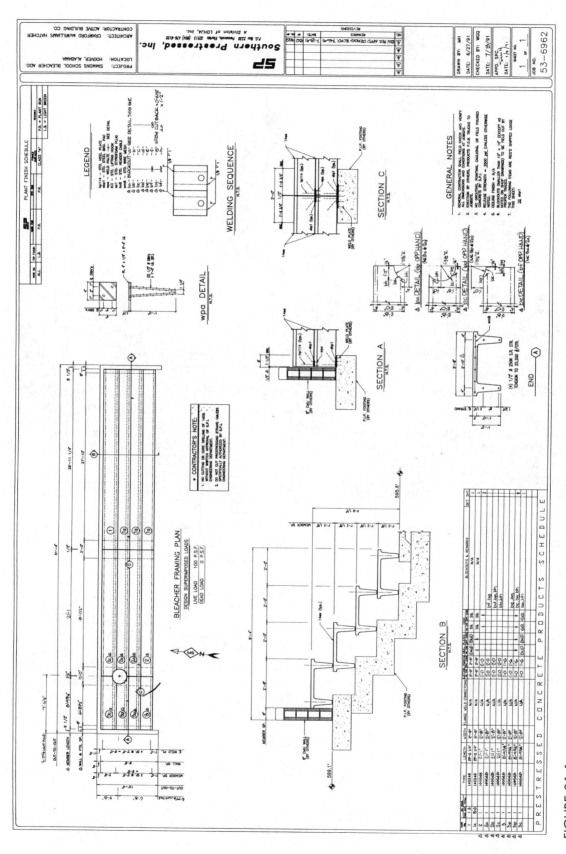

FIGURE 31.1 ■ Bleacher framing plan.

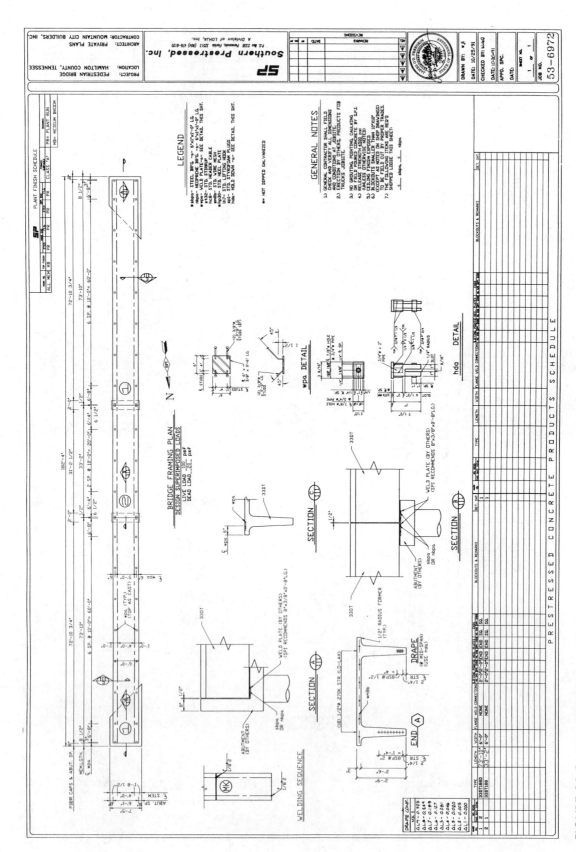

FIGURE 31.2 ■ Bridge framing plan.

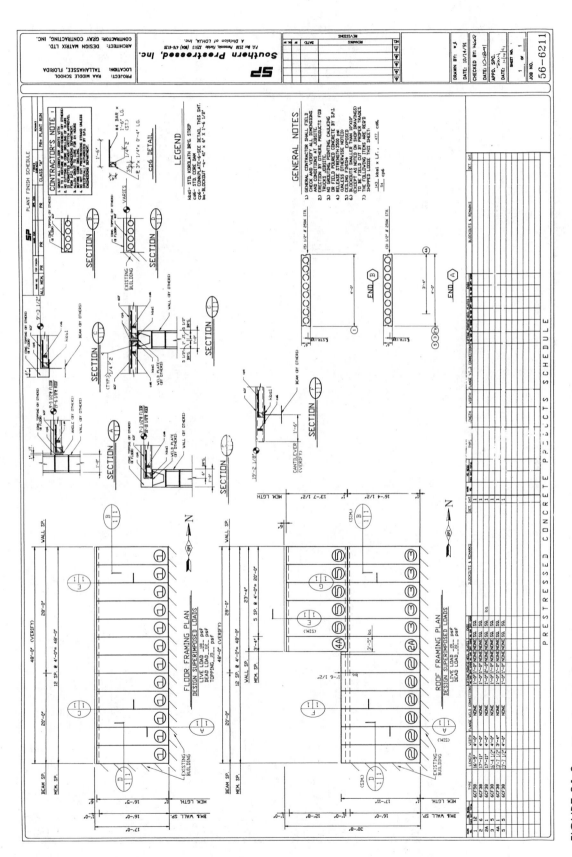

FIGURE 31.3 ■ Floor framing plan.

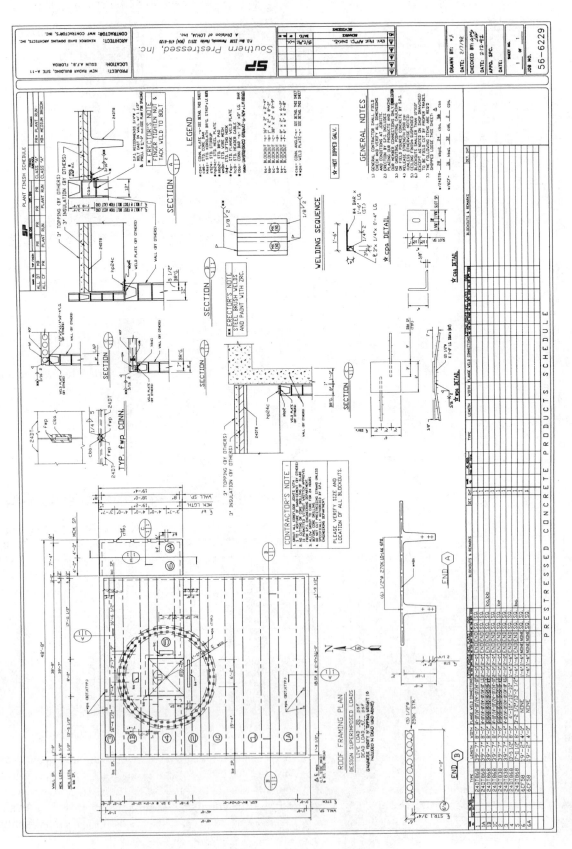

FIGURE 31.4 ■ Roof framing plan.

REINFORCEMENT ACTIVITY 5: BEAM DETAIL

Figure 31.5 is a shop drawing showing a beam fabrication detail. Reconstruct the drawing, making the following changes:

■ WAA should be placed at approximately equal intervals along the entire length of the beam (a total of 8 WAAs).

■ Use double lifting hooks at each end (two lh7) spaced 4″ apart.

REINFORCEMENT ACTIVITY 6: COLUMN DETAIL

Figure 31.6 is a shop drawing showing a column fabrication detail. Reconstruct the drawing, making the following changes: Change the depth of the beam to 1′-8″ and the width to 12″ and adjust the drawing accordingly.

REINFORCEMENT ACTIVITY 7: FLAT SLAB DETAIL

Figure 31.7 is a shop drawing showing a flat slab fabrication detail. Reconstruct the drawing, making the following changes: Change the width of the slab to 8′-6″, the thickness to 6″, and the length to 11′-2″ and adjust the drawing accordingly.

REINFORCEMENT ACTIVITY 8: DOUBLE TEE DETAIL

Figure 31.8 is a shop drawing showing a fabrication detail for a 10″ (depth) double tee. Reconstruct the drawing, making the following changes: Shorten the double tee by 1′-6″ and adjust all spacings and the drawing accordingly.

REINFORCEMENT ACTIVITY 9: COMPLETE SET OF PLANS

Figure 31.9 contains an architect's three-dimensional drawing of the structural skeleton for a multistory building. Assume a distance of 10′-4″ between stories and decide on your own length and width measurements and sizes of individual members. With only this information, develop a complete set of structural plans for the building including beam column, wall, and roof-framing plans; sections; connection details; and shop drawings for individual members. Note that the building contains prestressed beams, columns, and flat slabs.

REINFORCEMENT ACTIVITY 10: STEEL FABRICATION DETAILS

Figure 31.10 contains six beam fabrication details. Reconstruct the details, making the following changes:

■ B23, B25, and B26 should have the connection plates currently shown at one end on both ends.

■ Revise the bill of material as you revise the drawing.

REINFORCEMENT ACTIVITY 11: STEEL CONNECTOR DETAILS

Figure 31.11 contains details for several brackets and beam seats. Reconstruct the drawing making the following changes:

■ M13 Change the overall dimensions from 6″ to 8″ wide and the length from 8″ to 10″. Adjust all other dimensions accordingly.

■ M15 Add 2″ to all three overall dimensions and adjust the drawing accordingly.

REINFORCEMENT ACTIVITY 12: STEEL BEAM DETAILS

Figure 31.12 contains six beam fabrication details. Using the information that can be derived from this drawing, develop connection details for both ends of these beams. You may connect them to steel columns or other steel beams. You may select the size and types of the other members but may not alter the beams shown in this figure.

REINFORCEMENT ACTIVITY 13: PRE-ENGINEERED METAL BUILDINGS

One can find prefabricated metal buildings in almost any community. Find one in your community and approach the owner for permission to complete this activity. Develop a complete set of "as-built" drawings for the building including an anchor bolt plan, rigid-frame sections, all of the necessary framing plans, and the connection details.

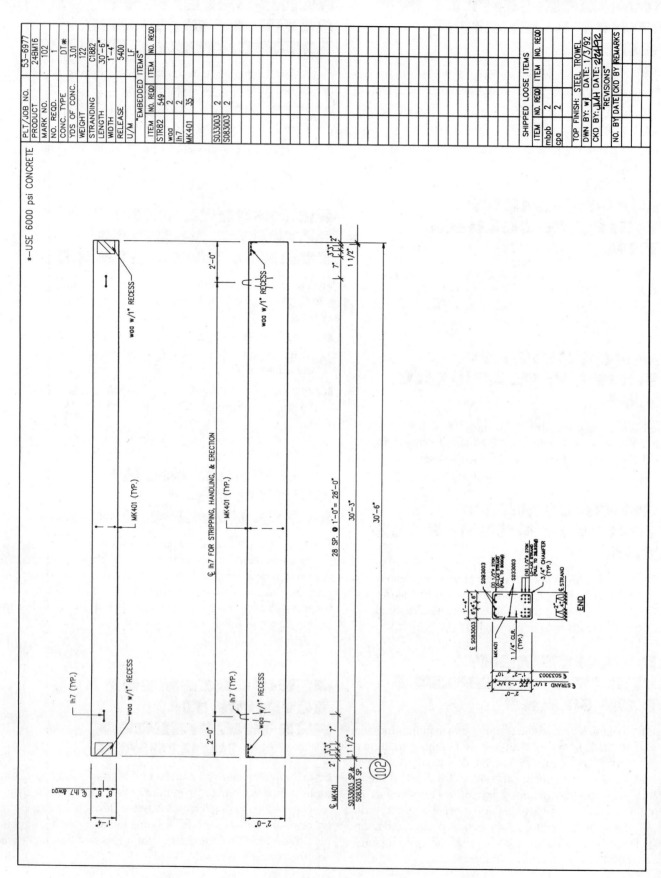

PLT/JOB NO.	53-6977
PRODUCT	24BM16
MARK NO.	102
NO. REQD.	1
CONC. TYPE	DT*
YDS OF CONC.	3.01
WEIGHT	122
STRANDING	C1882
LENGTH	30'-6"
WIDTH	1'-4"
RELEASE	5400
U/M	LF

"EMBEDDED ITEMS"			
ITEM	NO. REQD	ITEM	NO. REQD
STR82	549		
waa	2		
lh7	2		
MK401	35		
S033003	2		
S083003	2		

SHIPPED LOOSE ITEMS			
ITEM	NO. REQD	ITEM	NO. REQD
mbpb	2		
cpa	2		

TOP FINISH: STEEL TROWEL
DWN BY: wj   DATE: 1/3/92
CKD BY: JMH DATE: 2/4/92
"REVISIONS"
NO. BY DATE CKD BY REMARKS

*—USE 6000 psi CONCRETE

FIGURE 31.5 ■ Beam fabrication detail.

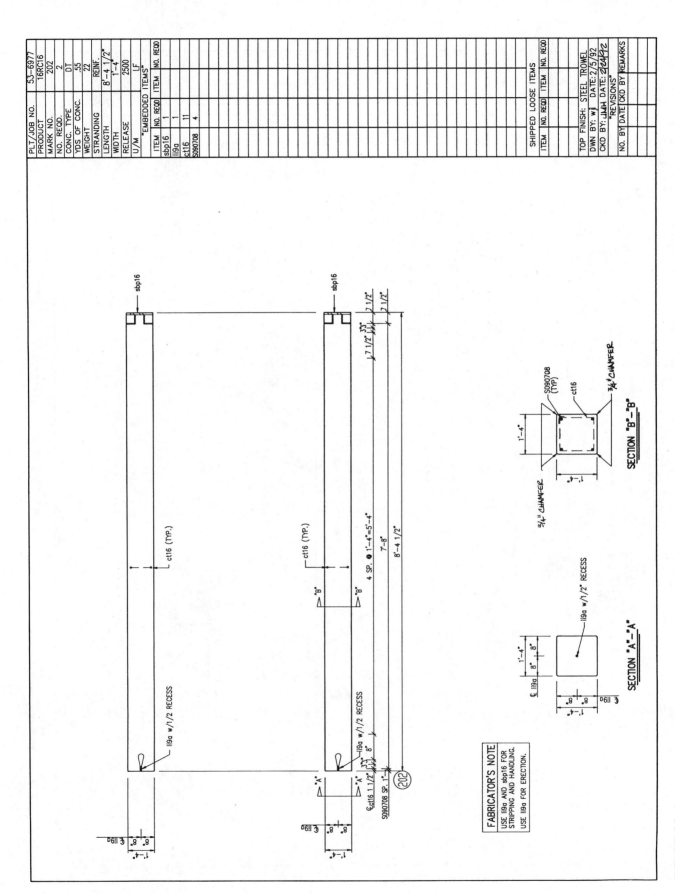

PLT/JOB NO.	53-6977		
PRODUCT	16RC16		
MARK NO.	202		
NO. REQD.	2		
CONC. TYPE	DT		
YDS OF CONC.	.55		
WEIGHT	22		
STRANDING	REINF.		
LENGTH	8'-4 1/2"		
WIDTH	1'-4"		
RELEASE	2500		
U/M	LF		

"EMBEDDED ITEMS"			
ITEM	NO. REQD	ITEM	NO. REQD
sbp16	1		
ll9a	1		
ct16	11		
S090708	4		

SHIPPED LOOSE ITEMS			
ITEM	NO. REQD	ITEM	NO. REQD

TOP FINISH: STEEL TROWEL
DWN BY: wj   DATE: 2/5/92
CKD BY: JMH   DATE: 2/24/92

"REVISIONS"			
NO.	BY DATE	CKD BY	REMARKS

FABRICATOR'S NOTE
USE ll9a AND sbp16 FOR
STRIPPING AND HANDLING.
USE ll9a FOR ERECTION.

FIGURE 31.6 ■ Column fabrication detail.

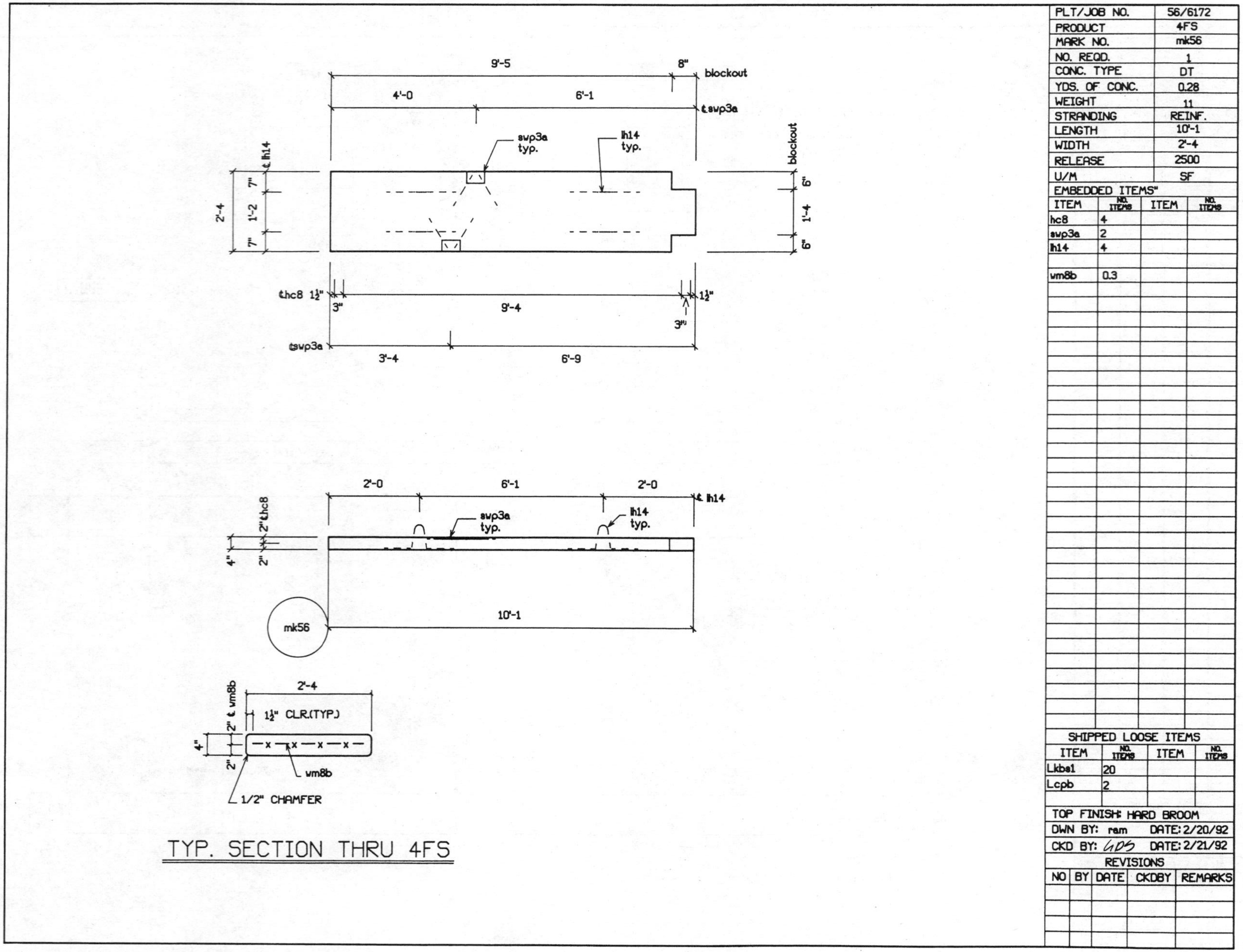
9'-5
8"
blockout
4'-0
6'-1
℄ swp3a
swp3a
typ.
lh14
typ.
℄ lh14
2'-4
7"
1'-2
7"
blockout
6"
1'-4
6"
℄hc8 1½"
3"
9'-4
1½"
3"
℄swp3a
3'-4
6'-9
2'-0
6'-1
2'-0
℄ lh14
2"℄hc8
4"
2"
mk56
10'-1
2'-4
2" ℄ vm8b
1½" CLR.(TYP.)
vm8b
1/2" CHAMFER
TYP. SECTION THRU 4FS
PLT/JOB NO. 56/6172
PRODUCT 4FS
MARK NO. mk56
NO. REQD. 1
CONC. TYPE DT
YDS. OF CONC. 0.28
WEIGHT 11
STRANDING REINF.
LENGTH 10'-1
WIDTH 2'-4
RELEASE 2500
U/M SF
EMBEDDED ITEMS"
ITEM NO. ITEMS ITEM NO. ITEMS
hc8 4
swp3a 2
lh14 4
vm8b 0.3
SHIPPED LOOSE ITEMS
ITEM NO. ITEMS ITEM NO. ITEMS
Lkbs1 20
Lcpb 2
TOP FINISH: HARD BROOM
DWN BY: ram DATE: 2/20/92
CKD BY: GDS DATE: 2/21/92
REVISIONS
NO BY DATE CKDBY REMARKS

FIGURE 31.7 ■ Slab fabrication detail.

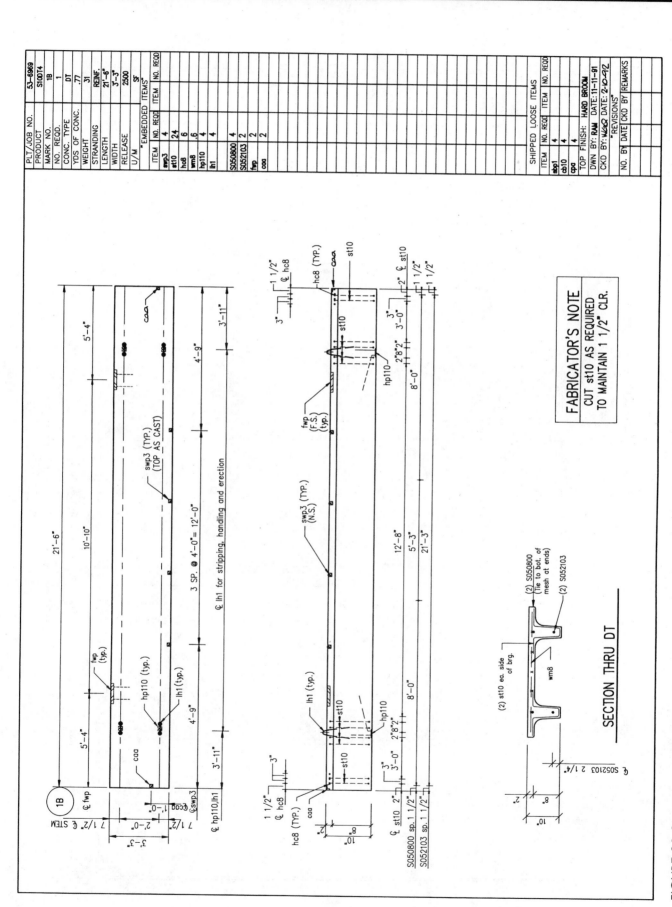

FIGURE 31.8 ■ Fabrication detail for 10" double tee.

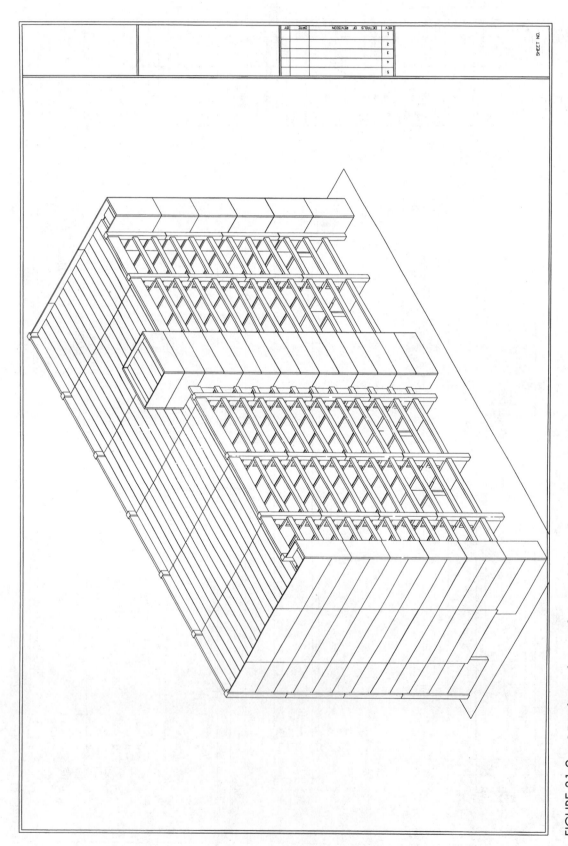

FIGURE 31.9 ■ 3-D architect's drawing of structural skeleton for a multistory building.

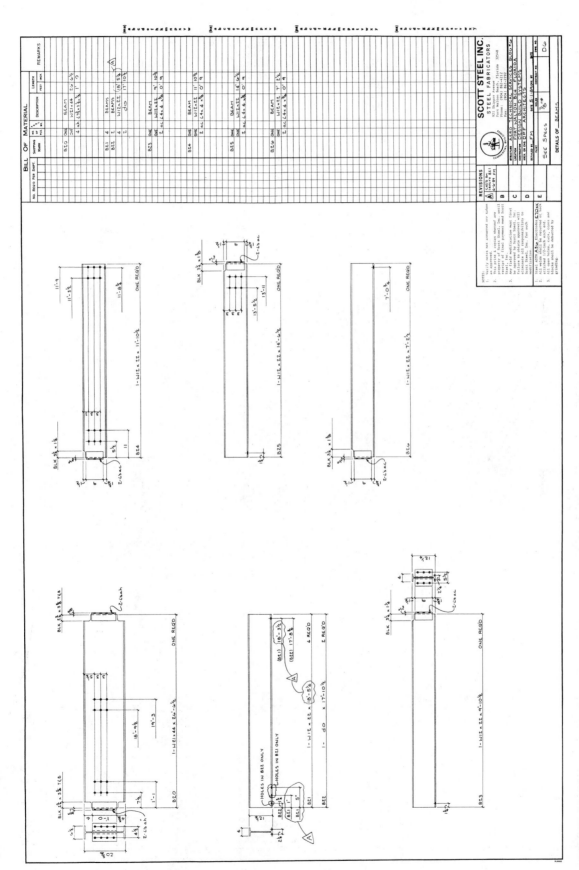

FIGURE 31.10 ■ Beam fabrication details.

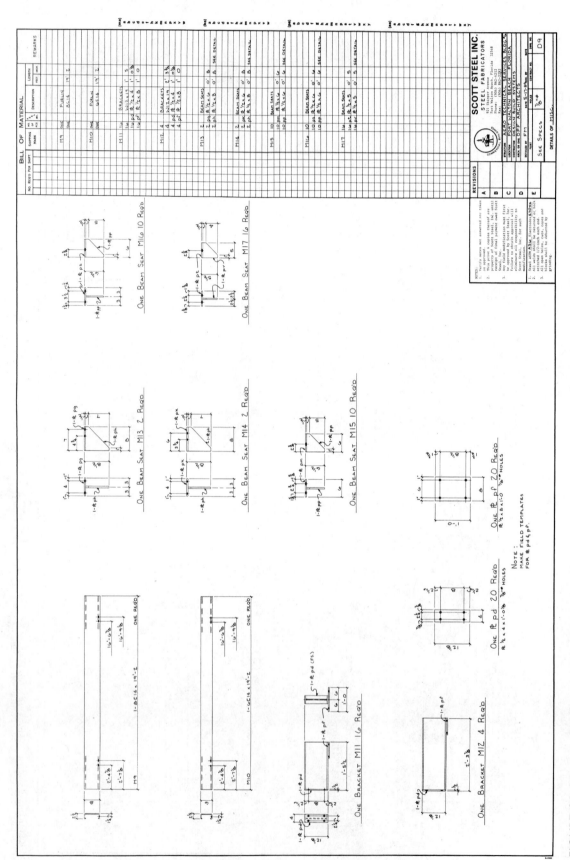

FIGURE 31.11 ■ Fabrication details for brackets and beam seats.

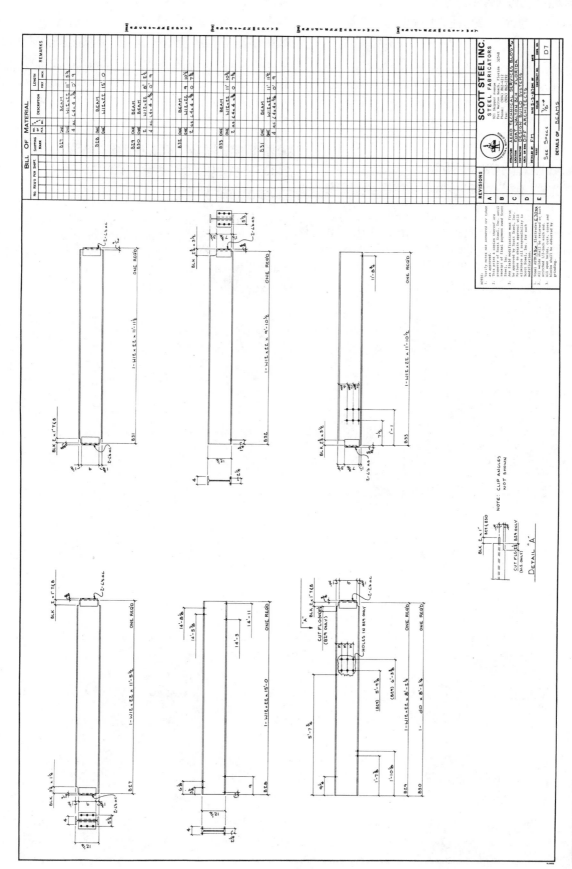

FIGURE 31.12 ■ Beam fabrication details.

REINFORCEMENT ACTIVITY 14: PROPERTY MAPS

Go to the office of the Clerk of the Circuit Court in your community and ask to see where subdivision plats are recorded. Obtain a copy of a small subdivision and the metes and bounds descriptions for all parcels in it. Using this information, reconstruct a drawing of the entire subdivision.

REINFORCEMENT ACTIVITY 15: PLOT PLANS

Locate the survey for the house you live in or one of a friend or relative. You might have to visit the office of the Clerk of the Circuit Court and secure a copy there. Develop a comprehensive plot plan for the house containing all necessary information.

REINFORCEMENT ACTIVITY 16: PLAN AND PROFILE DRAWING

Visit the office of the county or city engineer in your community and ask to see several plan and profile drawings of streets or roads. Using the data available from the county engineering personnel, reconstruct the drawing.

REINFORCEMENT ACTIVITY 17: PIPE DRAWING

Visit the office of the City Water and Sewer Department in your community and get permission to complete this activity. Gain access to a pump house and make a sketch of the piping system in it. Then use your sketch to create a single-line isometric drawing of the piping system.

GLOSSARY

Alloy Two or more metals, or metal combined with nonmetallic substances, to obtain a desired performance characteristic, such as hardness, elasticity, and corrosion resistance.

American standard beam Common name for an S-shape steel beam.

Anchor bolts Bolts used to secure building components to the foundation. In the case of primary framing, these bolts are embedded in the foundation and secured to the column baseplate.

Angle Structural steel shape resembling L. May be equal leg angle or unequal leg angle. Used in trusses and built-up girders.

Architecture The combined art and science of designing buildings and other structures for human use.

ASTM American Society for Testing and Materials.

Azimuth A horizontal direction measured in degrees from 0 to 360 and usually measured from north.

Balloon framing A type of wooden framing in which the studs extend in one piece from the sill to the double top plate. At one time popular for two-story dwellings, it is no longer widely used.

Bars Round, square, or rectangular steel rods used for reinforcing concrete or fabricating steel members.

Bay spacing The distance between primary framing members measured parallel to the ridge or eave. Interior bays are measured from centerline of frame to centerline of frame.

Beam A large structural member of concrete, steel, or wood used to support members over openings or from column to column.

Bill of materials A comprehensive list of all materials needed to fabricate and erect the structural members for a job.

Blueprints Common term for copies of original drawings. The term originated with a copying method seldom used any longer that actually produced a copy with white lines on a blue background. Most modern prints have a white background and dark blue or black lines.

Cadastral maps Maps that depict features in cities, towns, and counties. They are often used by county property appraisers for assessing taxes and by civil engineers for various types of municipal projects.

Chain A measuring tool used in surveying composed of links that were originally 66 feet long. The steel measuring tapes used by modern surveyors are sometimes called chains.

Channels A general term used to describe steel C and MC shapes.

Civil engineering The engineering discipline concerned with planning of roads, bridges, earthwork, maps, surveys, property descriptions, dams, canals, pipelines, and related projects.

Clear height Distance from the finished floor to the bottom of the rafter at the rafter-to-column connection.

Clear span Distance between columns.

COGO (coordinate geometry) A method used to create drawings in CAD that uses points or coordinates, geometric angles, bearings, and lengths.

Continuous beam frame A multiple-span structural frame consisting of straight or tapered solid-web sections whose exterior rafter-to-column moment connection stabilizes the structure.

Columns A vertical structural member usually attached to a footing and extending to the roof of the building. May be steel, concrete, or wood.

Concrete A mixture of cement, sand, aggregate, and water. It is usually reinforced with wire mesh and steel reinforcing bars when used in heavy construction.

Contour lines Lines on a map used to connect points of equal elevation and, thereby, to indicate the elevation of land.

Contractor The person who supplies the necessary work and materials and coordinates subcontractors in building a structure.

Cross section A full section cut at right angles to the longitudinal axis of a building.

C-shape or channel Structural steel shape which has a cross section resembling an opening bracket ([). Similar to W-shapes with half-width flanges on one side. Used in trusses and built-up girders.

Cut and fill A civil engineering term that describes moving earth from high areas to fill in low areas.

Cutting plane A hypothetical plane that cuts through a structure at designated locations to reveal inside conditions. Indicated on drawings with various types of arrows.

Details A small part of a structure drawn separated from the structure to accentuate certain information.

Eave height The vertical dimension from finished floor to eave.

Eave purlin A roof secondary framing member located at the eave and used for attaching roof and wall panels.

Eave strut A structural member at the eave to support roof panels and wall panels. It may also transmit wind forced from roof bracing to wall bracing.

Elevations An orthographic representation of a structure or part of a structure drawn on a flat, vertical plane as if the viewer's line of sight is perpendicular to the plane.

End post spacing Distance between centerlines of end posts.

Endwall An exterior wall which is perpendicular to the ridge and parallel to the gable of the building.

Engineering drawings Basic layout drawings of a structure used for design and engineering purposes.

Erection drawings Drawings prepared especially for use on the jobsite in erecting a building. Used primarily in steel and precast concrete construction to show how the building fits together and in what order each piece is to be erected.

Exterior bays Last frame spacing on either end of a building measured from the building line (outside face of girt) to the centerline of the first interior frame.

Extrusion A structural member formed by forcing a material, such as steel, through a hole of the desired cross section; refers to both the process and the final product.

Fabrication details Detailed drawings of individual structural members describing exactly how they are to be fabricated.

Flange On structural steel shapes, such as C-shapes, S-shapes, and W-shapes, the horizontal portions at the top and bottom which are perpendicular to the web.

Flame brace Flame used in roof and walls to transfer loads, such as wind loads, and seismic and crane thrusts to the foundation.

Footing An enlargement at the base of a column or bottom of a wall to distribute the load over a greater portion of ground and thereby prevent settling. Most footings are made of poured concrete.

Forge Process used in forming a metal structural member by heating and hammering to the desired shape.

Foundation The bottommost portion of a wall or that part of the wall that rests on the footing and upon which the rest of the wall is built.

Framing plan A plan view drawn to scale providing a bird's eye view of the structural components of a building. Columns, beams and girders, roof members, floor members, and wall members all require separate framing plans.

Geographic Information System (GIS) A geographical database that contains information on features under, on, and above the ground (e.g., topography, streets, utility lines, transmission lines, sewer lines, and land use).

Girder A large, horizontal support member similar to a beam. The terms are differentiated in some schools of thought as follows: beams span from column to column, while girders span from beam to beam.

Girt A secondary horizontal structural member attached to sidewall or endwall columns to which wall covering is attached and supported horizontally.

Grid survey The method of dividing a plot of land into a grid and taking elevations at each grid intersection.

Haunch The area of increased depth of the column or rafter member which is designed to account for the higher bending moments that occur at such places. Typically, this occurs at the rafter-to-column connection.

I-beam Common name for an S-shape steel beam.

Interpolation An inexact but acceptable method of inserting missing values between two known values. CAD technicians sometimes interpolate when placing contour lines between known points.

Iron A malleable (may be pressed and shaped without returning to its original form), ductile (may be stretched or hammered without breaking), metallic element. The main ingredient used in the production of steel. Once a common building material for bridges, it was gradually replaced by steel around the turn of the twentieth century.

Cast iron has a higher carbon content (2.0%–4.5%) and is less malleable (more brittle). It is shaped by pouring it in a fluid, molten state into molds. *Steel alloys* are next in decreasing order of carbon content (approximately 0.2%–2.0%), followed by *wrought iron,* which has less carbon content (approximately 0.2%). This makes wrought iron tough but more malleable; it is more easily shaped by heating and hammering (forging).

Joists Horizontal structural members which support the floors and/or roof of a building.

Longitudinal section A section cut through the length of a structure.

Lot and block A method of describing parcels of land that can be used only if a plat containing metes and bounds descriptions has been recorded with a local government entity. Each parcel is given a lot number within a given block. This method is typically used in subdivisions.

Magnetic declination Horizontal angle between the magnetic north meridian and the true north meridian.

Metes and bounds A method of describing parcels of property using bearings and distances from a known point of beginning (POB).

MOD 24 framing A wooden framing system that places joists, studs, rafters, and trusses in 24-inch modules.

Narrow flange beam An S-shape steel beam.

Pipe Hollow, cylindrical structural steel shape.

Plan and profile Drawing of a road, canal, utility line, or other feature that shows a plan view and a profile (section cut into the earth vertically). The profile view is typically projected directly down and drawn below the plan view.

Plat A drawing of a given piece of land from one individual parcel to a multiparcel subdivision.

Plate In structural steel, a flat steel piece rectangular in cross section. In wood construction, a term applied to 2 × 4 nailers placed on the sill and on top of the stud wall.

Plot plan A drawing of a piece of property that shows all features on the property (also known as a site plan).

Point of curve The point at which a straight line (such as the line representing the centerline of a highway) begins to curve.

Post-and-beam endframe A structural framing system utilized at the endwall which is composed of corner post, end post and rake beams.

Post tensioning A method of prestressing concrete by stressing the steel strands after the concrete has been poured and allowed to harden.

Precast concrete products Concrete members that are poured in forms at a plant or factory and allowed to harden. There are two types of precast products: prestressed products and reinforced products.

Prestressed concrete products Concrete products that are stressed before being erected in a job. This is accomplished by passing high-strength steel strands through the form and applying stress to the strands either before or after the concrete is poured.

Pretensioning Stressing the steel strands in a prestressed member before the concrete is poured into the form.

Purlin A secondary horizontal structural member attached to the primary frame which transfers the roof loads from the roof covering to the primary members.

Rafter A fabricated member, with parallel flanges, that extends from the haunch member to the frame ridge. Any beam, in general, used in a primary frame. Ridge apex of building.

Rebar Short term used for steel reinforcing bars used to reinforce concrete.

Rectangular system A system developed by the U.S. Bureau of Land Management used to describe parcels of land in a coordinate system within a given state.

Retaining wall Structural wall used to hold back earth or other materials. There are two types: gravity and cantilever.

Rigid frame A clear-span structural frame consisting of straight or tapered sections whose rafter-to-column connection stabilizes the frame with respect to imposed loads. This frame is designed in accordance with AISC Type I construction.

Rivet A metal fastener with a large head on one end used to connect multiple metal plates by passing the shank through aligned holes in the plates and hammering the plain end to form a second head.

Rod A square pole used by surveyors to take elevations when used in conjunction with a surveying device.

Rod bracing Rods are utilized in conjunction with purlins and girts to form a truss-type bracing system located in both roof and wall planes.

Rolled section A structural member formed by heating material, such as steel, and passing it through a series of rollers to achieve a desired shape.

Roof purlin A roof secondary member which is secured to frame rafters and supports the roof covering.

Roof system The exterior roof surface consisting of panels, closures, and attachments.

Section (In civil engineering). A portion of a township that contains 640 acres and is one mile square (except in areas where geographical features interfere). For example, many sections in the state of Florida contain less than 640 acres because of the state's odd shape and coastlines.

Section (In structural and pipe). A type of drawing used to clarify details of construction.

Shear walls Walls designed to resist lateral loading from winds, underground disturbances, or blasts.

Shop drawings Drawings prepared to guide shop personnel in the fabrication of structural members for a job. Usually include fabrication details and a bill of materials.

Sidewall An exterior wall which is parallel to the ridge and sidewall of the building.

Slab A flat concrete area usually reinforced with wire mesh and/or rebar(s).

Slope The degree of incline of a roof expressed as a ratio of the vertical rise to the horizontal run.

Specifications Written instructions accompanying the drawing containing information about materials, workmanship, style, and other pertinent information.

S-shape or narrow flange beam Structural steel shape which has a cross section resembling the letter *I* with sloped inner flange surfaces adjacent to the web. May be formed by extrusion or rolling. Designated by the prefix *S* followed by the depth in inches and the weight per linear foot in pounds, such as S6×10. Commonly called I-beam or American standard beam. Compare to W-shape.

Steel Any of a variety of iron-based metallic alloys having less carbon content than cast iron but more than wrought iron.

Stress Forces acting on structural members due to various types of loads. These forces are torsion, tension, compression, or shear.

Studs The primary vertical members of a wooden wall.

Subdivision A larger parcel of land that has been divided into smaller parcels. Many residential communities in the suburbs of larger cities are built on land that was subdivided.

Suspended floor A concrete floor system built above and off the ground.

Tilt-up walls Poured-in-place concrete walls that are poured in forms on the ground and then tilted up into place by cranes or hoists.

Townships Divisions of land described by parallels and meridians, each containing 36 square miles.

Truss A framed structure consisting of straight members jointed to form a pattern of interconnecting triangles, usually made of wood or metal.

Utilities Services to homes, businesses, government buildings, and nonprofit organizations such as electricity, water, sewer, cable, telephone, and Internet connections.

Wall girt A horizontal wall secondary member which is secured to columns and supports the wall covering.

Wall system The exterior wall surface consisting of panels, closures, and attachments.

Web On structural steel shapes, such as C-shapes, S-shapes, and W-shapes, the flat portion which is perpendicular to and joining the flanges. Also the system of members connecting the top and bottom chords of a truss.

Weld Joining two metal pieces by heating them and allowing them to flow together. Creating a bond by using another nonferrous metal which melts below 800° F is called *soldering*. Creating a bond by using another nonferrous metal which melts above 800° F is called *brazing*. The continuous deposit of fused metal created in these processes is called a *bead*. Other common fasteners used in metal structures include *rivets, threaded bolts,* and *pin/eyebar* connections.

Working drawings A set of drawings containing all information to complete a job from start to finish.

W-shape or wide flange beam Structural steel shape which has a cross section resembling an *H* with flat inner flange surfaces adjacent to the web. May be formed by extrusion or rolling. Designated by the prefix *W* followed by the depth in inches and the weight per linear foot in pounds, such as W18×40. Compare to S-shape.

APPENDIX A

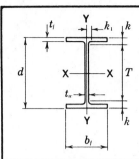

W SHAPES
Dimensions

Designation	Area A	Depth d		Web Thickness t_w		Web $\frac{t_w}{2}$	Flange Width b_f		Flange Thickness t_f		T	k	k_1
	In.²	In.		In.		In.	In.		In.		In.	In.	In.
W 36x300	88.3	36.74	36¾	0.945	15/16	½	16.655	16⅝	1.680	1 11/16	31⅛	2 13/16	1½
x280	82.4	36.52	36½	0.885	⅞	7/16	16.595	16⅝	1.570	1 9/16	31⅛	2 11/16	1½
x260	76.5	36.26	36¼	0.840	13/16	7/16	16.550	16½	1.440	1 7/16	31⅛	2 9/16	1½
x245	72.1	36.08	36⅛	0.800	13/16	7/16	16.510	16½	1.350	1⅜	31⅛	2½	1 7/16
x230	67.6	35.90	35⅞	0.760	¾	⅜	16.470	16½	1.260	1¼	31⅛	2⅜	1 7/16
W 36x210	61.8	36.69	36¾	0.830	13/16	7/16	12.180	12⅛	1.360	1⅜	32⅛	2 5/16	1¼
x194	57.0	36.49	36½	0.765	¾	⅜	12.115	12⅛	1.260	1¼	32⅛	2 3/16	1 3/16
x182	53.6	36.33	36⅜	0.725	¾	⅜	12.075	12⅛	1.180	1 3/16	32⅛	2⅛	1 3/16
x170	50.0	36.17	36⅛	0.680	11/16	⅜	12.030	12	1.100	1⅛	32⅛	2	1 3/16
x160	47.0	36.01	36	0.650	⅝	5/16	12.000	12	1.020	1	32⅛	1 15/16	1⅛
x150	44.2	35.85	35⅞	0.625	⅝	5/16	11.975	12	0.940	15/16	32⅛	1⅞	1⅛
x135	39.7	35.55	35½	0.600	⅝	5/16	11.950	12	0.790	13/16	32⅛	1 11/16	1⅛
W 33x241	70.9	34.18	34⅛	0.830	13/16	7/16	15.860	15⅞	1.400	1⅜	29¾	2 3/16	1 3/16
x221	65.0	33.93	33⅞	0.775	¾	⅜	15.805	15¾	1.275	1¼	29¾	2 1/16	1 3/16
x201	59.1	33.68	33⅝	0.715	11/16	⅜	15.745	15¾	1.150	1⅛	29¾	1 15/16	1⅛
W 33x152	44.7	33.49	33½	0.635	⅝	5/16	11.565	11⅝	1.055	1 1/16	29¾	1⅞	1⅛
x141	41.6	33.30	33¼	0.605	⅝	5/16	11.535	11½	0.960	15/16	29¾	1¾	1 1/16
x130	38.3	33.09	33⅛	0.580	9/16	5/16	11.510	11½	0.855	⅞	29¾	1 11/16	1 1/16
x118	34.7	32.86	32⅞	0.550	9/16	5/16	11.480	11½	0.740	¾	29¾	1 9/16	1 1/16
W 30x211	62.0	30.94	31	0.775	¾	⅜	15.105	15⅛	1.315	1 5/16	26¾	2⅛	1⅛
x191	56.1	30.68	30⅝	0.710	11/16	⅜	15.040	15	1.185	1 3/16	26¾	1 15/16	1 1/16
x173	50.8	30.44	30½	0.655	⅝	5/16	14.985	15	1.065	1 1/16	26¾	1⅞	1 1/16
W 30x132	38.9	30.31	30¼	0.615	⅝	5/16	10.545	10½	1.000	1	26¾	1¾	1 1/16
x124	36.5	30.17	30⅛	0.585	9/16	5/16	10.515	10½	0.930	15/16	26¾	1 11/16	1
x116	34.2	30.01	30	0.565	9/16	5/16	10.495	10½	0.850	⅞	26¾	1⅝	1
x108	31.7	29.83	29⅞	0.545	9/16	5/16	10.475	10½	0.760	¾	26¾	1 9/16	1
x 99	29.1	29.65	29⅝	0.520	½	¼	10.450	10½	0.670	11/16	26¾	1 7/16	1

AMERICAN INSTITUTE OF STEEL CONSTRUCTION

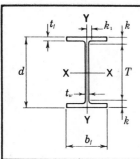

W SHAPES
Dimensions

Designation	Area A	Depth d		Web Thickness t_w		$\frac{t_w}{2}$	Flange Width b_f		Flange Thickness t_f		T	k	k_1
	In.²	In.		In.		In.	In.		In.		In.	In.	In.
W 27×178	52.3	27.81	27¾	0.725	¾	⅜	14.085	14⅛	1.190	1³/₁₆	24	1⅞	1¹/₁₆
×161	47.4	27.59	27⅝	0.660	¹¹/₁₆	⅜	14.020	14	1.080	1¹/₁₆	24	1¹³/₁₆	1
×146	42.9	27.38	27⅜	0.605	⅝	⁵/₁₆	13.965	14	0.975	1	24	1¹¹/₁₆	1
W 27×114	33.5	27.29	27¼	0.570	⁹/₁₆	⁵/₁₆	10.070	10⅛	0.930	¹⁵/₁₆	24	1⅝	¹⁵/₁₆
×102	30.0	27.09	27⅛	0.515	½	¼	10.015	10	0.830	¹³/₁₆	24	1⁹/₁₆	¹⁵/₁₆
× 94	27.7	26.92	26⅞	0.490	½	¼	9.990	10	0.745	¾	24	1⁷/₁₆	¹⁵/₁₆
× 84	24.8	26.71	26¾	0.460	⁷/₁₆	¼	9.960	10	0.640	⅝	24	1⅜	¹⁵/₁₆
W 24×162	47.7	25.00	25	0.705	¹¹/₁₆	⅜	12.955	13	1.220	1¼	21	2	1¹/₁₆
×146	43.0	24.74	24¾	0.650	⅝	⁵/₁₆	12.900	12⅞	1.090	1¹/₁₆	21	1⅞	1¹/₁₆
×131	38.5	24.48	24½	0.605	⅝	⁵/₁₆	12.855	12⅞	0.960	¹⁵/₁₆	21	1¾	1¹/₁₆
×117	34.4	24.26	24¼	0.550	⁹/₁₆	⁵/₁₆	12.800	12¾	0.850	⅞	21	1⅝	1
×104	30.6	24.06	24	0.500	½	¼	12.750	12¾	0.750	¾	21	1½	1
W 24× 94	27.7	24.31	24¼	0.515	½	¼	9.065	9⅛	0.875	⅞	21	1⅝	1
× 84	24.7	24.10	24⅛	0.470	½	¼	9.020	9	0.770	¾	21	1⁹/₁₆	¹⁵/₁₆
× 76	22.4	23.92	23⅞	0.440	⁷/₁₆	¼	8.990	9	0.680	¹¹/₁₆	21	1⁷/₁₆	¹⁵/₁₆
× 68	20.1	23.73	23¾	0.415	⁷/₁₆	¼	8.965	9	0.585	⁹/₁₆	21	1⅜	¹⁵/₁₆
W 24× 62	18.2	23.74	23¾	0.430	⁷/₁₆	¼	7.040	7	0.590	⁹/₁₆	21	1⅜	¹⁵/₁₆
× 55	16.2	23.57	23⅝	0.395	⅜	³/₁₆	7.005	7	0.505	½	21	1⁵/₁₆	¹⁵/₁₆
W 21×147	43.2	22.06	22	0.720	¾	⅜	12.510	12½	1.150	1⅛	18¼	1⅞	1¹/₁₆
×132	38.8	21.83	21⅞	0.650	⅝	⁵/₁₆	12.440	12½	1.035	1¹/₁₆	18¼	1¹³/₁₆	1
×122	35.9	21.68	21⅝	0.600	⅝	⁵/₁₆	12.390	12⅜	0.960	¹⁵/₁₆	18¼	1¹¹/₁₆	1
×111	32.7	21.51	21½	0.550	⁹/₁₆	⁵/₁₆	12.340	12⅜	0.875	⅞	18¼	1⅝	¹⁵/₁₆
×101	29.8	21.36	21⅜	0.500	½	¼	12.290	12¼	0.800	¹³/₁₆	18¼	1⁹/₁₆	¹⁵/₁₆
W 21× 93	27.3	21.62	21⅝	0.580	⁹/₁₆	⁵/₁₆	8.420	8⅜	0.930	¹⁵/₁₆	18¼	1¹¹/₁₆	1
× 83	24.3	21.43	21⅜	0.515	½	¼	8.355	8⅜	0.835	¹³/₁₆	18¼	1⁹/₁₆	¹⁵/₁₆
× 73	21.5	21.24	21¼	0.455	⁷/₁₆	¼	8.295	8¼	0.740	¾	18¼	1½	¹⁵/₁₆
× 68	20.0	21.13	21⅛	0.430	⁷/₁₆	¼	8.270	8¼	0.685	¹¹/₁₆	18¼	1⁷/₁₆	⅞
× 62	18.3	20.99	21	0.400	⅜	³/₁₆	8.240	8¼	0.615	⅝	18¼	1⅜	⅞
W 21× 57	16.7	21.06	21	0.405	⅜	³/₁₆	6.555	6½	0.650	⅝	18¼	1⅜	⅞
× 50	14.7	20.83	20⅞	0.380	⅜	³/₁₆	6.530	6½	0.535	⁹/₁₆	18¼	1⁵/₁₆	⅞
× 44	13.0	20.66	20⅝	0.350	⅜	³/₁₆	6.500	6½	0.450	⁷/₁₆	18¼	1³/₁₆	⅞

AMERICAN INSTITUTE OF STEEL CONSTRUCTION

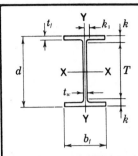

W SHAPES
Dimensions

Designation	Area A	Depth d		Web				Flange			Distance		
				Thickness t_w		$\frac{t_w}{2}$		Width b_f		Thickness t_f	T	k	k_1
	In.²	In.		In.		In.		In.		In.	In.	In.	In.
W 18x119	35.1	18.97	19	0.655	5/8	5/16		11.265	11¼	1.060	1 1/16	15½	1¾	15/16
x106	31.1	18.73	18¾	0.590	9/16	5/16		11.200	11¼	0.940	15/16	15½	1 5/8	15/16
x 97	28.5	18.59	18 5/8	0.535	9/16	5/16		11.145	11 1/8	0.870	7/8	15½	1 9/16	7/8
x 86	25.3	18.39	18 3/8	0.480	½	¼		11.090	11 1/8	0.770	¾	15½	1 7/16	7/8
x 76	22.3	18.21	18¼	0.425	7/16	¼		11.035	11	0.680	11/16	15½	1 3/8	13/16
W 18x 71	20.8	18.47	18½	0.495	½	¼		7.635	7 5/8	0.810	13/16	15½	1½	7/8
x 65	19.1	18.35	18 3/8	0.450	7/16	¼		7.590	7 5/8	0.750	¾	15½	1 7/16	7/8
x 60	17.6	18.24	18¼	0.415	7/16	¼		7.555	7½	0.695	11/16	15½	1 3/8	13/16
x 55	16.2	18.11	18 1/8	0.390	3/8	3/16		7.530	7½	0.630	5/8	15½	1 5/16	13/16
x 50	14.7	17.99	18	0.355	3/8	3/16		7.495	7½	0.570	9/16	15½	1¼	13/16
W 18x 46	13.5	18.06	18	0.360	3/8	3/16		6.060	6	0.605	5/8	15½	1¼	13/16
x 40	11.8	17.90	17 7/8	0.315	5/16	3/16		6.015	6	0.525	½	15½	1 3/16	13/16
x 35	10.3	17.70	17¾	0.300	5/16	3/16		6.000	6	0.425	7/16	15½	1 1/8	¾
W 16x100	29.4	16.97	17	0.585	9/16	5/16		10.425	10 3/8	0.985	1	13 5/8	1 11/16	15/16
x 89	26.2	16.75	16¾	0.525	½	¼		10.365	10 3/8	0.875	7/8	13 5/8	1 9/16	7/8
x 77	22.6	16.52	16½	0.455	7/16	¼		10.295	10¼	0.760	¾	13 5/8	1 7/16	7/8
x 67	19.7	16.33	16 3/8	0.395	3/8	3/16		10.235	10¼	0.665	11/16	13 5/8	1 3/8	13/16
W 16x 57	16.8	16.43	16 3/8	0.430	7/16	¼		7.120	7 1/8	0.715	11/16	13 5/8	1 3/8	7/8
x 50	14.7	16.26	16¼	0.380	3/8	3/16		7.070	7 1/8	0.630	5/8	13 5/8	1 5/16	13/16
x 45	13.3	16.13	16 1/8	0.345	3/8	3/16		7.035	7	0.565	9/16	13 5/8	1¼	13/16
x 40	11.8	16.01	16	0.305	5/16	3/16		6.995	7	0.505	½	13 5/8	1 3/16	13/16
x 36	10.6	15.86	15 7/8	0.295	5/16	3/16		6.985	7	0.430	7/16	13 5/8	1 1/8	¾
W 16x 31	9.12	15.88	15 7/8	0.275	¼	1/8		5.525	5½	0.440	7/16	13 5/8	1 1/8	¾
x 26	7.68	15.69	15¾	0.250	¼	1/8		5.500	5½	0.345	3/8	13 5/8	1 1/16	¾

AMERICAN INSTITUTE OF STEEL CONSTRUCTION

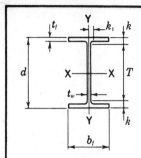

W SHAPES
Dimensions

Designation	Area A	Depth d		Web Thickness t_w		$\frac{t_w}{2}$	Flange Width b_f		Flange Thickness t_f		T	k	k_1
	In.²	In.		In.		In.	In.		In.		In.	In.	In.
W 14x730	215.0	22.42	$22\frac{3}{8}$	3.070	$3\frac{1}{16}$	$1\frac{9}{16}$	17.890	$17\frac{7}{8}$	4.910	$4\frac{15}{16}$	$11\frac{1}{4}$	$5\frac{9}{16}$	$2\frac{3}{16}$
x665	196.0	21.64	$21\frac{5}{8}$	2.830	$2\frac{13}{16}$	$1\frac{7}{16}$	17.650	$17\frac{5}{8}$	4.520	$4\frac{1}{2}$	$11\frac{1}{4}$	$5\frac{3}{16}$	$2\frac{1}{16}$
x605	178.0	20.92	$20\frac{7}{8}$	2.595	$2\frac{5}{8}$	$1\frac{5}{16}$	17.415	$17\frac{3}{8}$	4.160	$4\frac{3}{16}$	$11\frac{1}{4}$	$4\frac{13}{16}$	$1\frac{15}{16}$
x550	162.0	20.24	$20\frac{1}{4}$	2.380	$2\frac{3}{8}$	$1\frac{3}{16}$	17.200	$17\frac{1}{4}$	3.820	$3\frac{13}{16}$	$11\frac{1}{4}$	$4\frac{1}{2}$	$1\frac{13}{16}$
x500	147.0	19.60	$19\frac{5}{8}$	2.190	$2\frac{3}{16}$	$1\frac{1}{8}$	17.010	17	3.500	$3\frac{1}{2}$	$11\frac{1}{4}$	$4\frac{3}{16}$	$1\frac{3}{4}$
x455	134.0	19.02	19	2.015	2	1	16.835	$16\frac{7}{8}$	3.210	$3\frac{3}{16}$	$11\frac{1}{4}$	$3\frac{7}{8}$	$1\frac{5}{8}$
W 14x426	125.0	18.67	$18\frac{5}{8}$	1.875	$1\frac{7}{8}$	$\frac{15}{16}$	16.695	$16\frac{3}{4}$	3.035	$3\frac{1}{16}$	$11\frac{1}{4}$	$3\frac{11}{16}$	$1\frac{9}{16}$
x398	117.0	18.29	$18\frac{1}{4}$	1.770	$1\frac{3}{4}$	$\frac{7}{8}$	16.590	$16\frac{5}{8}$	2.845	$2\frac{7}{8}$	$11\frac{1}{4}$	$3\frac{1}{2}$	$1\frac{1}{2}$
x370	109.0	17.92	$17\frac{7}{8}$	1.655	$1\frac{5}{8}$	$\frac{13}{16}$	16.475	$16\frac{1}{2}$	2.660	$2\frac{11}{16}$	$11\frac{1}{4}$	$3\frac{5}{16}$	$1\frac{7}{16}$
x342	101.0	17.54	$17\frac{1}{2}$	1.540	$1\frac{9}{16}$	$\frac{13}{16}$	16.360	$16\frac{3}{8}$	2.470	$2\frac{1}{2}$	$11\frac{1}{4}$	$3\frac{1}{8}$	$1\frac{3}{8}$
x311	91.4	17.12	$17\frac{1}{8}$	1.410	$1\frac{7}{16}$	$\frac{3}{4}$	16.230	$16\frac{1}{4}$	2.260	$2\frac{1}{4}$	$11\frac{1}{4}$	$2\frac{15}{16}$	$1\frac{5}{16}$
x283	83.3	16.74	$16\frac{3}{4}$	1.290	$1\frac{5}{16}$	$\frac{11}{16}$	16.110	$16\frac{1}{8}$	2.070	$2\frac{1}{16}$	$11\frac{1}{4}$	$2\frac{3}{4}$	$1\frac{1}{4}$
x257	75.6	16.38	$16\frac{3}{8}$	1.175	$1\frac{3}{16}$	$\frac{5}{8}$	15.995	16	1.890	$1\frac{7}{8}$	$11\frac{1}{4}$	$2\frac{9}{16}$	$1\frac{3}{16}$
x233	68.5	16.04	16	1.070	$1\frac{1}{16}$	$\frac{9}{16}$	15.890	$15\frac{7}{8}$	1.720	$1\frac{3}{4}$	$11\frac{1}{4}$	$2\frac{3}{8}$	$1\frac{3}{16}$
x211	62.0	15.72	$15\frac{3}{4}$	0.980	1	$\frac{1}{2}$	15.800	$15\frac{3}{4}$	1.560	$1\frac{9}{16}$	$11\frac{1}{4}$	$2\frac{1}{4}$	$1\frac{1}{8}$
x193	56.8	15.48	$15\frac{1}{2}$	0.890	$\frac{7}{8}$	$\frac{7}{16}$	15.710	$15\frac{3}{4}$	1.440	$1\frac{7}{16}$	$11\frac{1}{4}$	$2\frac{1}{8}$	$1\frac{1}{16}$
x176	51.8	15.22	$15\frac{1}{4}$	0.830	$\frac{13}{16}$	$\frac{7}{16}$	15.650	$15\frac{5}{8}$	1.310	$1\frac{5}{16}$	$11\frac{1}{4}$	2	$1\frac{1}{16}$
x159	46.7	14.98	15	0.745	$\frac{3}{4}$	$\frac{3}{8}$	15.565	$15\frac{5}{8}$	1.190	$1\frac{3}{16}$	$11\frac{1}{4}$	$1\frac{7}{8}$	1
x145	42.7	14.78	$14\frac{3}{4}$	0.680	$\frac{11}{16}$	$\frac{3}{8}$	15.500	$15\frac{1}{2}$	1.090	$1\frac{1}{16}$	$11\frac{1}{4}$	$1\frac{3}{4}$	1

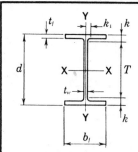

W SHAPES
Dimensions

Designation	Area A	Depth d	Web Thickness t_w	$\dfrac{t_w}{2}$	Flange Width b_f	Flange Thickness t_f	T	k	k_1
	In.²	In.	In.	In.	In.	In.	In.	In.	In.
W 14×132	38.8	14.66 14⅝	0.645 ⅝	⁵⁄₁₆	14.725 14¾	1.030 1	11¼	1¹¹⁄₁₆	¹⁵⁄₁₆
×120	35.3	14.48 14½	0.590 ⁹⁄₁₆	⁵⁄₁₆	14.670 14⅝	0.940 ¹⁵⁄₁₆	11¼	1⅝	¹⁵⁄₁₆
×109	32.0	14.32 14⅜	0.525 ½	¼	14.605 14⅝	0.860 ⅞	11¼	1⁹⁄₁₆	⅞
× 99	29.1	14.16 14⅛	0.485 ½	¼	14.565 14⅝	0.780 ¾	11¼	1⁷⁄₁₆	⅞
× 90	26.5	14.02 14	0.440 ⁷⁄₁₆	¼	14.520 14½	0.710 ¹¹⁄₁₆	11¼	1⅜	⅞
W 14× 82	24.1	14.31 14¼	0.510 ½	¼	10.130 10⅛	0.855 ⅞	11	1⅝	1
× 74	21.8	14.17 14⅛	0.450 ⁷⁄₁₆	¼	10.070 10⅛	0.785 ¹³⁄₁₆	11	1⁹⁄₁₆	¹⁵⁄₁₆
× 68	20.0	14.04 14	0.415 ⁷⁄₁₆	¼	10.035 10	0.720 ¾	11	1½	¹⁵⁄₁₆
× 61	17.9	13.89 13⅞	0.375 ⅜	³⁄₁₆	9.995 10	0.645 ⅝	11	1⁷⁄₁₆	¹⁵⁄₁₆
W 14× 53	15.6	13.92 13⅞	0.370 ⅜	³⁄₁₆	8.060 8	0.660 ¹¹⁄₁₆	11	1⁷⁄₁₆	¹⁵⁄₁₆
× 48	14.1	13.79 13¾	0.340 ⁵⁄₁₆	³⁄₁₆	8.030 8	0.595 ⅝	11	1⅜	⅞
× 43	12.6	13.66 13⅝	0.305 ⁵⁄₁₆	³⁄₁₆	7.995 8	0.530 ½	11	1⁵⁄₁₆	⅞
W 14× 38	11.2	14.10 14⅛	0.310 ⁵⁄₁₆	³⁄₁₆	6.770 6¾	0.515 ½	12	1¹⁄₁₆	⅝
× 34	10.0	13.98 14	0.285 ⁵⁄₁₆	³⁄₁₆	6.745 6¾	0.455 ⁷⁄₁₆	12	1	⅝
× 30	8.85	13.84 13⅞	0.270 ¼	⅛	6.730 6¾	0.385 ⅜	12	¹⁵⁄₁₆	⅝
W 14× 26	7.69	13.91 13⅞	0.255 ¼	⅛	5.025 5	0.420 ⁷⁄₁₆	12	¹⁵⁄₁₆	⁹⁄₁₆
× 22	6.49	13.74 13¾	0.230 ¼	⅛	5.000 5	0.335 ⁵⁄₁₆	12	⅞	⁹⁄₁₆

AMERICAN INSTITUTE OF STEEL CONSTRUCTION

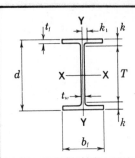

W SHAPES
Dimensions

Designation	Area A	Depth d		Web Thickness t_w		$\frac{t_w}{2}$	Flange Width b_f		Flange Thickness t_f		T	k	k_1
	In.^{2}	In.		In.		In.	In.		In.		In.	In.	In.
W 12x336	98.8	16.82	$16\frac{7}{8}$	1.775	$1\frac{3}{4}$	$\frac{7}{8}$	13.385	$13\frac{3}{8}$	2.955	$2\frac{15}{16}$	$9\frac{1}{2}$	$3\frac{11}{16}$	$1\frac{1}{2}$
x305	89.6	16.32	$16\frac{3}{8}$	1.625	$1\frac{5}{8}$	$\frac{13}{16}$	13.235	$13\frac{1}{4}$	2.705	$2\frac{11}{16}$	$9\frac{1}{2}$	$3\frac{7}{16}$	$1\frac{7}{16}$
x279	81.9	15.85	$15\frac{7}{8}$	1.530	$1\frac{1}{2}$	$\frac{3}{4}$	13.140	$13\frac{1}{8}$	2.470	$2\frac{1}{2}$	$9\frac{1}{2}$	$3\frac{3}{16}$	$1\frac{3}{8}$
x252	74.1	15.41	$15\frac{3}{8}$	1.395	$1\frac{3}{8}$	$\frac{11}{16}$	13.005	13	2.250	$2\frac{1}{4}$	$9\frac{1}{2}$	$2\frac{15}{16}$	$1\frac{5}{16}$
x230	67.7	15.05	15	1.285	$1\frac{5}{16}$	$\frac{11}{16}$	12.895	$12\frac{7}{8}$	2.070	$2\frac{1}{16}$	$9\frac{1}{2}$	$2\frac{3}{4}$	$1\frac{1}{4}$
x210	61.8	14.71	$14\frac{3}{4}$	1.180	$1\frac{3}{16}$	$\frac{5}{8}$	12.790	$12\frac{3}{4}$	1.900	$1\frac{7}{8}$	$9\frac{1}{2}$	$2\frac{5}{8}$	$1\frac{1}{4}$
x190	55.8	14.38	$14\frac{3}{8}$	1.060	$1\frac{1}{16}$	$\frac{9}{16}$	12.670	$12\frac{5}{8}$	1.735	$1\frac{3}{4}$	$9\frac{1}{2}$	$2\frac{7}{16}$	$1\frac{3}{16}$
x170	50.0	14.03	14	0.960	$\frac{15}{16}$	$\frac{1}{2}$	12.570	$12\frac{5}{8}$	1.560	$1\frac{9}{16}$	$9\frac{1}{2}$	$2\frac{1}{4}$	$1\frac{1}{8}$
x152	44.7	13.71	$13\frac{3}{4}$	0.870	$\frac{7}{8}$	$\frac{7}{16}$	12.480	$12\frac{1}{2}$	1.400	$1\frac{3}{8}$	$9\frac{1}{2}$	$2\frac{1}{8}$	$1\frac{1}{16}$
x136	39.9	13.41	$13\frac{3}{8}$	0.790	$\frac{13}{16}$	$\frac{7}{16}$	12.400	$12\frac{3}{8}$	1.250	$1\frac{1}{4}$	$9\frac{1}{2}$	$1\frac{15}{16}$	1
x120	35.3	13.12	$13\frac{1}{8}$	0.710	$\frac{11}{16}$	$\frac{3}{8}$	12.320	$12\frac{3}{8}$	1.105	$1\frac{1}{8}$	$9\frac{1}{2}$	$1\frac{13}{16}$	1
x106	31.2	12.89	$12\frac{7}{8}$	0.610	$\frac{5}{8}$	$\frac{5}{16}$	12.220	$12\frac{1}{4}$	0.990	1	$9\frac{1}{2}$	$1\frac{11}{16}$	$\frac{15}{16}$
x 96	28.2	12.71	$12\frac{3}{4}$	0.550	$\frac{9}{16}$	$\frac{5}{16}$	12.160	$12\frac{1}{8}$	0.900	$\frac{7}{8}$	$9\frac{1}{2}$	$1\frac{5}{8}$	$\frac{7}{8}$
x 87	25.6	12.53	$12\frac{1}{2}$	0.515	$\frac{1}{2}$	$\frac{1}{4}$	12.125	$12\frac{1}{8}$	0.810	$\frac{13}{16}$	$9\frac{1}{2}$	$1\frac{1}{2}$	$\frac{7}{8}$
x 79	23.2	12.38	$12\frac{3}{8}$	0.470	$\frac{1}{2}$	$\frac{1}{4}$	12.080	$12\frac{1}{8}$	0.735	$\frac{3}{4}$	$9\frac{1}{2}$	$1\frac{7}{16}$	$\frac{7}{8}$
x 72	21.1	12.25	$12\frac{1}{4}$	0.430	$\frac{7}{16}$	$\frac{1}{4}$	12.040	12	0.670	$\frac{11}{16}$	$9\frac{1}{2}$	$1\frac{3}{8}$	$\frac{7}{8}$
x 65	19.1	12.12	$12\frac{1}{8}$	0.390	$\frac{3}{8}$	$\frac{3}{16}$	12.000	12	0.605	$\frac{5}{8}$	$9\frac{1}{2}$	$1\frac{5}{16}$	$\frac{13}{16}$
W 12x 58	17.0	12.19	$12\frac{1}{4}$	0.360	$\frac{3}{8}$	$\frac{3}{16}$	10.010	10	0.640	$\frac{5}{8}$	$9\frac{1}{2}$	$1\frac{3}{8}$	$\frac{13}{16}$
x 53	15.6	12.06	12	0.345	$\frac{3}{8}$	$\frac{3}{16}$	9.995	10	0.575	$\frac{9}{16}$	$9\frac{1}{2}$	$1\frac{1}{4}$	$\frac{13}{16}$
W 12x 50	14.7	12.19	$12\frac{1}{4}$	0.370	$\frac{3}{8}$	$\frac{3}{16}$	8.080	$8\frac{1}{8}$	0.640	$\frac{5}{8}$	$9\frac{1}{2}$	$1\frac{3}{8}$	$\frac{13}{16}$
x 45	13.2	12.06	12	0.335	$\frac{5}{16}$	$\frac{3}{16}$	8.045	8	0.575	$\frac{9}{16}$	$9\frac{1}{2}$	$1\frac{1}{4}$	$\frac{13}{16}$
x 40	11.8	11.94	12	0.295	$\frac{5}{16}$	$\frac{3}{16}$	8.005	8	0.515	$\frac{1}{2}$	$9\frac{1}{2}$	$1\frac{1}{4}$	$\frac{3}{4}$
W 12x 35	10.3	12.50	$12\frac{1}{2}$	0.300	$\frac{5}{16}$	$\frac{3}{16}$	6.560	$6\frac{1}{2}$	0.520	$\frac{1}{2}$	$10\frac{1}{2}$	1	$\frac{9}{16}$
x 30	8.79	12.34	$12\frac{3}{8}$	0.260	$\frac{1}{4}$	$\frac{1}{8}$	6.520	$6\frac{1}{2}$	0.440	$\frac{7}{16}$	$10\frac{1}{2}$	$\frac{15}{16}$	$\frac{1}{2}$
x 26	7.65	12.22	$12\frac{1}{4}$	0.230	$\frac{1}{4}$	$\frac{1}{8}$	6.490	$6\frac{1}{2}$	0.380	$\frac{3}{8}$	$10\frac{1}{2}$	$\frac{7}{8}$	$\frac{1}{2}$
W 12x 22	6.48	12.31	$12\frac{1}{4}$	0.260	$\frac{1}{4}$	$\frac{1}{8}$	4.030	4	0.425	$\frac{7}{16}$	$10\frac{1}{2}$	$\frac{7}{8}$	$\frac{1}{2}$
x 19	5.57	12.16	$12\frac{1}{8}$	0.235	$\frac{1}{4}$	$\frac{1}{8}$	4.005	4	0.350	$\frac{3}{8}$	$10\frac{1}{2}$	$\frac{13}{16}$	$\frac{1}{2}$
x 16	4.71	11.99	12	0.220	$\frac{1}{4}$	$\frac{1}{8}$	3.990	4	0.265	$\frac{1}{4}$	$10\frac{1}{2}$	$\frac{3}{4}$	$\frac{1}{2}$
x 14	4.16	11.91	$11\frac{7}{8}$	0.200	$\frac{3}{16}$	$\frac{1}{8}$	3.970	4	0.225	$\frac{1}{4}$	$10\frac{1}{2}$	$\frac{11}{16}$	$\frac{1}{2}$

AMERICAN INSTITUTE OF STEEL CONSTRUCTION

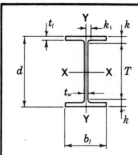

W SHAPES
Dimensions

Designation	Area A	Depth d		Web			Flange				Distance		
				Thickness t_w		$\frac{t_w}{2}$	Width b_f		Thickness t_f		T	k	k_1
	In.²	In.		In.		In.	In.		In.		In.	In.	In.
W 8x67	19.7	9.00	9	0.570	⁹/₁₆	⁵/₁₆	8.280	8¼	0.935	¹⁵/₁₆	6⅛	1⁷/₁₆	¹¹/₁₆
x58	17.1	8.75	8¾	0.510	½	¼	8.220	8¼	0.810	¹³/₁₆	6⅛	1⁵/₁₆	¹¹/₁₆
x48	14.1	8.50	8½	0.400	⅜	³/₁₆	8.110	8⅛	0.685	¹¹/₁₆	6⅛	1³/₁₆	⅝
x40	11.7	8.25	8¼	0.360	⅜	³/₁₆	8.070	8⅛	0.560	⁹/₁₆	6⅛	1¹/₁₆	⅝
x35	10.3	8.12	8⅛	0.310	⁵/₁₆	³/₁₆	8.020	8	0.495	½	6⅛	1	⁹/₁₆
x31	9.13	8.00	8	0.285	⁵/₁₆	³/₁₆	7.995	8	0.435	⁷/₁₆	6⅛	¹⁵/₁₆	⁹/₁₆
W 8x28	8.25	8.06	8	0.285	⁵/₁₆	³/₁₆	6.535	6½	0.465	⁷/₁₆	6⅛	¹⁵/₁₆	⁹/₁₆
x24	7.08	7.93	7⅞	0.245	¼	⅛	6.495	6½	0.400	⅜	6⅛	⅞	⁹/₁₆
W 8x21	6.16	8.28	8¼	0.250	¼	⅛	5.270	5¼	0.400	⅜	6⅝	¹³/₁₆	½
x18	5.26	8.14	8⅛	0.230	¼	⅛	5.250	5¼	0.330	⁵/₁₆	6⅝	¾	⁷/₁₆
W 8x15	4.44	8.11	8⅛	0.245	¼	⅛	4.015	4	0.315	⁵/₁₆	6⅝	¾	½
x13	3.84	7.99	8	0.230	¼	⅛	4.000	4	0.255	¼	6⅝	¹¹/₁₆	⁷/₁₆
x10	2.96	7.89	7⅞	0.170	³/₁₆	⅛	3.940	4	0.205	³/₁₆	6⅝	⅝	⁷/₁₆
W 6x25	7.34	6.38	6⅜	0.320	⁵/₁₆	³/₁₆	6.080	6⅛	0.455	⁷/₁₆	4¾	¹³/₁₆	⁷/₁₆
x20	5.87	6.20	6¼	0.260	¼	⅛	6.020	6	0.365	⅜	4¾	¾	⁷/₁₆
x15	4.43	5.99	6	0.230	¼	⅛	5.990	6	0.260	¼	4¾	⅝	⅜
W 6x16	4.74	6.28	6¼	0.260	¼	⅛	4.030	4	0.405	⅜	4¾	¾	⁷/₁₆
x12	3.55	6.03	6	0.230	¼	⅛	4.000	4	0.280	¼	4¾	⅝	⅜
x 9	2.68	5.90	5⅞	0.170	³/₁₆	⅛	3.940	4	0.215	³/₁₆	4¾	⁹/₁₆	⅜
W 5x19	5.54	5.15	5⅛	0.270	¼	⅛	5.030	5	0.430	⁷/₁₆	3½	¹³/₁₆	⁷/₁₆
x16	4.68	5.01	5	0.240	¼	⅛	5.000	5	0.360	⅜	3½	¾	⁷/₁₆
W 4x13	3.83	4.16	4⅛	0.280	¼	⅛	4.060	4	0.345	⅜	2¾	¹¹/₁₆	⁷/₁₆

AMERICAN INSTITUTE OF STEEL CONSTRUCTION

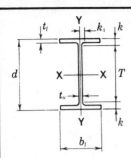

W SHAPES
Dimensions

Designation	Area A	Depth d		Web Thickness t_w		$\frac{t_w}{2}$	Flange Width b_f		Thickness t_f		Distance T	k	k_1
	In.²	In.		In.		In.	In.		In.		In.	In.	In.
W 10×112	32.9	11.36	11³⁄₈	0.755	³⁄₄	³⁄₈	10.415	10³⁄₈	1.250	1¹⁄₄	7⁵⁄₈	1⁷⁄₈	15⁄16
×100	29.4	11.10	11¹⁄₈	0.680	11⁄16	³⁄₈	10.340	10³⁄₈	1.120	1¹⁄₈	7⁵⁄₈	1³⁄₄	⁷⁄₈
× 88	25.9	10.84	10⁷⁄₈	0.605	⁵⁄₈	⁵⁄16	10.265	10¹⁄₄	0.990	1	7⁵⁄₈	1⁵⁄₈	13⁄16
× 77	22.6	10.60	10⁵⁄₈	0.530	¹⁄₂	¹⁄₄	10.190	10¹⁄₄	0.870	⁷⁄₈	7⁵⁄₈	1¹⁄₂	13⁄16
× 68	20.0	10.40	10³⁄₈	0.470	¹⁄₂	¹⁄₄	10.130	10¹⁄₈	0.770	³⁄₄	7⁵⁄₈	1³⁄₈	³⁄₄
× 60	17.6	10.22	10¹⁄₄	0.420	⁷⁄16	¹⁄₄	10.080	10¹⁄₈	0.680	11⁄16	7⁵⁄₈	1⁵⁄16	³⁄₄
× 54	15.8	10.09	10¹⁄₈	0.370	³⁄₈	³⁄16	10.030	10	0.615	⁵⁄₈	7⁵⁄₈	1¹⁄₄	11⁄16
× 49	14.4	9.98	10	0.340	⁵⁄16	³⁄16	10.000	10	0.560	⁹⁄16	7⁵⁄₈	1³⁄16	11⁄16
W 10× 45	13.3	10.10	10¹⁄₈	0.350	³⁄₈	³⁄16	8.020	8	0.620	⁵⁄₈	7⁵⁄₈	1¹⁄₄	11⁄16
× 39	11.5	9.92	9⁷⁄₈	0.315	⁵⁄16	³⁄16	7.985	8	0.530	¹⁄₂	7⁵⁄₈	1¹⁄₈	11⁄16
× 33	9.71	9.73	9³⁄₄	0.290	⁵⁄16	³⁄16	7.960	8	0.435	⁷⁄16	7⁵⁄₈	1¹⁄16	11⁄16
W 10× 30	8.84	10.47	10¹⁄₂	0.300	⁵⁄16	³⁄16	5.810	5³⁄₄	0.510	¹⁄₂	8⁵⁄₈	15⁄16	¹⁄₂
× 26	7.61	10.33	10³⁄₈	0.260	¹⁄₄	¹⁄₈	5.770	5³⁄₄	0.440	⁷⁄16	8⁵⁄₈	⁷⁄₈	¹⁄₂
× 22	6.49	10.17	10¹⁄₈	0.240	¹⁄₄	¹⁄₈	5.750	5³⁄₄	0.360	³⁄₈	8⁵⁄₈	³⁄₄	¹⁄₂
W 10× 19	5.62	10.24	10¹⁄₄	0.250	¹⁄₄	¹⁄₈	4.020	4	0.395	³⁄₈	8⁵⁄₈	13⁄16	¹⁄₂
× 17	4.99	10.11	10¹⁄₈	0.240	¹⁄₄	¹⁄₈	4.010	4	0.330	⁵⁄16	8⁵⁄₈	³⁄₄	¹⁄₂
× 15	4.41	9.99	10	0.230	¹⁄₄	¹⁄₈	4.000	4	0.270	¹⁄₄	8⁵⁄₈	11⁄16	⁷⁄16
× 12	3.54	9.87	9⁷⁄₈	0.190	³⁄16	¹⁄₈	3.960	4	0.210	³⁄16	8⁵⁄₈	⁵⁄₈	⁷⁄16

AMERICAN INSTITUTE OF STEEL CONSTRUCTION

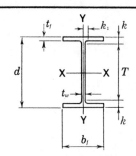

M SHAPES
Dimensions

Designation	Area A	Depth d		Web Thickness t_w		$\frac{t_w}{2}$	Flange Width b_f		Thickness t_f		Distance T	k	Grip	Max. Flge. Fastener
	In.²	In.		In.		In.	In.		In.		In.	In.	In.	In.
M 14x18	5.10	14.00	14	0.215	3/16	1/8	4.000	4	0.270	1/4	12³/4	5/8	1/4	3/4
M 12x11.8	3.47	12.00	12	0.177	3/16	1/8	3.065	3¹/8	0.225	1/4	10⁷/8	9/16	1/4	—
M 10x9	2.65	10.00	10	0.157	3/16	1/8	2.690	2³/4	0.206	3/16	8⁷/8	9/16	3/16	—
M 8x6.5	1.92	8.00	8	0.135	1/8	1/16	2.281	2¹/4	0.189	3/16	7	1/2	3/16	—
M 6x20	5.89	6.00	6	0.250	1/4	1/8	5.938	6	0.379	3/8	4¹/4	7/8	3/8	7/8
M 6x4.4	1.29	6.00	6	0.114	1/8	1/16	1.844	1⁷/8	0.171	3/16	5¹/8	7/16	3/16	—
M 5x18.9	5.55	5.00	5	0.316	5/16	3/16	5.003	5	0.416	7/16	3¹/4	7/8	7/16	7/8
M 4x13	3.81	4.00	4	0.254	1/4	1/8	3.940	4	0.371	3/8	2³/8	13/16	3/8	3/4

AMERICAN INSTITUTE OF STEEL CONSTRUCTION

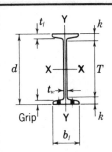

S SHAPES
Dimensions

Designation	Area A	Depth d		Web Thickness t_w		$\dfrac{t_w}{2}$	Flange Width b_f		Thickness t_f		Distance T	k	Grip	Max. Flge. Fastener
	In.²	In.		In.		In.	In.		In.		In.	In.	In.	In.
S 24x121	35.6	24.50	24½	0.800	13/16	7/16	8.050	8	1.090	1 1/16	20½	2	1 1/8	1
x106	31.2	24.50	24½	0.620	5/8	5/16	7.870	7 7/8	1.090	1 1/16	20½	2	1 1/8	1
S 24x100	29.3	24.00	24	0.745	3/4	3/8	7.245	7 1/4	0.870	7/8	20½	1 3/4	7/8	1
x90	26.5	24.00	24	0.625	5/8	5/16	7.125	7 1/8	0.870	7/8	20½	1 3/4	7/8	1
x80	23.5	24.00	24	0.500	1/2	1/4	7.000	7	0.870	7/8	20½	1 3/4	7/8	1
S 20x96	28.2	20.30	20 1/4	0.800	13/16	7/16	7.200	7 1/4	0.920	15/16	16 3/4	1 3/4	15/16	1
x86	25.3	20.30	20 1/4	0.660	11/16	3/8	7.060	7	0.920	15/16	16 3/4	1 3/4	15/16	1
S 20x75	22.0	20.00	20	0.635	5/8	5/16	6.385	6 3/8	0.795	13/16	16 3/4	1 5/8	13/16	7/8
x66	19.4	20.00	20	0.505	1/2	1/4	6.255	6 1/4	0.795	13/16	16 3/4	1 5/8	13/16	7/8
S 18x70	20.6	18.00	18	0.711	11/16	3/8	6.251	6 1/4	0.691	11/16	15	1 1/2	11/16	7/8
x54.7	16.1	18.00	18	0.461	7/16	1/4	6.001	6	0.691	11/16	15	1 1/2	11/16	7/8
S 15x50	14.7	15.00	15	0.550	9/16	5/16	5.640	5 5/8	0.622	5/8	12 1/4	1 3/8	9/16	3/4
x42.9	12.6	15.00	15	0.411	7/16	1/4	5.501	5 1/2	0.622	5/8	12 1/4	1 3/8	9/16	3/4
S 12x50	14.7	12.00	12	0.687	11/16	3/8	5.477	5 1/2	0.659	11/16	9 1/8	1 7/16	11/16	3/4
x40.8	12.0	12.00	12	0.462	7/16	1/4	5.252	5 1/4	0.659	11/16	9 1/8	1 7/16	5/8	3/4
S 12x35	10.3	12.00	12	0.428	7/16	1/4	5.078	5 1/8	0.544	9/16	9 5/8	1 3/16	1/2	3/4
x31.8	9.35	12.00	12	0.350	3/8	3/16	5.000	5	0.544	9/16	9 5/8	1 3/16	1/2	3/4
S 10x35	10.3	10.00	10	0.594	5/8	5/16	4.944	5	0.491	1/2	7 3/4	1 1/8	1/2	3/4
x25.4	7.46	10.00	10	0.311	5/16	3/16	4.661	4 5/8	0.491	1/2	7 3/4	1 1/8	1/2	3/4
S 8x23	6.77	8.00	8	0.441	7/16	1/4	4.171	4 1/8	0.426	7/16	6	1	7/16	3/4
x18.4	5.41	8.00	8	0.271	1/4	1/8	4.001	4	0.426	7/16	6	1	7/16	3/4
S 7x20	5.88	7.00	7	0.450	7/16	1/4	3.860	3 7/8	0.392	3/8	5 1/8	15/16	3/8	5/8
x15.3	4.50	7.00	7	0.252	1/4	1/8	3.662	3 5/8	0.392	3/8	5 1/8	15/16	3/8	5/8
S 6x17.25	5.07	6.00	6	0.465	7/16	1/4	3.565	3 5/8	0.359	3/8	4 1/4	7/8	3/8	5/8
x12.5	3.67	6.00	6	0.232	1/4	1/8	3.332	3 3/8	0.359	3/8	4 1/4	7/8	3/8	—
S 5x14.75	4.34	5.00	5	0.494	1/2	1/4	3.284	3 1/4	0.326	5/16	3 3/8	13/16	5/16	—
x10	2.94	5.00	5	0.214	3/16	1/8	3.004	3	0.326	5/16	3 3/8	13/16	5/16	—
S 4x9.5	2.79	4.00	4	0.326	5/16	3/16	2.796	2 3/4	0.293	5/16	2 1/2	3/4	5/16	—
x7.7	2.26	4.00	4	0.193	3/16	1/8	2.663	2 5/8	0.293	5/16	2 1/2	3/4	5/16	—
S 3x7.5	2.21	3.00	3	0.349	3/8	3/16	2.509	2 1/2	0.260	1/4	1 5/8	11/16	1/4	—
x5.7	1.67	3.00	3	0.170	3/16	1/8	2.330	2 3/8	0.260	1/4	1 5/8	11/16	1/4	—

AMERICAN INSTITUTE OF STEEL CONSTRUCTION

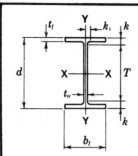

HP SHAPES
Dimensions

Designation	Area A	Depth d		Web		Flange				Distance		
				Thickness t_w		$\dfrac{t_w}{2}$	Width b_f		Thickness t_f		T	k	k_1
	In.²	In.		In.		In.	In.		In.		In.	In.	In.
HP 14x117	34.4	14.21	14¼	0.805	¹³/₁₆	⁷/₁₆	14.885	14⅞	0.805	¹³/₁₆	11¼	1½	1¹/₁₆
x102	30.0	14.01	14	0.705	¹¹/₁₆	⅜	14.785	14¾	0.705	¹¹/₁₆	11¼	1⅜	1
x 89	26.1	13.83	13⅞	0.615	⅝	⁵/₁₆	14.695	14¾	0.615	⅝	11¼	1⁵/₁₆	¹⁵/₁₆
x 73	21.4	13.61	13⅝	0.505	½	¼	14.585	14⅝	0.505	½	11¼	1³/₁₆	⅞
HP 13x100	29.4	13.15	13⅛	0.765	¾	⅜	13.205	13¼	0.765	¾	10¼	1⁷/₁₆	1
x 87	25.5	12.95	13	0.665	¹¹/₁₆	⅜	13.105	13⅛	0.665	¹¹/₁₆	10¼	1⅜	¹⁵/₁₆
x 73	21.6	12.75	12¾	0.565	⁹/₁₆	⁵/₁₆	13.005	13	0.565	⁹/₁₆	10¼	1¼	¹⁵/₁₆
x 60	17.5	12.54	12½	0.460	⁷/₁₆	¼	12.900	12⅞	0.460	⁷/₁₆	10¼	1⅛	⅞
HP 12x 84	24.6	12.28	12¼	0.685	¹¹/₁₆	⅜	12.295	12¼	0.685	¹¹/₁₆	9½	1⅜	1
x 74	21.8	12.13	12⅛	0.605	⅝	⁵/₁₆	12.215	12¼	0.610	⅝	9½	1⁵/₁₆	¹⁵/₁₆
x 63	18.4	11.94	12	0.515	½	¼	12.125	12⅛	0.515	½	9½	1¼	⅞
x 53	15.5	11.78	11¾	0.435	⁷/₁₆	¼	12.045	12	0.435	⁷/₁₆	9½	1⅛	⅞
HP 10x 57	16.8	9.99	10	0.565	⁹/₁₆	⁵/₁₆	10.225	10¼	0.565	⁹/₁₆	7⅝	1³/₁₆	¹³/₁₆
x 42	12.4	9.70	9¾	0.415	⁷/₁₆	¼	10.075	10⅛	0.420	⁷/₁₆	7⅝	1¹/₁₆	¾
HP 8x 36	10.6	8.02	8	0.445	⁷/₁₆	¼	8.155	8⅛	0.445	⁷/₁₆	6⅛	¹⁵/₁₆	⅝

AMERICAN INSTITUTE OF STEEL CONSTRUCTION

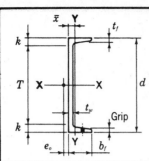

CHANNELS
AMERICAN STANDARD
Dimensions

Designation	Area A	Depth d	Web Thickness t_w		$\dfrac{t_w}{2}$	Flange Width b_f		Average thickness t_f		Distance T	k	Grip	Max. Flge. Fastener
	In.²	In.	In.		In.	In.		In.		In.	In.	In.	In.
C 15x50	14.7	15.00	0.716	11/16	3/8	3.716	3¾	0.650	5/8	12⅛	1 7/16	5/8	1
x40	11.8	15.00	0.520	½	¼	3.520	3½	0.650	5/8	12⅛	1 7/16	5/8	1
x33.9	9.96	15.00	0.400	3/8	3/16	3.400	3⅜	0.650	5/8	12⅛	1 7/16	5/8	1
C 12x30	8.82	12.00	0.510	½	¼	3.170	3⅛	0.501	½	9¾	1⅛	½	7/8
x25	7.35	12.00	0.387	3/8	3/16	3.047	3	0.501	½	9¾	1⅛	½	7/8
x20.7	6.09	12.00	0.282	5/16	⅛	2.942	3	0.501	½	9¾	1⅛	½	7/8
C 10x30	8.82	10.00	0.673	11/16	5/16	3.033	3	0.436	7/16	8	1	7/16	¾
x25	7.35	10.00	0.526	½	¼	2.886	2⅞	0.436	7/16	8	1	7/16	¾
x20	5.88	10.00	0.379	3/8	3/16	2.739	2¾	0.436	7/16	8	1	7/16	¾
x15.3	4.49	10.00	0.240	¼	⅛	2.600	2⅝	0.436	7/16	8	1	7/16	¾
C 9x20	5.88	9.00	0.448	7/16	¼	2.648	2⅝	0.413	7/16	7⅛	15/16	7/16	¾
x15	4.41	9.00	0.285	5/16	⅛	2.485	2½	0.413	7/16	7⅛	15/16	7/16	¾
x13.4	3.94	9.00	0.233	¼	⅛	2.433	2⅜	0.413	7/16	7⅛	15/16	7/16	¾
C 8x18.75	5.51	8.00	0.487	½	¼	2.527	2½	0.390	3/8	6⅛	15/16	3/8	¾
x13.75	4.04	8.00	0.303	5/16	⅛	2.343	2⅜	0.390	3/8	6⅛	15/16	3/8	¾
x11.5	3.38	8.00	0.220	¼	⅛	2.260	2¼	0.390	3/8	6⅛	15/16	3/8	¾
C 7x14.75	4.33	7.00	0.419	7/16	3/16	2.299	2¼	0.366	3/8	5¼	7/8	3/8	5/8
x12.25	3.60	7.00	0.314	5/16	3/16	2.194	2¼	0.366	3/8	5¼	7/8	3/8	5/8
x 9.8	2.87	7.00	0.210	3/16	⅛	2.090	2⅛	0.366	3/8	5¼	7/8	3/8	5/8
C 6x13	3.83	6.00	0.437	7/16	3/16	2.157	2⅛	0.343	5/16	4⅜	13/16	5/16	5/8
x10.5	3.09	6.00	0.314	5/16	3/16	2.034	2	0.343	5/16	4⅜	13/16	3/8	5/8
x 8.2	2.40	6.00	0.200	3/16	⅛	1.920	1⅞	0.343	5/16	4⅜	13/16	5/16	5/8
C 5x 9	2.64	5.00	0.325	5/16	3/16	1.885	1⅞	0.320	5/16	3½	¾	5/16	5/8
x 6.7	1.97	5.00	0.190	3/16	⅛	1.750	1¾	0.320	5/16	3½	¾	—	—
C 4x 7.25	2.13	4.00	0.321	5/16	3/16	1.721	1¾	0.296	5/16	2⅝	11/16	5/16	5/8
x 5.4	1.59	4.00	0.184	3/16	1/16	1.584	1⅝	0.296	5/16	2⅝	11/16	—	—
C 3x 6	1.76	3.00	0.356	3/8	3/16	1.596	1⅝	0.273	¼	1⅝	11/16	—	—
x 5	1.47	3.00	0.258	¼	⅛	1.498	1½	0.273	¼	1⅝	11/16	—	—
x 4.1	1.21	3.00	0.170	3/16	1/16	1.410	1⅜	0.273	¼	1⅝	11/16	—	—

AMERICAN INSTITUTE OF STEEL CONSTRUCTION

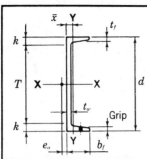

CHANNELS
MISCELLANEOUS
Dimensions

Designation	Area A	Depth d	Web Thickness t_w		$\frac{t_w}{2}$	Flange Width b_f		Flange Average thickness t_f		Distance T	Distance k	Grip	Max. Flge. Fastener
	In.²	In.	In.		In.	In.		In.		In.	In.	In.	In.
MC 18x58	17.1	18.00	0.700	¹¹/₁₆	³/₈	4.200	4¹/₄	0.625	⁵/₈	15¹/₄	1³/₈	⁵/₈	1
x51.9	15.3	18.00	0.600	⁵/₈	⁵/₁₆	4.100	4¹/₈	0.625	⁵/₈	15¹/₄	1³/₈	⁵/₈	1
x45.8	13.5	18.00	0.500	¹/₂	¹/₄	4.000	4	0.625	⁵/₈	15¹/₄	1³/₈	⁵/₈	1
x42.7	12.6	18.00	0.450	⁷/₁₆	¹/₄	3.950	4	0.625	⁵/₈	15¹/₄	1³/₈	⁵/₈	1
MC 13x50	14.7	13.00	0.787	¹³/₁₆	³/₈	4.412	4³/₈	0.610	⁵/₈	10¹/₄	1³/₈	⁵/₈	1
x40	11.8	13.00	0.560	⁹/₁₆	¹/₄	4.185	4¹/₈	0.610	⁵/₈	10¹/₄	1³/₈	⁹/₁₆	1
x35	10.3	13.00	0.447	⁷/₁₆	¹/₄	4.072	4¹/₈	0.610	⁵/₈	10¹/₄	1³/₈	⁹/₁₆	1
x31.8	9.35	13.00	0.375	³/₈	³/₁₆	4.000	4	0.610	⁵/₈	10¹/₄	1³/₈	⁹/₁₆	1
MC 12x50	14.7	12.00	0.835	¹³/₁₆	⁷/₁₆	4.135	4¹/₈	0.700	¹¹/₁₆	9³/₈	1⁵/₁₆	¹¹/₁₆	1
x45	13.2	12.00	0.712	¹¹/₁₆	³/₈	4.012	4	0.700	¹¹/₁₆	9³/₈	1⁵/₁₆	¹¹/₁₆	1
x40	11.8	12.00	0.590	⁹/₁₆	⁵/₁₆	3.890	3⁷/₈	0.700	¹¹/₁₆	9³/₈	1⁵/₁₆	¹¹/₁₆	1
x35	10.3	12.00	0.467	⁷/₁₆	¹/₄	3.767	3³/₄	0.700	¹¹/₁₆	9³/₈	1⁵/₁₆	¹¹/₁₆	1
MC 12x37	10.9	12.00	0.600	⁵/₈	⁵/₁₆	3.600	3⁵/₈	0.600	⁵/₈	9³/₈	1⁵/₁₆	⁵/₈	⁷/₈
x32.9	9.67	12.00	0.500	¹/₂	¹/₄	3.500	3¹/₂	0.600	⁵/₈	9³/₈	1⁵/₁₆	⁹/₁₆	⁷/₈
x30.9	9.07	12.00	0.450	⁷/₁₆	¹/₄	3.450	3¹/₂	0.600	⁵/₈	9³/₈	1⁵/₁₆	⁹/₁₆	⁷/₈
MC 12x10.6	3.10	12.00	0.190	³/₁₆	¹/₈	1.500	1¹/₂	0.309	⁵/₁₆	10⁵/₈	¹¹/₁₆	—	—
MC 10x41.1	12.1	10.00	0.796	¹³/₁₆	³/₈	4.321	4³/₈	0.575	⁹/₁₆	7¹/₂	1¹/₄	⁹/₁₆	⁷/₈
x33.6	9.87	10.00	0.575	⁹/₁₆	⁵/₁₆	4.100	4¹/₈	0.575	⁹/₁₆	7¹/₂	1¹/₄	⁹/₁₆	⁷/₈
x28.5	8.37	10.00	0.425	⁷/₁₆	³/₁₆	3.950	4	0.575	⁹/₁₆	7¹/₂	1¹/₄	⁹/₁₆	⁷/₈
MC 10x28.3	8.32	10.00	0.477	¹/₂	¹/₄	3.502	3¹/₂	0.575	⁹/₁₆	7¹/₂	1¹/₄	⁹/₁₆	⁷/₈
x25.3	7.43	10.00	0.425	⁷/₁₆	³/₁₆	3.550	3¹/₂	0.500	¹/₂	7³/₄	1¹/₈	¹/₂	⁷/₈
x24.9	7.32	10.00	0.377	³/₈	³/₁₆	3.402	3³/₈	0.575	⁹/₁₆	7¹/₂	1¹/₄	⁹/₁₆	⁷/₈
x21.9	6.43	10.00	0.325	⁵/₁₆	³/₁₆	3.450	3¹/₂	0.500	¹/₂	7³/₄	1¹/₈	¹/₂	⁷/₈
MC 10x 8.4	2.46	10.00	0.170	³/₁₆	¹/₁₆	1.500	1¹/₂	0.280	¹/₄	8⁵/₈	¹¹/₁₆	—	—
MC 10x 6.5	1.91	10.00	0.152	¹/₈	¹/₁₆	1.127	1¹/₈	0.202	³/₁₆	9¹/₈	⁷/₁₆	—	—

AMERICAN INSTITUTE OF STEEL CONSTRUCTION

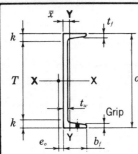

CHANNELS
MISCELLANEOUS
Dimensions

Designation	Area A	Depth d	Web Thickness t_w		$\dfrac{t_w}{2}$	Flange Width b_f		Flange Average thickness t_f		Distance T	Distance k	Grip	Max. Flge. Fastener
	In.²	In.	In.		In.	In.		In.		In.	In.	In.	In.
MC 9x25.4	7.47	9.00	0.450	7/16	1/4	3.500	3 1/2	0.550	9/16	6 5/8	1 3/16	9/16	7/8
x23.9	7.02	9.00	0.400	3/8	3/16	3.450	3 1/2	0.550	9/16	6 5/8	1 3/16	9/16	7/8
MC 8x22.8	6.70	8.00	0.427	7/16	3/16	3.502	3 1/2	0.525	1/2	5 5/8	1 3/16	1/2	7/8
x21.4	6.28	8.00	0.375	3/8	3/16	3.450	3 1/2	0.525	1/2	5 5/8	1 3/16	1/2	7/8
MC 8x20	5.88	8.00	0.400	3/8	3/16	3.025	3	0.500	1/2	5 3/4	1 1/8	1/2	7/8
x18.7	5.50	8.00	0.353	3/8	3/16	2.978	3	0.500	1/2	5 3/4	1 1/8	1/2	7/8
MC 8x 8.5	2.50	8.00	0.179	3/16	1/16	1.874	1 7/8	0.311	5/16	6 1/2	3/4	5/16	5/8
MC 7x22.7	6.67	7.00	0.503	1/2	1/4	3.603	3 5/8	0.500	1/2	4 3/4	1 1/8	1/2	7/8
x19.1	5.61	7.00	0.352	3/8	3/16	3.452	3 1/2	0.500	1/2	4 3/4	1 1/8	1/2	7/8
MC 7x17.6	5.17	7.00	0.375	3/8	3/16	3.000	3	0.475	1/2	4 7/8	1 1/16	1/2	3/4
MC 6x18	5.29	6.00	0.379	3/8	3/16	3.504	3 1/2	0.475	1/2	3 7/8	1 1/16	1/2	7/8
x15.3	4.50	6.00	0.340	5/16	3/16	3.500	3 1/2	0.385	3/8	4 1/4	7/8	3/8	7/8
MC 6x16.3	4.79	6.00	0.375	3/8	3/16	3.000	3	0.475	1/2	3 7/8	1 1/16	1/2	3/4
x15.1	4.44	6.00	0.316	5/16	3/16	2.941	3	0.475	1/2	3 7/8	1 1/16	1/2	3/4
MC 6x12	3.53	6.00	0.310	5/16	1/8	2.497	2 1/2	0.375	3/8	4 3/8	13/16	3/8	5/8

AMERICAN INSTITUTE OF STEEL CONSTRUCTION

WELDED JOINTS
Standard Symbols

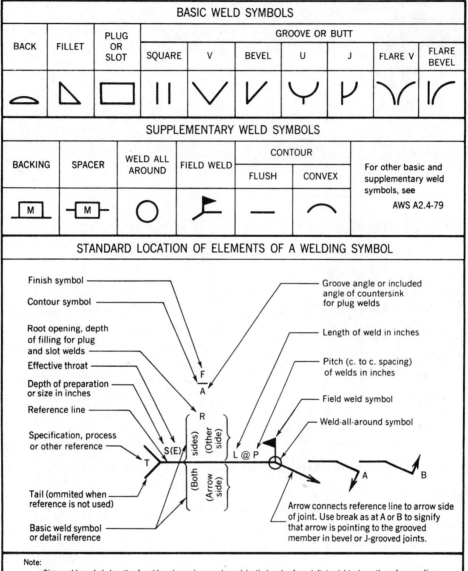

BASIC WELD SYMBOLS									
BACK	FILLET	PLUG OR SLOT	GROOVE OR BUTT						
			SQUARE	V	BEVEL	U	J	FLARE V	FLARE BEVEL

SUPPLEMENTARY WELD SYMBOLS						
BACKING	SPACER	WELD ALL AROUND	FIELD WELD	CONTOUR		For other basic and supplementary weld symbols, see AWS A2.4-79
				FLUSH	CONVEX	

STANDARD LOCATION OF ELEMENTS OF A WELDING SYMBOL

Finish symbol

Contour symbol

Root opening, depth of filling for plug and slot welds

Effective throat

Depth of preparation or size in inches

Reference line

Specification, process or other reference

Tail (ommited when reference is not used)

Basic weld symbol or detail reference

Groove angle or included angle of countersink for plug welds

Length of weld in inches

Pitch (c. to c. spacing) of welds in inches

Field weld symbol

Weld-all-around symbol

Arrow connects reference line to arrow side of joint. Use break as at A or B to signify that arrow is pointing to the grooved member in bevel or J-grooved joints.

Note:

Size, weld symbol, length of weld and spacing must read in that order from left to right along the reference line. Neither orientation of reference line nor location of the arrow alter this rule.

The perpendicular leg of △, V, V, ⊮ weld symbols must be at left.

Arrow and Other Side welds are of the same size unless otherwise shown. Dimensions of fillet welds must be shown on both the Arrow Side and the Other Side Symbol.

The point of the field weld symbol must point toward the tail.

Symbols apply between abrupt changes in direction of welding unless governed by the "all around" symbol or otherwise dimensioned.

These symbols do not explicitly provide for the case that frequently occurs in structural work, where duplicate material (such as stiffeners) occurs on the far side of a web or gusset plate. The fabricating industry has adopted this convention: that when the billing of the detail material discloses the existence of a member on the far side as well as on the near side, the welding shown for the near side shall be duplicated on the far side.

APPENDIX B

USE OF THE SPAN TABLES

Spans for floor and ceiling joists are calculated on the basis of the modulus of elasticity (E) with the required fiber bending stress (F_b) listed below each span. Spans for rafters are calculated on the basis of fiber bending stress (F_b) with the required modulus of elasticity (E) listed below each span. Use of the tables is illustrated in the following examples:

Example 1—Floor Joists. Assume a required span of 12'–9", a live load of 40 psf and joists spaced 16 inches on centers. Table J–1 shows that a grade of 2 × 8 having an E value of 1,600,000 psi and an F_b value of 1,250 psi would have a span of 12'–10", which satisfies the condition.

Example 2—Rafters. Assume a horizontal projection span of 13'–0", a live load of 20 psf, dead load of 7 psf, no attached ceiling, and rafters spaced 16 inches on centers. Table R–13 shows that a 2 × 6 having an F_b value of 1,200 psi and an E value of 940,000 psi would have a span of 13'–0" of horizontal projection. Conversion of horizontal to sloping distance is shown in the chart, Conversion Diagram for Rafters.

Since many combinations of size, spacing, and E and F_b values are possible, it is recommended that the user examine the tables to determine which combination fits her particular case most effectively.

The spans for nominal 2 × 5 joists or rafters are 82 percent of the spans tabulated for the same spacing of nominal 2 × 6 joists or rafters. For each joist or rafter spacing, the required values of F_b or E for 2 × 5s are the same as the tabulated values of 2 × 6s.

ACKNOWLEDGMENT

The material contained in this appendix is taken from: *Span Tables for Joists and Rafters* (American Softwood Lumber, Standard Sizes PS 20–70), National Forest Products Association, 1619 Massachusetts Avenue, N.W., Washington, DC 20036.

TABLE J-1
FLOOR JOISTS
40 Lbs. Per Sq. Ft. Live Load
(All rooms except those used for sleeping areas and attic floors.)

DESIGN CRITERIA:
Deflection - For 40 lbs. per sq. ft. live load.
 Limited to span in inches divided by 360.
Strength - Live Load of 40 lbs. per sq. ft. plus
dead load of 10 lbs. per sq. ft. determines the
required fiber stress value.

Modulus of Elasticity, "E", in 1,000,000 psi

Each cell shows span (feet-inches) over the required extreme fiber stress in bending, "Fb", in pounds per square inch.

JOIST SIZE (IN)	SPACING (IN)	0.4	0.5	0.6	0.7	0.8	0.9	1.0	1.1	1.2	1.3	1.4	1.5	1.6	1.7	1.8	1.9	2.0	2.2	2.4
2x6	12.0	6-9 / 450	7-3 / 520	7-9 / 590	8-2 / 660	8-6 / 720	8-10 / 780	9-2 / 830	9-6 / 890	9-9 / 940	10-0 / 990	10-3 / 1040	10-6 / 1090	10-9 / 1140	10-11 / 1190	11-2 / 1230	11-4 / 1280	11-7 / 1320	11-11 / 1410	12-3 / 1490
	13.7	6-6 / 470	7-0 / 550	7-5 / 620	7-9 / 690	8-2 / 750	8-6 / 810	8-9 / 870	9-1 / 930	9-4 / 980	9-7 / 1040	9-10 / 1090	10-0 / 1140	10-3 / 1190	10-6 / 1240	10-8 / 1290	10-10 / 1340	11-1 / 1380	11-5 / 1470	11-9 / 1560
	16.0	6-2 / 500	6-7 / 580	7-0 / 650	7-5 / 720	7-9 / 790	8-0 / 860	8-4 / 920	8-7 / 980	8-10 / 1040	9-1 / 1090	9-4 / 1150	9-6 / 1200	9-9 / 1250	9-11 / 1310	10-2 / 1360	10-4 / 1410	10-6 / 1460	10-10 / 1550	11-2 / 1640
	19.2	5-9 / 530	6-3 / 610	6-7 / 690	7-0 / 770	7-3 / 840	7-7 / 910	7-10 / 970	8-1 / 1040	8-4 / 1100	8-7 / 1160	8-9 / 1220	9-0 / 1280	9-2 / 1330	9-4 / 1390	9-6 / 1440	9-8 / 1500	9-10 / 1550	10-2 / 1650	10-6 / 1750
	24.0	5-4 / 570	5-9 / 660	6-2 / 750	6-6 / 830	6-9 / 900	7-0 / 980	7-3 / 1050	7-6 / 1120	7-9 / 1190	7-11 / 1250	8-2 / 1310	8-4 / 1380	8-6 / 1440	8-8 / 1500	8-10 / 1550	9-0 / 1610	9-2 / 1670	9-6 / 1780	9-9 / 1880
	32.0					6-2 / 1010	6-5 / 1090	6-7 / 1150	6-10 / 1230	7-0 / 1300	7-3 / 1390	7-5 / 1450	7-7 / 1520	7-9 / 1590	7-11 / 1660	8-0 / 1690	8-2 / 1760	8-4 / 1840	8-7 / 1950	8-10 / 2060
2x8	12.0	8-11 / 450	9-7 / 520	10-2 / 590	10-9 / 660	11-3 / 720	11-8 / 780	12-1 / 830	12-6 / 890	12-10 / 940	13-2 / 990	13-6 / 1040	13-10 / 1090	14-2 / 1140	14-5 / 1190	14-8 / 1230	15-0 / 1280	15-3 / 1320	15-9 / 1410	16-2 / 1490
	13.7	8-6 / 470	9-2 / 550	9-9 / 620	10-3 / 690	10-9 / 750	11-2 / 810	11-7 / 870	11-11 / 930	12-3 / 980	12-7 / 1040	12-11 / 1090	13-3 / 1140	13-6 / 1190	13-10 / 1240	14-1 / 1290	14-4 / 1340	14-7 / 1380	15-0 / 1470	15-6 / 1560
	16.0	8-1 / 500	8-9 / 580	9-3 / 650	9-9 / 720	10-2 / 790	10-7 / 850	11-0 / 920	11-4 / 980	11-8 / 1040	12-0 / 1090	12-3 / 1150	12-7 / 1200	12-10 / 1250	13-1 / 1310	13-4 / 1360	13-7 / 1410	13-10 / 1460	14-3 / 1550	14-8 / 1640
	19.2	7-7 / 530	8-2 / 610	8-9 / 690	9-2 / 770	9-7 / 840	10-0 / 910	10-4 / 970	10-8 / 1040	11-0 / 1100	11-3 / 1160	11-7 / 1220	11-10 / 1280	12-1 / 1330	12-4 / 1390	12-7* / 1440	12-10 / 1500	13-0 / 1550	13-5 / 1650	13-10 / 1750
	24.0	7-1 / 570	7-7 / 660	8-1 / 750	8-6 / 830	8-11 / 900	9-3 / 980	9-7 / 1050	9-11 / 1120	10-2 / 1190	10-6 / 1250	10-9 / 1310	11-0 / 1380	11-3 / 1440	11-5 / 1500	11-8 / 1550	11-11 / 1610	12-1 / 1670	12-6 / 1780	12-10 / 1880
	32.0					8-1 / 990	8-5 / 1080	8-9 / 1170	9-0 / 1230	9-3 / 1300	9-6 / 1370	9-9 / 1450	10-0 / 1520	10-2 / 1570	10-5 / 1650	10-7 / 1700	10-10 / 1790	11-0 / 1840	11-4 / 1950	11-8 / 2070
2x10	12.0	11-4 / 450	12-3 / 520	13-0 / 590	13-8 / 660	14-4 / 720	14-11 / 780	15-5 / 830	15-11 / 890	16-5 / 940	16-10 / 990	17-3 / 1040	17-8 / 1090	18-0 / 1140	18-5 / 1190	18-9 / 1230	19-1 / 1280	19-5 / 1320	20-1 / 1410	20-8 / 1490
	13.7	10-10 / 470	11-8 / 550	12-5 / 620	13-1 / 690	13-8 / 750	14-3 / 810	14-9 / 870	15-3 / 930	15-8 / 980	16-1 / 1040	16-6 / 1090	16-11 / 1140	17-3 / 1190	17-7 / 1240	17-11 / 1290	18-3 / 1340	18-7 / 1380	19-2 / 1470	19-9 / 1560
	16.0	10-4 / 500	11-1 / 580	11-10 / 650	12-5 / 720	13-0 / 790	13-6 / 850	14-0 / 920	14-6 / 980	14-11 / 1040	15-3 / 1090	15-8 / 1150	16-0 / 1200	16-5 / 1250	16-9 / 1310	17-0 / 1360	17-4 / 1410	17-8 / 1460	18-3 / 1550	18-9 / 1640
	19.2	9-9 / 530	10-6 / 610	11-1 / 690	11-8 / 770	12-3 / 840	12-9 / 910	13-2 / 970	13-7 / 1040	14-0 / 1100	14-5 / 1160	14-9 / 1220	15-1 / 1280	15-5 / 1330	15-9 / 1390	16-0 / 1440	16-4 / 1500	16-7 / 1550	17-2 / 1650	17-8 / 1750
	24.0	9-0 / 570	9-9 / 660	10-4 / 750	10-10 / 830	11-4 / 900	11-10 / 980	12-3 / 1050	12-8 / 1120	13-0 / 1190	13-4 / 1250	13-8 / 1310	14-0 / 1380	14-4 / 1440	14-7 / 1500	14-11 / 1550	15-2 / 1610	15-5 / 1670	15-11 / 1780	16-5 / 1880
	32.0					10-4 / 1000	10-9 / 1080	11-1 / 1150	11-6 / 1240	11-10 / 1310	12-2 / 1380	12-5 / 1440	12-9 / 1520	13-0 / 1580	13-3 / 1640	13-6 / 1700	13-9 / 1770	14-0 / 1830	14-6 / 1970	14-11 / 2080
2x12	12.0	13-10 / 450	14-11 / 520	15-10 / 590	16-8 / 660	17-5 / 720	18-1 / 780	18-9 / 830	19-4 / 890	19-11 / 940	20-6 / 990	21-0 / 1040	21-6 / 1090	21-11 / 1140	22-5 / 1190	22-10 / 1230	23-3 / 1280	23-7 / 1320	24-5 / 1410	25-1 / 1490
	13.7	13-3 / 470	14-3 / 550	15-2 / 620	15-11 / 690	16-8 / 750	17-4 / 810	17-11 / 870	18-6 / 930	19-1 / 980	19-7 / 1040	20-1 / 1090	20-6 / 1140	21-0 / 1190	21-5 / 1240	21-10 / 1290	22-3 / 1340	22-7 / 1380	23-4 / 1470	24-0 / 1560
	16.0	12-7 / 500	13-6 / 580	14-4 / 650	15-2 / 720	15-10 / 790	16-5 / 860	17-0 / 920	17-7 / 980	18-1 / 1040	18-7 / 1090	19-1 / 1150	19-6 / 1200	19-11 / 1250	20-4 / 1310	20-9 / 1360	21-1 / 1410	21-6 / 1460	22-2 / 1550	22-10 / 1640
	19.2	11-10 / 530	12-9 / 610	13-6 / 690	14-3 / 770	14-11 / 840	15-6 / 910	16-0 / 970	16-7 / 1040	17-0 / 1100	17-6 / 1160	17-11 / 1220	18-4 / 1280	18-9 / 1330	19-2 / 1390	19-6 / 1440	19-10 / 1500	20-2 / 1550	20-10 / 1650	21-6 / 1750
	24.0	11-0 / 570	11-10 / 660	12-7 / 750	13-3 / 830	13-10 / 900	14-4 / 980	14-11 / 1050	15-4 / 1120	15-10 / 1190	16-3 / 1250	16-8 / 1310	17-0 / 1380	17-5 / 1440	17-9 / 1500	18-1 / 1550	18-5 / 1610	18-9 / 1670	19-4 / 1780	19-11 / 1880
	32.0					12-7 / 1000	13-1 / 1080	13-6 / 1150	13-11 / 1220	14-4 / 1300	14-9 / 1380	15-2 / 1450	15-6 / 1520	15-10 / 1580	16-2 / 1650	16-5 / 1700	16-9 / 1770	17-0 / 1830	17-7 / 1950	18-1 / 2070

Note: The required extreme fiber stress in bending, "F$_b$", in pounds per square inch is shown below each span.

TABLE R-13
MEDIUM OR HIGH SLOPE RAFTERS
No Ceiling Load
Slope over 3 in 12
Live Load - 20 lb. per. sq. ft.
(Light roof covering)

DESIGN CRITERIA:

Strength - 7 lbs. per sq. ft. dead load plus 20 lbs. per sq. ft. live load determines required fiber stress.

Deflection - For 20 lbs. per sq. ft. live load. Limited to span in inches divided by 180.

RAFTER SIZE (IN)	SPACING (IN)	Extreme Fiber Stress in Bending, "Fb" (psi).											
		200	300	400	500	600	700	800	900	1000	1100	1200	1300
2x4	12.0	3-11 0.07	4-9 0.14	5-6 0.21	6-2 0.29	6-9 0.38	7-3 0.49	7-9 0.59	8-3 0.71	8-8 0.83	9-1 0.96	9-6 1.09	9-11 1.23
	13.7	3-8 0.07	4-5 0.13	5-2 0.20	5-9 0.27	6-4 0.36	6-10 0.45	7-3 0.55	7-9 0.66	8-2 0.77	8-6 0.89	8-11 1.02	9-3 1.15
	16.0	3-4 0.06	4-1 0.12	4-9 0.18	5-4 0.25	5-10 0.33	6-4 0.42	6-9 0.51	7-2 0.61	7-6 0.72	7-11 0.83	8-3 0.94	8-7 1.06
	19.2	3-1 0.06	3-9 0.11	4-4 0.17	4-10 0.23	5-4 0.30	5-9 0.38	6-2 0.47	6-6 0.56	6-10 0.65	7-3 0.76	7-6 0.86	7-10 0.97
	24.0	2-9 0.05	3-4 0.10	3-11 0.15	4-4 0.21	4-9 0.27	5-2 0.34	5-6 0.42	5-10 0.50	6-2 0.59	6-5 0.68	6-9 0.77	7-0 0.87
2x6	12.0	6-1 0.07	7-6 0.14	8-8 0.21	9-8 0.29	10-7 0.38	11-5 0.49	12-3 0.59	13-0 0.71	13-8 0.83	14-4 0.96	15-0 1.09	15-7 1.23
	13.7	5-9 0.07	7-0 0.13	8-1 0.20	9-0 0.27	9-11 0.36	10-8 0.45	11-5 0.55	12-2 0.66	12-9 0.77	13-5 0.89	14-0 1.02	14-7 1.15
	16.0	5-4 0.06	6-6 0.12	7-6 0.18	8-4 0.25	9-2 0.33	9-11 0.42	10-7 0.51	11-3 0.61	11-10 0.72	12-5 0.83	13-0 0.94	13-6 1.06
	19.2	4-10 0.06	5-11 0.11	6-10 0.17	7-8 0.23	8-4 0.30	9-0 0.38	9-8 0.47	10-3 0.56	10-10 0.65	11-4 0.76	11-10 0.86	12-4 0.97
	24.0	4-4 0.05	5-4 0.10	6-1 0.15	6-10 0.21	7-6 0.27	8-1 0.34	8-8 0.42	9-2 0.50	9-8 0.59	10-2 0.68	10-7 0.77	11-0 0.87
2x8	12.0	8-1 0.07	9-10 0.14	11-5 0.21	12-9 0.29	13-11 0.38	15-1 0.49	16-1 0.59	17-1 0.71	18-0 0.83	18-11 0.96	19-9 1.09	20-6 1.23
	13.7	7-6 0.07	9-3 0.13	10-8 0.20	11-11 0.27	13-1 0.36	14-1 0.45	15-1 0.55	16-0 0.66	16-10 0.77	17-8 0.89	18-5 1.02	19-3 1.15
	16.0	7-0 0.06	8-7 0.12	9-10 0.18	11-0 0.25	12-1 0.33	13-1 0.42	13-11 0.51	14-10 0.61	15-7 0.72	16-4 0.83	17-1 0.94	17-9 1.06
	19.2	6-4 0.06	7-10 0.11	9-0 0.17	10-1 0.23	11-0 0.30	11-11 0.38	12-9 0.47	13-6 0.56	14-3 0.65	14-11 0.76	15-7 0.86	16-3 0.97
	24.0	5-8 0.05	7-0 0.10	8-1 0.15	9-0 0.21	9-10 0.27	10-8 0.34	11-5 0.42	12-1 0.50	12-9 0.59	13-4 0.68	13-11 0.77	14-6 0.87
2x10	12.0	10-3 0.07	12-7 0.14	14-6 0.21	16-3 0.29	17-10 0.38	19-3 0.49	20-7 0.59	21-10 0.71	23-0 0.83	24-1 0.96	25-2 1.09	26-2 1.23
	13.7	9-7 0.07	11-9 0.13	13-7 0.20	15-2 0.27	16-8 0.36	18-0 0.45	19-3 0.55	20-5 0.66	21-6 0.77	22-7 0.89	23-7 1.02	24-6 1.15
	16.0	8-11 0.06	10-11 0.12	12-7 0.18	14-1 0.25	15-5 0.33	16-8 0.42	17-10 0.51	18-11 0.61	19-11 0.72	20-10 0.83	21-10 0.94	22-8 1.06
	19.2	8-2 0.06	9-11 0.11	11-6 0.17	12-10 0.23	14-1 0.30	15-2 0.38	16-3 0.47	17-3 0.56	18-2 0.65	19-1 0.76	19-11 0.86	20-9 0.97
	24.0	7-3 0.05	8-11 0.10	10-3 0.15	11-6 0.21	12-7 0.27	13-7 0.34	14-6 0.42	15-5 0.50	16-3 0.59	17-1 0.68	17-10 0.77	18-6 0.87

Note: The required modulus of elasticity, "E", in 1,000,000 pounds per square inch is shown below each span.

TABLE R-13 (cont.)

RAFTERS: Spans are measured along the horizontal projection and loads are considered as applied on the horizontal projection.

1400	1500	1600	1700	1800	1900	2000	2100	2200	2400	2700	RAFTER SPACING (IN)	SIZE (IN)
10-3 / 1.37	10-8 / 1.52	11-0 / 1.68	11-4 / 1.84	11-8 / 2.00	12-0 / 2.17	12-4 / 2.34	12-7 / 2.52				12.0	
9-7 / 1.28	10-0 / 1.42	10-3 / 1.57	10-7 / 1.72	10-11 / 1.87	11-3 / 2.03	11-6 / 2.19	11-9 / 2.36	12-1 / 2.53			13.7	
8-11 / 1.19	9-3 / 1.32	9-6 / 1.45	9-10 / 1.59	10-1 / 1.73	10-5 / 1.88	10-8 / 2.03	10-11 / 2.18	11-2 / 2.34			16.0	2x4.
8-2 / 1.08	8-5 / 1.20	8-8 / 1.33	9-0 / 1.45	9-3 / 1.58	9-6 / 1.71	9-9 / 1.85	10-0 / 1.99	10-2 / 2.14	10-8 / 2.43		19.2	
7-3 / 0.97	7-6 / 1.08	7-9 / 1.19	8-0 / 1.30	8-3 / 1.41	8-6 / 1.53	8-8 / 1.66	8-11 / 1.78	9-1 / 1.91	9-6 / 2.18	10-1 / 2.60	24.0	
16-2 / 1.37	16-9 / 1.52	17-3 / 1.68	17-10 / 1.84	18-4 / 2.00	18-10 / 2.17	19-4 / 2.34	19-10 / 2.52				12.0	
15-1 / 1.28	15-8 / 1.42	16-2 / 1.57	16-8 / 1.72	17-2 / 1.87	17-7 / 2.03	18-1 / 2.19	18-6 / 2.36	19-0 / 2.53			13.7	
14-0 / 1.19	14-6 / 1.32	15-0 / 1.45	15-5 / 1.59	15-11 / 1.73	16-4 / 1.88	16-9 / 2.03	17-2 / 2.18	17-7 / 2.34			16.0	2x6
12-9 / 1.08	13-3 / 1.20	13-8 / 1.33	14-1 / 1.45	14-6 / 1.58	14-11 / 1.71	15-3 / 1.85	15-8 / 1.99	16-0 / 2.14	16-9 / 2.43		19.2	
11-5 / 0.97	11-10 / 1.08	12-3 / 1.19	12-7 / 1.30	13-0 / 1.41	13-4 / 1.53	13-8 / 1.66	14-0 / 1.78	14-4 / 1.91	15-0 / 2.18	15-11 / 2.60	24.0	
21-4 / 1.37	22-1 / 1.52	22-9 / 1.68	23-6 / 1.84	24-2 / 2.00	24-10 / 2.17	25-6 / 2.34	26-1 / 2.52				12.0	
19-11 / 1.28	20-8 / 1.42	21-4 / 1.57	22-0 / 1.72	22-7 / 1.87	23-3 / 2.03	23-10 / 2.19	24-5 / 2.36	25-0 / 2.53			13.7	
18-5 / 1.19	19-1 / 1.32	19-9 / 1.45	20-4 / 1.59	20-11 / 1.73	21-6 / 1.88	22-1 / 2.03	22-7 / 2.18	23-2 / 2.34			16.0	2x8
16-10 / 1.08	17-5 / 1.20	18-0 / 1.33	18-7 / 1.45	19-1 / 1.58	19-8 / 1.71	20-2 / 1.85	20-8 / 1.99	21-1 / 2.14	22-1 / 2.43		19.2	
15-1 / 0.97	15-7 / 1.08	16-1 / 1.19	16-7 / 1.30	17-1 / 1.41	17-7 / 1.53	18-0 / 1.66	18-5 / 1.78	18-11 / 1.91	19-9 / 2.18	20-11 / 2.60	24.0	
27-2 / 1.37	28-2 / 1.52	29-1 / 1.68	30-0 / 1.84	30-10 / 2.00	31-8 / 2.17	32-6 / 2.34	33-4 / 2.52				12.0	
25-5 / 1.28	26-4 / 1.42	27-2 / 1.57	28-0 / 1.72	28-10 / 1.87	29-8 / 2.03	30-5 / 2.19	31-2 / 2.36	31-11 / 2.53			13.7	
23-7 / 1.19	24-5 / 1.32	25-2 / 1.45	25-11 / 1.59	26-8 / 1.73	27-5 / 1.88	28-2 / 2.03	28-10 / 2.18	29-6 / 2.34			16.0	2x10
21-6 / 1.08	22-3 / 1.20	23-0 / 1.33	23-8 / 1.45	24-5 / 1.58	25-1 / 1.71	25-8 / 1.85	26-4 / 1.99	26-11 / 2.14	28-2 / 2.43		19.2	
19-3 / 0.97	19-11 / 1.08	20-7 / 1.19	21-2 / 1.30	21-10 / 1.41	22-5 / 1.53	23-0 / 1.66	23-7 / 1.78	24-1 / 1.91	25-2 / 2.18	26-8 / 2.60	24.0	

Column header group: Extreme Fiber Stress in Bending, "F_b" (psi).

Note: The required modulus of elasticity, "E", in 1,000,000 pounds per square inch is shown below each span.

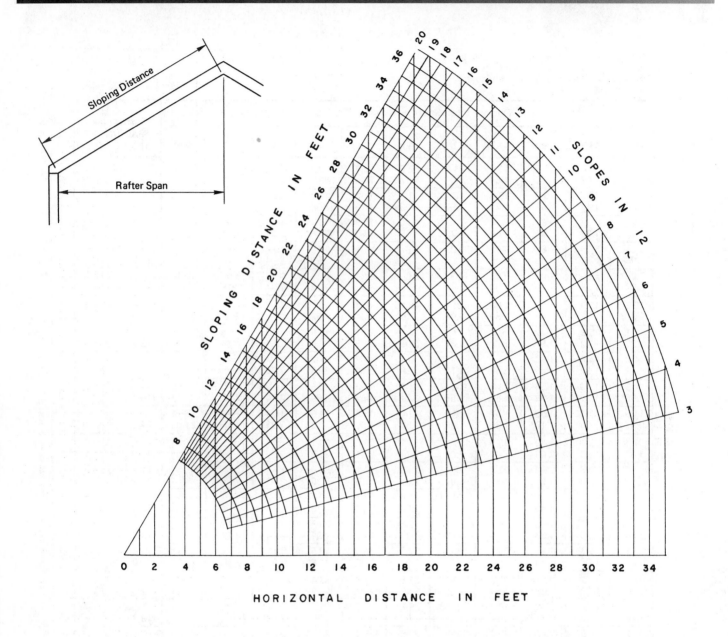

TABLE J-3
CEILING JOISTS
20 Lbs. Per Sq. Ft. Live Load
(Limited attic storage where development of future rooms is not possible)
(Plaster Ceiling)

DESIGN CRITERIA:
Deflection - For 20 lbs. per sq. ft. live load.
  Limited to span in inches divided by 360.
Strength - Live load of 20 lbs. per sq. ft. plus
  dead load of 10 lbs. per sq. ft. determines
  required fiber stress value.

Modulus of Elasticity, "E", in 1,000,000 psi

Each cell shows span (feet-inches) and, below it, the required extreme fiber stress in bending, F_b, in pounds per square inch.

Joist Size (IN)	Spacing (IN)	0.4	0.5	0.6	0.7	0.8	0.9	1.0	1.1	1.2	1.3	1.4	1.5	1.6	1.7	1.8	1.9	2.0	2.2	2.4
2x4	12.0	5-5 / 430	5-10 / 500	6-2 / 560	6-6 / 630	6-10 / 680	7-1 / 740	7-4 / 790	7-7 / 850	7-10 / 900	8-0 / 950	8-3 / 990	8-5 / 1040	8-7 / 1090	8-9 / 1130	8-11 / 1170	9-1 / 1220	9-3 / 1260	9-7 / 1340	9-10 / 1420
	13.7	5-2 / 450	5-7 / 520	5-11 / 590	6-3 / 650	6-6 / 720	6-9 / 770	7-0 / 830	7-3 / 880	7-6 / 940	7-8 / 990	7-10 / 1040	8-1 / 1090	8-3 / 1140	8-5 / 1180	8-7 / 1230	8-8 / 1270	8-10 / 1320	9-2 / 1400	9-5 / 1490
	16.0	4-11 / 470	5-4 / 550	5-8 / 620	5-11 / 690	6-2 / 750	6-5 / 810	6-8 / 870	6-11 / 930	7-1 / 990	7-3 / 1040	7-6 / 1090	7-8 / 1140	7-10 / 1200	8-0 / 1240	8-1 / 1290	8-3 / 1340	8-5 / 1390	8-8 / 1480	8-11 / 1570
	19.2	4-8 / 500	5-0 / 580	5-4 / 660	5-7 / 730	5-10 / 800	6-1 / 870	6-3 / 930	6-6 / 990	6-8 / 1050	6-10 / 1110	7-0 / 1160	7-2 / 1220	7-4 / 1270	7-6 / 1320	7-8 / 1370	7-9 / 1420	7-11 / 1470	8-2 / 1570	8-5 / 1660
	24.0	4-4 / 540	4-8 / 630	4-11 / 710	5-2 / 790	5-5 / 860	5-8 / 930	5-10 / 1000	6-0 / 1070	6-2 / 1130	6-4 / 1190	6-6 / 1250	6-8 / 1310	6-10 / 1370	7-0 / 1420	7-1 / 1480	7-3 / 1530	7-4 / 1590	7-7 / 1690	7-10 / 1790
2x6	12.0	8-6 / 430	9-2 / 500	9-9 / 560	10-3 / 630	10-9 / 680	11-2 / 740	11-7 / 790	11-11 / 850	12-3 / 900	12-7 / 950	12-11 / 990	13-3 / 1040	13-6 / 1090	13-9 / 1130	14-1 / 1170	14-4 / 1220	14-7 / 1260	15-0 / 1340	15-6 / 1420
	13.7	8-2 / 450	8-9 / 520	9-4 / 590	9-10 / 650	10-3 / 720	10-8 / 770	11-1 / 830	11-5 / 880	11-9 / 940	12-1 / 990	12-4 / 1040	12-8 / 1090	12-11 / 1140	13-2 / 1180	13-5 / 1230	13-8 / 1270	13-11 / 1320	14-4 / 1400	14-9 / 1490
	16.0	7-9 / 470	8-4 / 550	8-10 / 620	9-4 / 690	9-9 / 750	10-2 / 810	10-6 / 870	10-10 / 930	11-2 / 990	11-5 / 1040	11-9 / 1090	12-0 / 1140	12-3 / 1200	12-6 / 1240	12-9 / 1290	13-0 / 1340	13-3 / 1390	13-8 / 1480	14-1 / 1570
	19.2	7-3 / 500	7-10 / 580	8-4 / 660	8-9 / 730	9-2 / 800	9-6 / 870	9-10 / 930	10-2 / 990	10-6 / 1050	10-9 / 1110	11-1 / 1160	11-4 / 1220	11-7 / 1270	11-9 / 1320	12-0 / 1370	12-3 / 1420	12-5 / 1470	12-10 / 1570	13-3 / 1660
	24.0	6-9 / 540	7-3 / 630	7-9 / 710	8-2 / 790	8-6 / 860	8-10 / 930	9-2 / 1000	9-6 / 1070	9-9 / 1130	10-0 / 1190	10-3 / 1250	10-6 / 1310	10-9 / 1370	10-11 / 1420	11-2 / 1480	11-4 / 1530	11-7 / 1590	11-11 / 1690	12-3 / 1790
2x8	12.0	11-3 / 430	12-1 / 500	12-10 / 560	13-6 / 630	14-2 / 680	14-8 / 740	15-3 / 790	15-9 / 850	16-2 / 900	16-7 / 950	17-0 / 990	17-5 / 1040	17-10 / 1090	18-2 / 1130	18-6 / 1170	18-10 / 1220	19-2 / 1260	19-10 / 1340	20-5 / 1420
	13.7	10-9 / 450	11-7 / 520	12-3 / 590	12-11 / 650	13-6 / 720	14-1 / 770	14-7 / 830	15-0 / 880	15-6 / 940	15-11 / 990	16-3 / 1040	16-8 / 1090	17-0 / 1140	17-5 / 1180	17-9 / 1230	18-0 / 1270	18-4 / 1320	18-11 / 1400	19-6 / 1490
	16.0	10-2 / 470	11-0 / 550	11-8 / 620	12-3 / 690	12-10 / 750	13-4 / 810	13-10 / 870	14-3 / 930	14-8 / 990	15-1 / 1040	15-6 / 1090	15-10 / 1140	16-2 / 1200	16-6 / 1240	16-10 / 1290	17-2 / 1340	17-5 / 1390	18-0 / 1480	18-6 / 1570
	19.2	9-7 / 500	10-4 / 580	11-0 / 660	11-7 / 730	12-1 / 800	12-7 / 870	13-0 / 930	13-5 / 990	13-10 / 1050	14-2 / 1110	14-7 / 1160	14-11 / 1220	15-3 / 1270	15-6 / 1320	15-10 / 1370	16-1 / 1420	16-5 / 1470	16-11 / 1570	17-5 / 1660
	24.0	8-11 / 540	9-7 / 630	10-2 / 710	10-9 / 790	11-3 / 860	11-8 / 930	12-1 / 1000	12-6 / 1070	12-10 / 1130	13-2 / 1190	13-6 / 1250	13-10 / 1310	14-2 / 1370	14-5 / 1420	14-8 / 1480	15-0 / 1530	15-3 / 1590	15-9 / 1690	16-2 / 1790
2x10	12.0	14-4 / 430	15-5 / 500	16-5 / 560	17-3 / 630	18-0 / 680	18-9 / 740	19-5 / 790	20-1 / 850	20-8 / 900	21-2 / 950	21-9 / 990	22-3 / 1040	22-9 / 1090	23-2 / 1130	23-8 / 1170	24-1 / 1220	24-6 / 1260	25-3 / 1340	26-0 / 1420
	13.7	13-8 / 450	14-9 / 520	15-8 / 590	16-6 / 650	17-3 / 720	17-11 / 770	18-7 / 830	19-2 / 880	19-9 / 940	20-3 / 990	20-9 / 1040	21-3 / 1090	21-9 / 1140	22-2 / 1180	22-7 / 1230	23-0 / 1270	23-5 / 1320	24-2 / 1400	24-10 / 1490
	16.0	13-0 / 470	14-0 / 550	14-11 / 620	15-8 / 690	16-5 / 750	17-0 / 810	17-8 / 870	18-3 / 930	18-9 / 990	19-3 / 1040	19-9 / 1090	20-2 / 1140	20-8 / 1200	21-1 / 1240	21-6 / 1290	21-10 / 1340	22-3 / 1390	22-11 / 1480	23-8 / 1570
	19.2	12-3 / 500	13-2 / 580	14-0 / 660	14-9 / 730	15-5 / 800	16-0 / 870	16-7 / 930	17-2 / 990	17-8 / 1050	18-1 / 1110	18-7 / 1160	19-0 / 1220	19-5 / 1270	19-10 / 1320	20-2 / 1370	20-7 / 1420	20-11 / 1470	21-7 / 1570	22-3 / 1660
	24.0	11-4 / 540	12-3 / 630	13-0 / 710	13-8 / 790	14-4 / 860	14-11 / 930	15-5 / 1000	15-11 / 1070	16-5 / 1130	16-10 / 1190	17-3 / 1250	17-8 / 1310	18-0 / 1370	18-5 / 1420	18-9 / 1480	19-1 / 1530	19-5 / 1590	20-1 / 1690	20-8 / 1790

Note: The required extreme fiber stress in bending, "F_b", in pounds per square inch is shown below each span.

TABLE J-4
CEILING JOISTS
20 Lbs. Per Sq. Ft. Live Load
(Limited attic storage where development of future rooms is not possible)
(Drywall Ceiling)

DESIGN CRITERIA:
Deflection - For 20 lbs. per sq. ft. live load.
 Limited to span in inches divided by 240.
Strength - live load of 20 lbs. per sq. ft. plus dead load of 10 lbs. per sq. ft. determines required fiber stress value.

Each cell shows the allowable span (ft-in) above the required extreme fiber stress in bending, F_b, in pounds per square inch.

Joist Size	Spacing (in)	Modulus of Elasticity, "E", in 1,000,000 psi

Joist Size	Spacing (IN)	0.4	0.5	0.6	0.7	0.8	0.9	1.0	1.1	1.2	1.3	1.4	1.5	1.6	1.7	1.8	1.9	2.0	2.2	2.4
2x4	12.0	6-2 / 560	6-8 / 660	7-1 / 740	7-6 / 820	7-10 / 900	8-1 / 970	8-5 / 1040	8-8 / 1110	8-11 / 1170	9-2 / 1240	9-5 / 1300	9-8 / 1360	9-10 / 1420	10-0 / 1480	10-3 / 1540	10-5 / 1600	10-7 / 1650	10-11 / 1760	11-3 / 1860
	13.7	5-11 / 590	6-5 / 690	6-9 / 770	7-2 / 860	7-6 / 940	7-9 / 1010	8-1 / 1090	8-4 / 1160	8-7 / 1230	8-9 / 1300	9-0 / 1360	9-3 / 1420	9-5 / 1490	9-7 / 1550	9-9 / 1610	10-0 / 1670	10-2 / 1730	10-6 / 1840	10-9 / 1950
	16.0	5-8 / 620	6-1 / 720	6-5 / 810	6-9 / 900	7-1 / 990	7-5 / 1070	7-8 / 1140	7-11 / 1220	8-1 / 1290	8-4 / 1360	8-7 / 1430	8-9 / 1500	8-11 / 1570	9-1 / 1630	9-4 / 1690	9-6 / 1760	9-8 / 1820	9-11 / 1940	10-3 / 2050
	19.2	5-4 / 660	5-9 / 770	6-1 / 870	6-5 / 960	6-8 / 1050	6-11 / 1130	7-2 / 1220	7-5 / 1300	7-8 / 1370	7-10 / 1450	8-1 / 1520	8-3 / 1590	8-5 / 1660	8-7 / 1730	8-9 / 1800	8-11 / 1870	9-1 / 1930	9-4 / 2060	9-8 / 2180
	24.0	4-11 / 710	5-4 / 830	5-8 / 930	5-11 / 1030	6-2 / 1130	6-5 / 1220	6-8 / 1310	6-11 / 1400	7-1 / 1480	7-3 / 1560	7-6 / 1640	7-8 / 1720	7-10 / 1790	8-0 / 1870	8-1 / 1940	8-3 / 2010	8-5 / 2080	8-8 / 2220	8-11 / 2350
2x6	12.0	9-9 / 560	10-6 / 660	11-2 / 740	11-9 / 820	12-3 / 900	12-9 / 970	13-3 / 1040	13-8 / 1110	14-1 / 1170	14-5 / 1240	14-9 / 1300	15-2 / 1360	15-6 / 1420	15-9 / 1480	16-1 / 1540	16-4 / 1600	16-8 / 1650	17-2 / 1760	17-8 / 1860
	13.7	9-4 / 590	10-0 / 690	10-8 / 770	11-3 / 860	11-9 / 940	12-3 / 1010	12-8 / 1090	13-1 / 1160	13-5 / 1230	13-10 / 1300	14-2 / 1360	14-6 / 1420	14-9 / 1490	15-1 / 1550	15-5 / 1610	15-8 / 1670	15-11 / 1730	16-5 / 1840	16-11 / 1950
	16.0	8-10 / 620	9-6 / 720	10-2 / 810	10-8 / 900	11-2 / 990	11-7 / 1070	12-0 / 1140	12-5 / 1220	12-9 / 1290	13-1 / 1360	13-5 / 1430	13-9 / 1500	14-1 / 1570	14-4 / 1630	14-7 / 1690	14-11 / 1760	15-2 / 1820	15-7 / 1940	16-1 / 2050
	19.2	8-4 / 660	9-0 / 770	9-6 / 870	10-0 / 960	10-6 / 1050	10-11 / 1130	11-4 / 1220	11-8 / 1300	12-0 / 1370	12-4 / 1450	12-8 / 1520	12-11 / 1590	13-3 / 1660	13-6 / 1730	13-9 / 1800	14-0 / 1870	14-3 / 1930	14-8 / 2060	15-2 / 2180
	24.0	7-9 / 710	8-4 / 830	8-10 / 930	9-4 / 1030	9-9 / 1130	10-2 / 1220	10-6 / 1310	10-10 / 1400	11-2 / 1480	11-5 / 1560	11-9 / 1640	12-0 / 1720	12-3 / 1790	12-6 / 1870	12-9 / 1940	13-0 / 2010	13-3 / 2080	13-8 / 2220	14-1 / 2350
2x8	12.0	12-10 / 560	13-10 / 660	14-8 / 740	15-6 / 820	16-2 / 900	16-10 / 970	17-5 / 1040	18-0 / 1110	18-6 / 1170	19-0 / 1240	19-6 / 1300	19-11 / 1360	20-5 / 1420	20-10 / 1480	21-2 / 1540	21-7 / 1600	21-11 / 1650	22-8 / 1760	23-4 / 1860
	13.7	12-3 / 590	13-3 / 690	14-1 / 770	14-10 / 860	15-6 / 940	16-1 / 1010	16-8 / 1090	17-2 / 1160	17-9 / 1230	18-2 / 1300	18-8 / 1360	19-1 / 1420	19-6 / 1490	19-11 / 1550	20-3 / 1610	20-8 / 1670	21-0 / 1730	21-8 / 1840	22-4 / 1950
	16.0	11-8 / 620	12-7 / 720	13-4 / 810	14-1 / 900	14-8 / 990	15-3 / 1070	15-10 / 1140	16-4 / 1220	16-10 / 1290	17-3 / 1360	17-9 / 1430	18-2 / 1500	18-6 / 1570	18-11 / 1630	19-3 / 1690	19-7 / 1760	19-11 / 1820	20-7 / 1940	21-2 / 2050
	19.2	11-0 / 660	11-10 / 770	12-7 / 870	13-3 / 960	13-10 / 1050	14-5 / 1130	14-11 / 1220	15-5 / 1300	15-10 / 1370	16-3 / 1450	16-8 / 1520	17-1 / 1590	17-5 / 1660	17-9 / 1730	18-2 / 1800	18-5 / 1870	18-9 / 1930	19-5 / 2060	19-11 / 2180
	24.0	10-2 / 710	11-0 / 830	11-8 / 930	12-3 / 1030	12-10 / 1130	13-4 / 1220	13-10 / 1310	14-3 / 1400	14-8 / 1480	15-1 / 1560	15-6 / 1640	15-10 / 1720	16-2 / 1790	16-6 / 1870	16-10 / 1940	17-2 / 2010	17-5 / 2080	18-0 / 2220	18-6 / 2350
2x10	12.0	16-5 / 560	17-8 / 660	18-9 / 740	19-9 / 820	20-8 / 900	21-6 / 970	22-3 / 1040	22-11 / 1110	23-8 / 1170	24-3 / 1240	24-10 / 1300	25-5 / 1360	26-0 / 1420	26-6 / 1480	27-1 / 1540	27-6 / 1600	28-0 / 1650	28-11 / 1760	29-9 / 1860
	13.7	15-8 / 590	16-11 / 690	17-11 / 770	18-11 / 860	19-9 / 940	20-6 / 1010	21-3 / 1090	21-11 / 1160	22-7 / 1230	23-3 / 1300	23-9 / 1360	24-4 / 1420	24-10 / 1490	25-5 / 1550	25-10 / 1610	26-4 / 1670	26-10 / 1730	27-8 / 1840	28-6 / 1950
	16.0	14-11 / 620	16-0 / 720	17-0 / 810	17-11 / 900	18-9 / 990	19-6 / 1070	20-2 / 1140	20-10 / 1220	21-6 / 1290	22-1 / 1360	22-7 / 1430	23-2 / 1500	23-8 / 1570	24-1 / 1630	24-7 / 1690	25-0 / 1760	25-5 / 1820	26-3 / 1940	27-1 / 2050
	19.2	14-0 / 660	15-1 / 770	16-0 / 870	16-11 / 960	17-8 / 1050	18-4 / 1130	19-0 / 1220	19-7 / 1300	20-2 / 1370	20-9 / 1450	21-3 / 1520	21-9 / 1590	22-3 / 1660	22-8 / 1730	23-2 / 1800	23-7 / 1870	23-11 / 1930	24-9 / 2060	25-5 / 2180
	24.0	13-0 / 710	14-0 / 830	14-11 / 930	15-8 / 1030	16-5 / 1130	17-0 / 1220	17-8 / 1310	18-3 / 1400	18-9 / 1480	19-3 / 1560	19-9 / 1640	20-2 / 1720	20-8 / 1790	21-1 / 1870	21-6 / 1940	21-10 / 2010	22-3 / 2080	22-11 / 2220	23-8 / 2350

Note: The required extreme fiber stress in bending, "F_b", in pounds per square inch is shown below each span.

DESIGN CRITERIA:
Deflection - For 10 lbs. per sq. ft. live load.
  Limited to span in inches divided by 360.
Strength - live load of 10 lbs. per sq. ft. plus
  dead load of 5 lbs. per sq. ft. determines
  required fiber stress value.

TABLE J-5
CEILING JOISTS
10 Lbs. Per Sq. Ft. Live Load
(No attic storage and roof slope not steeper than 3 in 12)
(Plaster Ceiling)

Modulus of Elasticity, "E", in 1,000,000 psi

JOIST SIZE	SPACING (IN)	0.4	0.5	0.6	0.7	0.8	0.9	1.0	1.1	1.2	1.3	1.4	1.5	1.6	1.7	1.8	1.9	2.0	2.2	2.4
2x4	12.0	6-10 / 340	7-4 / 400	7-10 / 450	8-3 / 500	8-7 / 540	8-11 / 590	9-3 / 630	9-7 / 670	9-10 / 710	10-1 / 750	10-4 / 790	10-7 / 830	10-10 / 860	11-1 / 900	11-3 / 930	11-6 / 970	11-8 / 1000	12-1 / 1070	12-5 / 1130
	13.7	6-6 / 360	7-0 / 410	7-6 / 470	7-10 / 520	8-3 / 570	8-7 / 610	8-10 / 660	9-2 / 700	9-5 / 740	9-8 / 780	9-11 / 820	10-2 / 860	10-4 / 900	10-7 / 940	10-9 / 970	11-0 / 1010	11-2 / 1050	11-6 / 1110	11-10 / 1180
	16.0	6-2 / 380	6-8 / 440	7-1 / 490	7-6 / 550	7-10 / 600	8-1 / 650	8-5 / 690	8-8 / 740	8-11 / 780	9-2 / 830	9-5 / 870	9-8 / 910	9-10 / 950	10-0 / 990	10-3 / 1030	10-5 / 1060	10-7 / 1100	10-11 / 1170	11-3 / 1240
	19.2	5-10 / 400	6-3 / 460	6-8 / 520	7-0 / 580	7-4 / 630	7-8 / 690	7-11 / 740	8-2 / 790	8-5 / 830	8-8 / 880	8-10 / 920	9-1 / 970	9-3 / 1010	9-5 / 1050	9-8 / 1090	9-10 / 1130	10-0 / 1170	10-4 / 1250	10-7 / 1320
	24.0	5-5 / 430	5-10 / 500	6-2 / 560	6-6 / 630	6-10 / 680	7-1 / 740	7-4 / 790	7-7 / 850	7-10 / 900	8-0 / 950	8-3 / 990	8-5 / 1040	8-7 / 1090	8-9 / 1130	8-11 / 1170	9-1 / 1220	9-3 / 1260	9-7 / 1340	9-10 / 1420
2x6	12.0	10-9 / 340	11-7 / 400	12-3 / 450	12-11 / 500	13-6 / 540	14-1 / 590	14-7 / 630	15-0 / 670	15-6 / 710	15-11 / 750	16-3 / 790	16-8 / 830	17-0 / 860	17-4 / 900	17-8 / 930	18-0 / 970	18-4 / 1000	18-11 / 1070	19-6 / 1130
	13.7	10-3 / 360	11-1 / 410	11-9 / 470	12-4 / 520	12-11 / 570	13-5 / 610	13-11 / 660	14-4 / 700	14-9 / 740	15-2 / 780	15-7 / 820	15-11 / 860	16-3 / 900	16-7 / 940	16-11 / 970	17-3 / 1010	17-6 / 1050	18-1 / 1110	18-8 / 1180
	16.0	9-9 / 380	10-6 / 440	11-2 / 490	11-9 / 550	12-3 / 600	12-9 / 650	13-3 / 690	13-8 / 740	14-1 / 780	14-5 / 830	14-9 / 870	15-2 / 910	15-6 / 950	15-9 / 990	16-1 / 1030	16-4 / 1060	16-8 / 1100	17-2 / 1170	17-8 / 1240
	19.2	9-2 / 400	9-10 / 460	10-6 / 520	11-1 / 580	11-7 / 630	12-0 / 690	12-5 / 740	12-10 / 790	13-3 / 830	13-7 / 880	13-11 / 920	14-3 / 970	14-7 / 1010	14-10 / 1050	15-2 / 1090	15-5 / 1130	15-8 / 1170	16-2 / 1250	16-8 / 1320
	24.0	8-6 / 430	9-2 / 500	9-9 / 560	10-3 / 630	10-9 / 680	11-2 / 740	11-7 / 790	11-11 / 850	12-3 / 900	12-7 / 950	12-11 / 990	13-3 / 1040	13-6 / 1090	13-9 / 1130	14-1 / 1170	14-4 / 1220	14-7 / 1260	15-0 / 1340	15-6 / 1420
2x8	12.0	14-2 / 340	15-3 / 400	16-2 / 450	17-0 / 500	17-10 / 540	18-6 / 590	19-2 / 630	19-10 / 670	20-5 / 710	20-11 / 750	21-5 / 790	21-11 / 830	22-5 / 860	22-11 / 900	23-4 / 930	23-9 / 970	24-2 / 1000	24-11 / 1070	25-8 / 1130
	13.7	13-6 / 360	14-7 / 410	15-6 / 470	16-3 / 520	17-0 / 570	17-9 / 610	18-4 / 660	18-11 / 700	19-6 / 740	20-0 / 780	20-6 / 820	21-0 / 860	21-5 / 900	21-11 / 940	22-4 / 970	22-9 / 1010	23-1 / 1050	23-10 / 1110	24-7 / 1180
	16.0	12-10 / 380	13-10 / 440	14-8 / 490	15-6 / 550	16-2 / 600	16-10 / 650	17-5 / 690	18-0 / 740	18-6 / 780	19-0 / 830	19-6 / 870	19-11 / 910	20-5 / 950	20-10 / 990	21-2 / 1030	21-7 / 1060	21-11 / 1100	22-8 / 1170	23-4 / 1240
	19.2	12-1 / 400	13-0 / 460	13-10 / 520	14-7 / 580	15-3 / 630	15-10 / 690	16-5 / 740	16-11 / 790	17-5 / 830	17-11 / 880	18-4 / 920	18-9 / 970	19-2 / 1010	19-7 / 1050	19-11 / 1090	20-4 / 1130	20-8 / 1170	21-4 / 1250	21-11 / 1320
	24.0	11-3 / 430	12-1 / 500	12-10 / 560	13-6 / 630	14-2 / 680	14-8 / 740	15-3 / 790	15-9 / 850	16-2 / 900	16-7 / 950	17-0 / 990	17-5 / 1040	17-10 / 1090	18-2 / 1130	18-6 / 1170	18-10 / 1220	19-2 / 1260	19-10 / 1340	20-5 / 1420
2x10	12.0	18-0 / 340	19-5 / 400	20-8 / 450	21-9 / 500	22-9 / 540	23-8 / 590	24-6 / 630	25-3 / 670	26-0 / 710	26-9 / 750	27-5 / 790	28-0 / 830	28-7 / 860	29-2 / 900	29-9 / 930	30-4 / 970	30-10 / 1000	31-10 / 1070	32-9 / 1130
	13.7	17-3 / 360	18-7 / 410	19-9 / 470	20-9 / 520	21-9 / 570	22-7 / 610	23-5 / 660	24-2 / 700	24-10 / 740	25-7 / 780	26-2 / 820	26-10 / 860	27-5 / 900	27-11 / 940	28-6 / 970	29-0 / 1010	29-6 / 1050	30-5 / 1110	31-4 / 1180
	16.0	16-5 / 380	17-8 / 440	18-9 / 490	19-9 / 550	20-8 / 600	21-6 / 650	22-3 / 690	22-11 / 740	23-8 / 780	24-3 / 830	24-10 / 870	25-5 / 910	26-0 / 950	26-6 / 990	27-1 / 1030	27-6 / 1060	28-0 / 1100	28-11 / 1170	29-9 / 1240
	19.2	15-5 / 400	16-7 / 460	17-8 / 520	18-7 / 580	19-5 / 630	20-2 / 690	20-11 / 740	21-7 / 790	22-3 / 830	22-10 / 880	23-5 / 920	23-11 / 970	24-6 / 1010	25-0 / 1050	25-5 / 1090	25-11 / 1130	26-4 / 1170	27-3 / 1250	28-0 / 1320
	24.0	14-4 / 430	15-5 / 500	16-5 / 560	17-3 / 630	18-0 / 680	18-9 / 740	19-5 / 790	20-1 / 850	20-8 / 900	21-2 / 950	21-9 / 990	22-3 / 1040	22-9 / 1090	23-2 / 1130	23-8 / 1170	24-1 / 1220	24-6 / 1260	25-3 / 1340	26-0 / 1420

Note: The required extreme fiber stress in bending, "F_b" in pounds per square inch is shown below each span.

438 ■ *Appendix B*

TABLE J-6
CEILING JOISTS
10 Lbs. Per Sq. Ft. Live Load
(No attic storage and roof slope not steeper than 3 in 12)
(Drywall Ceiling)

DESIGN CRITERIA:
Deflection - For 10 lbs. per sq. ft. live load.
Limited to span in inches divided by 240.
Strength - live load of 10 lbs. per sq. ft. plus
dead load of 5 lbs. per sq. ft. determines
required fiber stress value.

JOIST SIZE (IN)	SPACING (IN)	Modulus of Elasticity, "E", in 1,000,000 psi																		
		0.4	0.5	0.6	0.7	0.8	0.9	1.0	1.1	1.2	1.3	1.4	1.5	1.6	1.7	1.8	1.9	2.0	2.2	2.4
2x4	12.0	7-10 / 450	8-5 / 520	8-11 / 590	9-5 / 650	9-10 / 710	10-3 / 770	10-7 / 830	10-11 / 880	11-3 / 930	11-7 / 980	11-10 / 1030	12-2 / 1080	12-5 / 1130	12-8 / 1180	12-11 / 1220	13-2 / 1270	13-4 / 1310	13-9 / 1400	14-2 / 1480
	13.7	7-6 / 470	8-1 / 540	8-7 / 610	9-0 / 680	9-5 / 740	9-9 / 800	10-2 / 860	10-6 / 920	10-9 / 970	11-1 / 1030	11-4 / 1080	11-7 / 1130	11-10 / 1180	12-1 / 1230	12-4 / 1280	12-7 / 1320	12-9 / 1370	13-2 / 1460	13-7 / 1550
	16.0	7-1 / 490	7-8 / 570	8-1 / 650	8-7 / 720	8-11 / 780	9-4 / 850	9-8 / 910	9-11 / 970	10-3 / 1030	10-6 / 1080	10-9 / 1140	11-0 / 1190	11-3 / 1240	11-6 / 1290	11-9 / 1340	11-11 / 1390	12-2 / 1440	12-6 / 1540	12-11 / 1630
	19.2	6-8 / 520	7-2 / 610	7-8 / 690	8-1 / 760	8-5 / 830	8-9 / 900	9-1 / 970	9-4 / 1030	9-8 / 1090	9-11 / 1150	10-2 / 1210	10-4 / 1270	10-7 / 1320	10-10 / 1380	11-0 / 1430	11-3 / 1480	11-5 / 1530	11-9 / 1630	12-2 / 1730
	24.0	6-2 / 560	6-8 / 660	7-1 / 740	7-6 / 820	7-10 / 900	8-1 / 970	8-5 / 1040	8-8 / 1110	8-11 / 1170	9-2 / 1240	9-5 / 1300	9-8 / 1360	9-10 / 1420	10-0 / 1480	10-3 / 1540	10-5 / 1600	10-7 / 1650	10-11 / 1760	11-3 / 1860
2x6	12.0	12-3 / 450	13-3 / 520	14-1 / 590	14-9 / 650	15-6 / 710	16-1 / 770	16-8 / 830	17-2 / 880	17-8 / 930	18-2 / 980	18-8 / 1030	19-1 / 1080	19-6 / 1130	19-11 / 1180	20-3 / 1220	20-8 / 1270	21-0 / 1310	21-8 / 1400	22-4 / 1480
	13.7	11-9 / 470	12-8 / 540	13-5 / 610	14-2 / 680	14-9 / 740	15-5 / 800	15-11 / 860	16-5 / 920	16-11 / 970	17-5 / 1030	17-10 / 1080	18-3 / 1130	18-8 / 1180	19-0 / 1230	19-5 / 1280	19-9 / 1320	20-1 / 1370	20-9 / 1460	21-4 / 1550
	16.0	11-2 / 490	12-0 / 570	12-9 / 650	13-5 / 720	14-1 / 780	14-7 / 850	15-2 / 910	15-7 / 970	16-1 / 1030	16-6 / 1080	16-11 / 1140	17-4 / 1190	17-8 / 1240	18-1 / 1290	18-5 / 1340	18-9 / 1390	19-1 / 1440	19-8 / 1540	20-3 / 1630
	19.2	10-6 / 520	11-4 / 610	12-0 / 690	12-8 / 760	13-3 / 830	13-9 / 900	14-3 / 970	14-8 / 1030	15-2 / 1090	15-7 / 1150	15-11 / 1210	16-4 / 1270	16-8 / 1320	17-0 / 1380	17-4 / 1430	17-8 / 1480	17-11 / 1530	18-6 / 1630	19-1 / 1730
	24.0	9-9 / 560	10-6 / 660	11-2 / 740	11-9 / 820	12-3 / 900	12-9 / 970	13-3 / 1040	13-8 / 1110	14-1 / 1170	14-5 / 1240	14-9 / 1300	15-2 / 1360	15-6 / 1420	15-9 / 1480	16-1 / 1540	16-4 / 1600	16-8 / 1650	17-2 / 1760	17-8 / 1860
2x8	12.0	16-2 / 450	17-5 / 520	18-6 / 590	19-6 / 650	20-5 / 710	21-2 / 770	21-11 / 830	22-8 / 880	23-4 / 930	24-0 / 980	24-7 / 1030	25-2 / 1080	25-8 / 1130	26-2 / 1180	26-9 / 1220	27-2 / 1270	27-8 / 1310	28-7 / 1400	29-5 / 1480
	13.7	15-6 / 470	16-8 / 540	17-9 / 610	18-8 / 680	19-6 / 740	20-3 / 800	21-0 / 860	21-8 / 920	22-4 / 970	22-11 / 1030	23-6 / 1080	24-0 / 1130	24-7 / 1180	25-1 / 1230	25-7 / 1280	26-0 / 1320	26-6 / 1370	27-4 / 1460	28-1 / 1550
	16.0	14-8 / 490	15-10 / 570	16-10 / 650	17-9 / 720	18-6 / 780	19-3 / 850	19-11 / 910	20-7 / 970	21-2 / 1030	21-9 / 1080	22-4 / 1140	22-10 / 1190	23-4 / 1240	23-10 / 1290	24-3 / 1340	24-8 / 1390	25-2 / 1440	25-11 / 1540	26-9 / 1630
	19.2	13-10 / 520	14-11 / 610	15-10 / 690	16-8 / 760	17-5 / 830	18-2 / 900	18-9 / 970	19-5 / 1030	19-11 / 1090	20-6 / 1150	21-0 / 1210	21-6 / 1270	21-11 / 1320	22-5 / 1380	22-10 / 1430	23-3 / 1480	23-8 / 1530	24-5 / 1630	25-2 / 1730
	24.0	12-10 / 560	13-10 / 660	14-8 / 740	15-6 / 820	16-2 / 900	16-10 / 970	17-5 / 1040	18-0 / 1110	18-6 / 1170	19-0 / 1240	19-6 / 1300	19-11 / 1360	20-5 / 1420	20-10 / 1480	21-2 / 1540	21-7 / 1600	21-11 / 1650	22-8 / 1760	23-4 / 1860
2x10	12.0	20-8 / 450	22-3 / 520	23-8 / 590	24-10 / 650	26-0 / 710	27-1 / 770	28-0 / 830	28-11 / 880	29-9 / 930	30-7 / 980	31-4 / 1030	32-1 / 1080	32-9 / 1130	33-5 / 1180	34-1 / 1220	34-8 / 1270	35-4 / 1310	36-5 / 1400	37-6 / 1480
	13.7	19-9 / 470	21-3 / 540	22-7 / 610	23-9 / 680	24-10 / 740	25-10 / 800	26-10 / 860	27-8 / 920	28-6 / 970	29-3 / 1030	30-0 / 1080	30-8 / 1130	31-4 / 1180	32-0 / 1230	32-7 / 1280	33-2 / 1320	33-9 / 1370	34-10 / 1460	35-10 / 1550
	16.0	18-9 / 490	20-2 / 570	21-6 / 650	22-7 / 720	23-8 / 780	24-7 / 850	25-5 / 910	26-3 / 970	27-1 / 1030	27-9 / 1080	28-6 / 1140	29-2 / 1190	29-9 / 1240	30-5 / 1290	31-0 / 1340	31-6 / 1390	32-1 / 1440	33-1 / 1540	34-1 / 1630
	19.2	17-8 / 520	19-0 / 610	20-2 / 690	21-3 / 760	22-3 / 830	23-2 / 900	23-11 / 970	24-9 / 1030	25-5 / 1090	26-2 / 1150	26-10 / 1210	27-5 / 1270	28-0 / 1320	28-7 / 1380	29-2 / 1430	29-8 / 1480	30-2 / 1530	31-2 / 1630	32-1 / 1730
	24.0	16-5 / 560	17-8 / 660	18-9 / 740	19-9 / 820	20-8 / 900	21-6 / 970	22-3 / 1040	22-11 / 1110	23-8 / 1170	24-3 / 1240	24-10 / 1300	25-5 / 1360	26-0 / 1420	26-6 / 1480	27-1 / 1540	27-6 / 1600	28-0 / 1650	28-11 / 1760	29-9 / 1860

Note: The required extreme fiber stress in bending, "F_b", in pounds per square inch is shown below each span.

INDEX

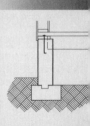